Dissertation

Besov Regularity of Stochastic Partial Differential Equations on Bounded Lipschitz Domains

Petru A. Cioica

2013

Bibliographic information published by the Deutsche Nationalbibliothek

The Deutsche Nationalbibliothek lists this publication in the Deutsche Nationalbibliografie; detailed bibliographic data are available in the Internet at http://dnb.d-nb.de .

ISBN 978-3-8325-3920-7

Logos Verlag Berlin GmbH
Comeniushof, Gubener Str. 47,
10243 Berlin
Tel.: +49 (0)30 42 85 10 90
Fax: +49 (0)30 42 85 10 92
INTERNET: http://www.logos-verlag.de

Besov Regularity of Stochastic Partial Differential Equations on Bounded Lipschitz Domains

Dissertation

zur

Erlangung des akademischen Grades

Doktor der Naturwissenschaften

(Dr. rer. nat.)

vorgelegt

dem Fachbereich Mathematik und Informatik

der

Philipps–Universität Marburg

von

Petru A. Cioica

geboren am 22. März 1983

in Cluj-Napoca/Klausenburg/Kolozsvár (Rumänien)

Vom Fachbereich Mathematik und Informatik
der Philipps-Universität Marburg (Hochschulkennziffer: 1180)
als Dissertation angenommen am: 20. Dezember 2013

Erstgutachter: **Prof. Dr. Stephan Dahlke**, Philipps-Universität Marburg

Zweitgutachter: **Prof. Dr. René L. Schilling**, Technische Universität Dresden

Drittgutachter: **Prof. Dr. Stig Larsson**, Chalmers University of Technology, Göteborg, Schweden

Tag der mündlichen Prüfung: 17. Februar 2014

For Christine.

In memory of my mother, Carmen Luminița Cioica (†2010).

Contents

Acknowledgement

During my work on this thesis, I have been supported by many people and institutions, whom I want to express my gratitude at this point. I want to start with my advisor, Prof. Dr. Stephan Dahlke. I am deeply grateful, Stephan, that you accepted me as your PhD student. Thank you for providing an excellent work and study environment and for supporting me in all respects, in particular, when applying for fellowships and grants, when demanding more background from different mathematical areas, or when any self-doubt came up. Next, I thank Prof. Dr. René L. Schilling for agreeing to overview this thesis and for the constant support whenever needed. Thank you also for several invitations to Dresden, I enjoyed the atmosphere at the 'Institut für Mathematische Stochastik' very much. My special thanks goes to Junior-Prof. Dr. Felix Lindner. Thank you, Felix, for many fruitful discussions on SPDEs and related topics, for a lot of good advice, for your encouragement and your kind hospitality during my visits to Dresden. Thank you also for reading very carefully parts of this manuscript.

Over the last years, I have had the pleasure to participate at several meetings of the project "Adaptive Wavelet Methods for SPDEs", which is funded by the German Research Council (DFG) and is part of the DFG-Priority Program 1324 "Mathematical methods for extracting quantifiable information from complex systems" (DFG-SPP 1324). I want to thank all the members of this project and their affiliates for many fruitful discussions. Besides the already mentioned persons, these are: Prof. Dr. Klaus Ritter and his assistants Nicolas Döhring and Dr. Tiange Xu from Kaiserslautern, Stefan Kinzel from Marburg and Junior-Prof. Dr. Thorsten Raasch from Mainz.

During one of my visits to Dresden, I have had the pleasure to meet Prof. Dr. Kyeong-Hun Kim and Prof. Dr. Kijung Lee, who have been in Dresden as Fellows of the DFG-SPP 1324. A very fruitful collaboration started and I want to thank you both, Kyeong-Hun and Kijung, for teaching me many details on the regularity theory of SPDEs in weighted Sobolev spaces and the techniques used in the analytic approach. Thank you also for inviting me to Korea, where I have had the opportunity to attend a very well organized summer school on "Stochastic Partial Differential Equations and Related Fields". I enjoyed the time at this summer school and at your department very much.

I also want to thank Dr. Sonja G. Cox for accepting to visit Marburg two years ago and provide an insight into recent results concerning numerical methods for SPDEs in Banach spaces. Thank you, Sonja, for explaining different aspects from the semigroup approach to SPDEs in Banach spaces to me. Thank you also for a copy of your very nice thesis and the delicious Swiss chocolate.

While working on this thesis, I have received great support from my colleagues from the Workgroup Numerics at Philipps-Universität Marburg. Thank you, guys, for daily lunch and coffee breaks, for helping me whenever needed, in particular, during the last year. I also want to thank Mrs Jutta Happel for being a perfectly organized and very kind secretary.

My PhD studies have been mainly funded by a doctoral scholarship of the Philipps-Universität Marburg. In this context I want to thank the selection panel for their confidence, as well as Dr. Susanne Igler, Dr. Katja Heitmann and Dr. Ute Kämper for excellent assistance. Over the last years, I have had the opportunity to attend many workshops and conferences, and to intensify my collaborations while visiting different mathematical departments. Financial support by the DFG-SPP 1324 for the travelling costs is gratefully acknowledged.

At this point, I also want to thank my father for his constant encouragement. I enjoy it very much, that you are (almost) always in good humour again. Last but not least, I want to thank you, Christine, for your love, your patience and for very precious moments 'after maths'.

Chapter 1

Introduction

This thesis is concerned with the regularity of (semi-)linear second order parabolic stochastic partial differential equations (SPDEs, for short) of Itô type on bounded Lipschitz domains. They have the following form:

$$
\left.
\begin{aligned}
\mathrm{d}u &= \left(\sum_{i,j=1}^{d} a^{ij} u_{x^i x^j} + \sum_{i=1}^{d} b^i u_{x^i} + cu + f + L(u) \right) \mathrm{d}t \\
&\quad + \sum_{k=1}^{\infty} \left(\sum_{i=1}^{d} \sigma^{ik} u_{x^i} + \mu^k u + g^k + (\Lambda(u))^k \right) \mathrm{d}w_t^k \qquad \text{on } \Omega \times [0,T] \times \mathcal{O}, \\
u &= 0 \qquad \text{on } \Omega \times (0,T] \times \partial\mathcal{O}, \\
u(0) &= u_0 \qquad \text{on } \Omega \times \mathcal{O}.
\end{aligned}
\right\} \quad (1.1)
$$

Here, and in the rest of this thesis, $\mathcal{O}$ is a bounded Lipschitz domain in $\mathbb{R}^d$ ($d \geq 2$) and $T \in (0,\infty)$ denotes a finite time horizon. Moreover, $(w_t^k)_{t \in [0,T]}$, $k \in \mathbb{N}$, is a sequence of independent real valued standard Brownian motions with respect to a normal filtration $(\mathcal{F}_t)_{t \in [0,T]}$ on a complete probability space $(\Omega, \mathcal{F}, \mathbb{P})$ and $\mathrm{d}u$ denotes Itô's stochastic differential with respect to the time $t \in [0,T]$. The coefficients a^{ij}, b^i, c, σ^{ik}, and μ^k with $i,j \in \{1,\ldots,d\}$ and $k \in \mathbb{N} := \{1,2,\ldots\}$, are real valued functions on $\Omega \times [0,T] \times \mathcal{O}$ and fulfil certain conditions which will be specified later on in Chapter 3, see Assumption 3.1. The non-linearities L and Λ are assumed to be Lipschitz continuous in suitable spaces, see Chapter 5, in particular Assumption 5.9, for details. In this thesis we take a functional analytic point of view, meaning that the solution u is not considered as a real valued function depending on $(\omega, t, x) \in \Omega \times [0,T] \times \mathcal{O}$ but as a function on $\Omega \times [0,T]$ taking values in the space $\mathcal{D}'(\mathcal{O})$ of real valued distributions on $\mathcal{O}$.

The most prominent equation of the type (1.1) is the stochastic heat equation with additive or multiplicative noise. More general equations of the form (1.1) with finitely many $(w_t^k)_{t \in [0,T]}$, $k \in \{1,\ldots,N\}$, appear in the context of non-linear filtering problems, see, e.g., [80, Section 8.1] and [107]. Choosing infinitely many Brownian motions $(w_t^k)_{t \in [0,T]}$, $k \in \mathbb{N}$, allows us to consider equations driven by space-time white noise, cf. [80, Section 8.3]. These equations are suggested, for instance, as mathematical models for reaction diffusion systems corrupted by noise, see [32, Section 0.7] and the references therein, in particular, [9]. In general, the question whether a unique solution to Eq. (1.1) exits is well-studied. However, in the majority of cases, this solution can not be specified. Thus, in order to make equations of the form (1.1) ready to use as mathematical models in applications, the solution has to be constructively approximated. Therefore, efficient numerical methods are needed. Usually, their performance depends on the regularity or smoothness of the solutions to the considered SPDEs in specific scales of function spaces. As we will elaborate later on in detail, the scale

$$B^{\alpha}_{\tau,\tau}(\mathcal{O}), \quad \frac{1}{\tau} = \frac{\alpha}{d} + \frac{1}{p}, \quad \alpha > 0, \tag{$*$}$$

of Besov spaces ($p \geq 2$ being fixed) plays an outstanding role in this context. We refer to Subsection 2.3.4 for the definition of Besov spaces.

In this thesis we analyse the regularity of SPDEs of the form (1.1) in the scale $(*)$. We will be mainly concerned with the following two tasks:

(T1) **Spatial regularity.** We use the scale $(*)$ to measure the smoothness of the solution u with respect to the space coordinates. That is, we ask for an $\alpha^* > 0$ as high as possible, such that for all $0 < \alpha < \alpha^*$ and $1/\tau = \alpha/d + 1/p$, the solution u is contained in the space of (equivalence classes of) p-integrable $B^{\alpha}_{\tau,\tau}(\mathcal{O})$-valued stochastic processes.

(T2) **Space time regularity.** Under the assumption that the solution u is a $B^{\alpha}_{\tau,\tau}(\mathcal{O})$-valued stochastic process with α and τ as in $(*)$, we analyse the Hölder regularity of the paths of this process.

Before we continue our exposition, we motivate our analysis by elaborating in detail the importance of the topics (T1) and (T2). In particular, we will emphasize their link with the convergence analysis of certain numerical methods.

1.1   Motivation

Our motivation to study the regularity of SPDEs in the scale $(*)$ of Besov spaces is closely related to the theme of *adaptive numerical wavelet methods*. Since this topic is not a common prerequisite in the stochastic analysis community we give a rather detailed exposition aiming to point out the significance of our results from the point of view of numerical analysis. However, we will not be too rigorous in a formal sense, but rather try to emphasize some key principles and basic results from the theory of numerical methods and non-linear approximation which motivate our analysis. For an in-depth treatment of these topics we refer to the monograph [27] on numerical wavelet methods and to the survey article [46] on non-linear approximation theory, see also [37].

Usually, the term *wavelet* is used for the elements of a specific kind of basis for the space $L_2(\mathcal{O})$ of quadratically Lebesgue-integrable functions on a domain $\mathcal{O} \subseteq \mathbb{R}^d$, which allows the decomposition of functions into components corresponding to different scales of resolution [33]. Such a basis is typically constructed by means of a *multiresolution analysis* (MRA, for short), i.e., a sequence $(S_j)_{j \geq j_0}$ of closed linear subspaces of $L_2(\mathcal{O})$ with

$$S_j \subset S_{j+1} \quad \text{for all } j \geq j_0, \quad \text{and} \quad \overline{\left(\bigcup_{j \geq j_0} S_j \right)}^{\|\cdot\|_{L_2(\mathcal{O})}} = L_2(\mathcal{O}).$$

The latter means that the union of all S_j, $j \geq j_0$, is dense in $L_2(\mathcal{O})$. The MRA is designed in such a way that for each $j \geq j_0$, the space S_j is spanned by a Riesz basis $\{\phi_\lambda : \lambda \in \Delta_j\}$ of so-called *scaling functions*. Furthermore, the complement of S_j in S_{j+1} is spanned by another Riesz basis $\{\psi_\lambda : \lambda \in \nabla_j\}$ of so-called *wavelets*. Following the notation from [27] we write $\nabla_{j_0-1} := \Delta_{j_0}$ and denote the scaling functions spanning S_{j_0} also by ψ_λ, $\lambda \in \nabla_{j_0-1}$. Then, setting $\nabla := \cup_{j \geq j_0-1} \nabla_j$, we call

$$\{\psi_\lambda : \lambda \in \nabla\} := \bigcup_{j \geq j_0-1} \{\psi_\lambda : \lambda \in \nabla_j\}$$

a *wavelet Riesz basis* of $L_2(\mathcal{O})$. The index $\lambda \in \nabla$ typically encodes several types of information, namely the *scale level* $j + 1 \geq j_0$, if $\lambda \in \nabla_j$, the *spatial location*, and also the *type* of the wavelet. For constructions of wavelet bases for diverse shapes of bounded domains including polygonal and polyhedral domains we refer to [42–44] or [19, 20], see also [27, Section 1.2] for a detailed discussion. Typically, the elements of a wavelet basis are local in the sense that they have compact supports and the size of the supports decays exponentially with the scale. Furthermore, they fulfil appropriate smoothness assumptions and have vanishing moments up to a prescribed order. These properties yield the following facts [33]:

- Weighted sequence norms of wavelet coefficients are equivalent to Lebesgue, Sobolev and Besov norms (for a certain range of regularity and integrability parameters, depending in particular on the smoothness of the wavelets).

- The representation of a wide class of operators in the wavelet basis is nearly diagonal.

- The vanishing moments of wavelets remove the smooth part of a function.

Due to these features, wavelets become a powerful tool for solving operator equations. Let us discuss this topic with the help of a classical example. We write $\mathring{W}_2^1(\mathcal{O})$ for the closure of the space $\mathcal{C}_0^\infty(\mathcal{O})$ of infinitely differentiable functions with compact support on $\mathcal{O}$ in the $L_2(\mathcal{O})$-Sobolev space of order one, which we denote by $W_2^1(\mathcal{O})$; see Subsection 2.3.1 for a precise definition of Sobolev spaces. Let $a : \mathring{W}_2^1(\mathcal{O}) \times \mathring{W}_2^1(\mathcal{O}) \to \mathbb{R}$ be a continuous, symmetric and elliptic bilinear form, so that, in particular, there exists a finite constant $C > 0$, such that

$$\frac{1}{C} \|u\|^2_{\mathring{W}_2^1(\mathcal{O})} \leq a(u, u) \leq C \|u\|^2_{\mathring{W}_2^1(\mathcal{O})}, \qquad u \in \mathring{W}_2^1(\mathcal{O}). \tag{1.2}$$

It defines an isomorphism

$$A : \mathring{W}_2^1(\mathcal{O}) \to W_2^{-1}(\mathcal{O})$$
$$u \mapsto a(u, \cdot),$$

where $W_2^{-1}(\mathcal{O})$ denotes the dual of $\mathring{W}_2^1(\mathcal{O})$. Thus, for $f \in W_2^{-1}(\mathcal{O})$, the equation

$$Au = f, \tag{1.3}$$

has a unique solution $u \in \mathring{W}_2^1(\mathcal{O})$, which is simultaneously the unique solution of the variational problem

$$a(u, v) = f(v), \qquad v \in \mathring{W}_2^1(\mathcal{O}). \tag{1.4}$$

However, in general this solution is not known explicitly. Therefore, in order to use (1.3) as a mathematical model in real-life applications, the solution has to be constructively approximated. To this end, Eq. (1.4) is discretized. One classical way to discretize this equation is to employ a Galerkin method. That is, we choose an increasing sequence $(V_m)_{m \in \mathcal{J}}$ with $\mathcal{J} \subseteq \mathbb{N}_0$ of subspaces of $\mathring{W}_2^1(\mathcal{O})$ and determine the solutions $u_m \in V_m$ to the variational problems

$$a(u_m, v_m) = f(v_m), \qquad v_m \in V_m, \tag{1.5}$$

successively for $m \in \mathcal{J}$. The index m denotes the number of degrees of freedom (here: scaling functions and wavelets) spanning the subspace V_m. We distinguish two kinds of numerical methods, depending on the way the refinement from a space V_m to its successor $V_{m'}$, $m, m' \in \mathcal{J}$, is performed. In our context 'refinement' means to add wavelets to the basis functions used to approximate the current approximative solution u_m. On the one hand, we can develop a *uniform method* which is based on the underlying MRA and set

$$V_{m(j)} := S_j, \qquad j \geq j_0,$$

where $m(j) = \left| \cup_{i \geq j_0 - 1}^j \nabla_i \right| \in \mathbb{N}$ for $j \geq j_0$ (usually, on bounded domains the cardinality of ∇_j behaves like 2^{jd}). This method is called 'uniform', since when passing from $V_{m(j)}$ to $V_{m(j+1)}$ we add all the wavelets at the scale level $j+1$, i.e., we choose a finer resolution uniformly on the entire domain. On the other hand, since the approximation u_m might be already sufficiently accurate in some regions of the domain, it is reasonable to look for a self-regulating updating strategy and try to refine the resolution only at that parts where the accuracy is not yet satisfactory. Such an *adaptive method*, executes the following steps successively for $m \in \mathcal{J}$:

1. Solve Eq. (1.5) in V_m.

2. Estimate the local error of $u - u_m$ in a suitable norm $\|\cdot\|_E$.

3. Refine where it is necessary.

Of course, for the second step one needs *a posteriori error estimators*, since the solution u is not known exactly. These estimators should rely on local error indicators, so that they provide information about the way the space V_m has to be refined in the subsequent step.

However, one is faced with at least three major difficulties on the way to a fully-fledged adaptive method. Firstly, the design of local error estimators as they are needed for adaptive strategies is not a trivial task. A second difficulty is the convergence proof for adaptive strategies and the estimation of their convergence rates. Thirdly, their implementation turns out to be much more difficult than the implementation of uniform counterparts. Thus, before we decide to work on the development of an adaptive method, we need to check whether adaptivity really pays, in the sense that there is any chance to obtain a higher convergence rate than by uniform alternatives. A numerical method is said to have *convergence rate* $s > 0$ in the Banach space $(E, \|\cdot\|_E)$, if there exists a constant $C \in (0, \infty)$, which does not depend on the number of degrees of freedom $m \in \mathcal{J}$ needed to describe the approximative solution $u_m \in V_m$, such that

$$\|u - u_m\|_E \leq C \, m^{-s}, \qquad m \in \mathcal{J},$$

where $u \in E$ denotes the exact solution of the given problem. The benchmark for any numerical approximation method based on $\{\psi_\lambda : \lambda \in \nabla\}$ is the rate of the *best m-term (wavelet) approximation error*

$$\sigma_{m,E}(u) := \inf_{u_m \in \widetilde{\Sigma}_m} \|u - u_m\|_E,$$

where

$$\widetilde{\Sigma}_m := \left\{ \sum_{\lambda \in \Lambda} c_\lambda \psi_\lambda : \Lambda \subset \nabla, \, |\Lambda| = m, \, c_\lambda \in \mathbb{R}, \, \lambda \in \Lambda \right\}$$

is the space of *m-term approximations* from $\{\psi_\lambda : \lambda \in \nabla\}$, $m \in \mathbb{N}$. As it is easy to see, $\widetilde{\Sigma}_m$ is not a linear space: The sum of two functions, each of which uses m basis elements, might make use of up to $2m$ basis elements and is therefore usually not contained in $\widetilde{\Sigma}_m$. This is why m-term approximation is referred to as a *non-linear approximation method*. Obviously, the convergence rate of any numerical method based on $\{\psi_\lambda : \lambda \in \nabla\}$ is dominated by the decay rate of the best m-term approximation error $\sigma_{m,E}(u)$, $m \in \mathbb{N}$. Since, in general, the solution u is not known, we will not be able to find approximations u_m, $m \in \mathbb{N}$, reproducing the errors $\sigma_{m,E}(u)$, $m \in \mathbb{N}$. However, what we can aim for is to develop a numerical method which has the same convergence rate as the best m-term approximation error.

If the convergence rate of a uniform method meets the benchmark, then working on the development of adaptive algorithms is superfluously. However, if the converse is true, i.e., if the rate of best m-term approximation is strictly higher than the convergence rate of uniform methods, the development of adaptive methods is completely justified. Since the error of uniform numerical methods based on $V_{m(j)} = S_j$, with $m(j) = \left|\cup_{i\geq j_0-1}^{j} \nabla_i\right| \in \mathbb{N}$, $j \geq j_0$, is dominated by

$$e_{m,E}(u) := \inf_{u_m \in V_m} \|u - u_m\|_E, \quad m = m(j), \quad j \geq j_0,$$

this means: Adaptivity pays, only if there exists an $\alpha > 0$ and a corresponding constant C, which does not depend on $m \in \mathbb{N}$, such that[1]

$$\sigma_{m,E}(u) \leq C\,m^{-\alpha/d}, \qquad m \in \mathbb{N}, \tag{1.6}$$

holds for the solution $u \in E$, and, simultaneously,

$$\alpha > s_{\max}(u) := \sup\left\{s \geq 0 : \forall j \in \mathbb{N} : e_{m(j),E}(u) \leq C\,m(j)^{-s/d}, C \text{ independent of } j\right\}. \tag{1.7}$$

The question whether (1.6) and (1.7) with $E = L_p(\mathcal{O})$ are simultaneously fulfilled, where $p \in (1,\infty)$, can be decided after a rigorous regularity analysis of the target function u. On the one hand, it is well-known that—under certain technical assumptions on the wavelet basis, which can be found, e.g., in [27, Chapter 3 and 4]—the decay rate of $e_m(u) := e_{m,L_p(\mathcal{O})}(u)$ is linked with the $L_p(\mathcal{O})$-Sobolev regularity of the target function. That is, there exists an upper bound $\tilde{s} \in \mathbb{N}$, depending on the smoothness and polynomial exactness of the wavelet basis, such that, for all $s \in (0, \tilde{s})$,

$$u \in W_p^s(\mathcal{O}) \quad \text{implies} \quad e_m(u) \leq C\,m^{-s/d}, \quad m = m(j), \quad j \geq j_0, \tag{1.8}$$

with a constant $C \in (0,\infty)$ which does not depend on m. As mentioned in the introduction of [27, Chapter 3], statements similar to (1.8) also hold for approximation methods based on finite elements instead of wavelets—of course, with adjusted spaces V_m, $m \in \mathcal{J}$ (see also the standard literature on finite elements like [21] or [100]). One can also show the following converse of (1.8): The existence of a constant $C \in (0,\infty)$ such that

$$e_m(u) \leq C\,m^{-s/d}, \quad m = m(j), \quad \text{for all } j \geq j_0, \quad \text{implies} \quad u \in W_p^{s'}(\mathcal{O}), \quad s' < s.$$

In particular, if $u \notin W_p^s(\mathcal{O})$ for some $s \in (0,\infty)$, then $s_{\max}(u) \leq s$ with $s_{\max}(u)$ as defined in (1.7). This yields

$$s_{\max}(u) = s_{\max}^{\text{Sob}}(u) := \sup\left\{s \geq 0 : u \in W_p^s(\mathcal{O})\right\}. \tag{1.9}$$

On the other hand, the convergence rate of the best m-term wavelet approximation error $\sigma_m(u) = \sigma_{m,L_p(\mathcal{O})}(u)$, $m \in \mathbb{N}$, is governed by the smoothness of u in the so-called $(L_p(\mathcal{O})$-)non-linear approximation scale

$$B_{\tau,\tau}^\alpha(\mathcal{O}), \quad \frac{1}{\tau} = \frac{\alpha}{d} + \frac{1}{p}, \quad \alpha > 0, \tag{$*$}$$

of Besov spaces. That is, for all $\alpha \in (0, \tilde{s})$,

$$u \in B_{\tau,\tau}^\alpha(\mathcal{O}), \quad \frac{1}{\tau} = \frac{\alpha}{d} + \frac{1}{p} \quad \text{implies} \quad \sigma_m(u) \leq C\,m^{-\alpha/d}, \quad m \in \mathbb{N}.$$

Therefore, if

$$u \in B_{\tau,\tau}^\alpha(\mathcal{O}), \quad \frac{1}{\tau} = \frac{\alpha}{d} + \frac{1}{p} \quad \text{with} \quad \alpha > s_{\max}^{\text{Sob}}(u), \tag{1.10}$$

---

[1] The factor $1/d$ in the exponent is just a useful convention in order to match with the results presented below.

then (1.6) and (1.7) are simultaneously satisfied with $s_{\max}(u) = s_{\max}^{\text{Sob}}(u)$. In this case, the decay rate of the best m-term wavelet approximation error is higher than the convergence rate of the uniform wavelet method presented above. Thus, by our expositions above, if (1.10) is fulfilled, working on the development of adaptive wavelet methods is completely justified.

For deterministic elliptic equations it could be already shown that, indeed, adaptivity pays: The results from [34–36,38,40] together with [57,58] show that solutions of elliptic equations on non-smooth domains generically behave like described by (1.10). Simultaneously, for this class of equations, there exist adaptive wavelet methods which realise the convergence rate of the best m-term approximation error in a Hilbert space setting ($p = 2$), see, e.g., [28,39]. The error is measured in the *energy norm* induced by the equation, which is, in general, equivalent to a suitable Sobolev norm. In our example from above, the energy norm is given by $\|\cdot\|_a := \sqrt{a(\cdot,\cdot)}$ and it is equivalent to the $L_2(\mathcal{O})$-Sobolev norm of order one in $\mathring{W}_2^1(\mathcal{O})$ by (1.2). There also exist optimal adaptive wavelet algorithms for more general deterministic equations, see, e.g., [29,112], this list being by no means complete.

Our analysis is motivated by the question whether these results can be extended to solutions of SPDEs of the form (1.1). We tackle and solve the tasks (T1) and (T2) with the following scopes:

ad (T1). **Spatial regularity.** By analysing the spatial regularity of the solution process u in the scale $(*)$ of Besov spaces we aim to clarify whether $u = u(\omega, t, \cdot)$ fulfils

$$u \in L_p(\Omega \times [0,T]; B_{\tau,\tau}^{\alpha}(\mathcal{O})), \quad \frac{1}{\tau} = \frac{\alpha}{d} + \frac{1}{p}, \quad \text{with} \quad \alpha > \tilde{s}_{\max}^{\text{Sob}}(u), \tag{1.11}$$

where

$$\tilde{s}_{\max}^{\text{Sob}}(u) := \sup\left\{ s \geq 0 : u \in L_p(\Omega \times [0,T]; W_p^s(\mathcal{O})) \right\}. \tag{1.12}$$

If so, the decay rate of the best m-term wavelet approximation error for the solution to the considered SPDE with respect to the space variables is higher than the convergence rates of uniform wavelet based alternatives. In this case, the attempt to develop numerical wavelet methods for SPDEs working adaptively in space direction is completely justified.

ad (T2). **Space time regularity.** If our analysis of the spatial regularity shows that, indeed, adaptivity with respect to the space coordinates pays, the next reasonable step is to develop a space time scheme for the pathwise approximation of the solutions to SPDEs of the type (1.1), which works adaptively in space direction. To this end, variants of Rothe's method suggest themselves. That is, the equations is first dicretized in time. Then, since for stability reasons one has to take an implicit scheme, in each time step an elliptic subproblem has to be solved. To this end, optimal adaptive solvers of the type mentioned above have to be employed. At the end, we need to estimate the overall error of such a scheme. We conjecture that our analysis of the Hölder regularity of the paths of the solution, considered as a stochastic process taking values in the Besov spaces from the non-linear approximation scale $(*)$, can be used for estimating the overall error of spatially adaptive variants of Rothe's method. Such an analysis has been started in [23], see also [77], but is still in its infancy.

1.2 Overview of the relevant regularity theory

In order to relate our results to the current state of research, we give a brief overview of the regularity theory which is relevant for our analysis. We begin with the significant achievements obtained from the analytic and from the semigroup approach to SPDEs. Then we discuss what is known about the regularity of (deterministic and stochastic) equations in the non-linear approximation scale ($*$). In contrast to the rest of this thesis, in this section, we do not assume that $\mathcal{O} \subset \mathbb{R}^d$ is bounded or Lipschitz.

The *analytic approach* of N.V. Krylov provides a quite complete and satisfactory L_p-theory ($p \geq 2$) for (semi-linear) parabolic SPDEs of second order on the whole space $\mathbb{R}^d$, see in particular [79,80]. Roughly speaking, the main results concerning the spatial regularity are of the form: If the free term f in Eq. (1.1)—with $\mathcal{O} = \mathbb{R}^d$ and without the boundary condition—takes values in the space $H_p^\gamma(\mathbb{R}^d)$ of Bessel potentials, and the $g = (g^k)_{k \in \mathbb{N}}$ take values in the corresponding space $H_p^{\gamma+1}(\mathbb{R}^d; \ell_2)$, then there exists a unique solution of this equation with values in $H_p^{\gamma+2}(\mathbb{R}^d)$. Thus, the spaces of Bessel potentials are suitable for the regularity analysis of SPDEs on the whole space $\mathbb{R}^d$. Recall that, for $\gamma \in \mathbb{N}$, $H_p^\gamma(\mathbb{R}^d)$ coincides with $W_p^\gamma(\mathbb{R}^d)$, the L_p-Sobolev space of order γ, see, e.g., [84, Theorem 13.3.12]. A precise definition of the spaces of Bessel potentials and their counterparts $H_p^\gamma(\mathbb{R}^d; \ell_2)$ for ℓ_2-valued functions can be found in Subsection 2.3.2.

On domains $\mathcal{O} \subset \mathbb{R}^d$ with non-empty boundary $\partial\mathcal{O}$ one is faced with (at least) two additional difficulties in order to obtain a similar theory. On the one hand, because of the behaviour of the infinitesimal differences of the driving noise, the second derivatives of the solution to Eq. (1.1) may blow up near the boundary. Then, the solution process fails to take values in $W_2^2(\mathcal{O})$. This may happen, even if the domain and the data of the equation are smooth, see, e.g., [78]. On the other hand, if the boundary of the domain is not very smooth, the singularities may become even worse caused by the influence of the shape of the boundary, see [91]. A natural way to deal with these difficulties is to consider the solution $(u(t))_{t \in [0,T]}$ as a stochastic process taking values in suitable *weighted* Sobolev spaces. These spaces allow to include solutions for which the higher-order derivatives might explode near the boundary, since this behaviour is compensated by the weight. This approach has been initiated and developed by Krylov and collaborators: first as an L_2-theory for general smooth domains [78], then as an L_p-theory ($p \geq 2$) for the half space [85,86] and subsequently also for general smooth domains [72,76]. Recently, an L_p-theory ($p \geq 2$) for SPDEs on more general bounded domains admitting Hardy's inequality, such as bounded Lipschitz domains, has been established by K.-H. Kim in [75]. The results in those publications are proven for linear equations of the form (1.1) with $L = \Lambda = 0$.

The weighted Sobolev spaces $H_{p,\theta}^\gamma(\mathcal{O}) \subset \mathcal{D}'(\mathcal{O})$ used in the theory described above are of the following form: For integer $\gamma \in \mathbb{N}$ and $\theta \in \mathbb{R}$, they consist of all measurable functions having finite norm

$$u \mapsto \left(\sum_{|\alpha| \leq \gamma} \int_{\mathcal{O}} \left| \rho_{\mathcal{O}}(x)^{|\alpha|} D^\alpha u(x) \right|^p \rho_{\mathcal{O}}(x)^{\theta-d} \, \mathrm{d}x \right)^{1/p},$$

where $\rho_{\mathcal{O}}(x)$ denotes the distance of a point $x \in \mathcal{O}$ to the boundary $\partial\mathcal{O}$ of the domain. For non-integer $\gamma > 0$ they can be characterized as complex interpolation spaces and for $\gamma < 0$ the usual duality relation holds. A precise definition can be found in Subsection 2.3.3. It turns out that this is a suitable scale to study the regularity of second-order (semi-)linear parabolic SPDEs on domains in the following sense: If the free terms f and $g = (g^k)_{k \in \mathbb{N}}$ in the equation have spatial weighted Sobolev regularity γ and $\gamma + 1$, respectively, and the initial condition u_0 is smooth enough, then the solution has spatial weighted Sobolev regularity $\gamma + 2$ (with properly chosen weight parameters $\theta \in \mathbb{R}$ on the different parts of the equation). Hence, the spatial regularity of the solution in the scale $H_{p,\theta}^\gamma(\mathcal{O})$, $\gamma > 0$, increases with the weighted Sobolev regularity of the free terms f and g of the equation. Furthermore, the weighted Sobolev norm of the solution process can be estimated from above by the corresponding weighted Sobolev norms of f, g, and u_0.

Another way to analyse the regularity of solutions of Eq. (1.1) is the *semigroup approach*. Developed mainly by G. Da Prato and J. Zabczyk in a Hilbert space framework [32], it has been generalized by Brzeźniak to M-type 2 Banach spaces [15, 16] and by J.M.A.M. van Neerven, M.C. Veraar and L. Weis to UMD Banach spaces [121, 122]—'UMD' stands for 'unconditional martingale differences'. In this approach, infinite dimensional ordinary stochastic differential equations (SDEs, for short) of the form

$$\left.\begin{aligned} \mathrm{d}U(t) + AU(t)\,\mathrm{d}t &= F(t, U(t))\,\mathrm{d}t + \Sigma(t, U(t))\,\mathrm{d}W_H(t), \qquad t \in [0, T], \\ U(0) &= u_0, \end{aligned}\right\} \tag{1.13}$$

are considered. The operator A is the infinitesimal generator of a strongly continuous analytic semigroup on a suitable Banach space E (usually $L_p(\mathcal{O})$ with $p \geq 2$), and Eq. (1.13) is interpreted as an abstract Cauchy problem. Roughly speaking, typical results are of the following form: If $(-A)$ has a 'good' H^∞-functional calculus (in the sense of McIntosh, see Section 2.4 for details) and the coefficients and non-linearities of the equations are smooth enough (where the smoothness is measured in domains of fractional powers of the leading operator), then there exists a unique strong solution in the space

$$L_q(\Omega \times (0, T); D(A)) \cap L_q(\Omega; C([0, T]; (E, D(A))_{1-\frac{1}{q}, q})).$$

Here, $D(A)$ denotes the domain of the operator A in the Banach space E, whereas $(E, D(A))_{1-\frac{1}{q}, q}$ is a real interpolation space.

For many prominent examples the domain of the leading operator A can be characterized in terms of well-studied function spaces, so that the abstract results of [121, 122] pave the way to a powerful regularity theory for SPDEs. In contrast to the theory of Krylov and collaborators, which relies mainly on hard PDE techniques, in this approach (almost) everything stands and falls with the 'good' H^∞-functional calculus of $(-A)$. To mention an example, the Dirichlet-Laplacian $\Delta_{p,w}^D$ on $L_p(\mathcal{O})$ ($p \geq 2$) has an H^∞-calculus which is good enough, provided the boundary $\partial\mathcal{O}$ of the domain is sufficiently regular—in general, $\mathcal{C}^2$ is assumed. In this case, $D(\Delta_{p,w}^D) = W_p^2(\mathcal{O}) \cap \mathring{W}_p^1(\mathcal{O})$, where $\mathring{W}_p^1(\mathcal{O})$ denotes the closure of $\mathcal{C}_0^\infty(\mathcal{O})$ in $W_p^1(\mathcal{O})$. Using these facts and the abstract theory from [121], one obtains an $L_q(L_p)$-theory for the heat equation on bounded smooth domains. It is worth noting that similar results hold also for more general second order elliptic operators, if the boundary of the domain $\mathcal{O}$ is smooth enough. Hence, equations of the form (1.1), which are analysed in the analytic approach, also fit into this framework. However, we would like to mention that in the semigroup approach certain compatibility conditions between the noise term and the leading operator A have to be fulfilled. This makes the admissible class of noises smaller compared to those that can be treated with the analytic approach, see, e.g., the discussion in [121, Section 7.4]. On the plus side, one obtains $L_q(L_p)$-regularity results with different integrability parameters q and p in time and space—even the case $q < p$ is possible. With the techniques used by Krylov and collaborators, such results could not yet been proven. Also, in the semigroup framework one can treat more general second ($2m$-th) order parabolic equations with Dirichlet and Neumann boundary conditions, stochastic Navier-Stokes equations and other important classes of equations (see, e.g., the examples presented in [121]).

In this thesis, we are explicitly interested in domains with non-smooth boundary, in particular, we focus on general bounded Lipschitz domains $\mathcal{O} \subset \mathbb{R}^d$. This covers nearly all domains of practical interest. However, the characterization of the domain of the Dirichlet-Laplacian in terms of Sobolev spaces presented above, fails to be true if the boundary of the domain $\mathcal{O}$ is assumed to be only Lipschitz. Indeed, it has been proven in [57, 58] for polygonal and polyhedral domains, and in [67] for general bounded Lipschitz domains, that $W_2^2(\mathcal{O}) \cap \mathring{W}_2^1(\mathcal{O}) \subsetneq D(\Delta_{p,w}^D)$. Moreover, to the best of our knowledge, in the case of general bounded Lipschitz domains, a characterization of $D(\Delta^D_{p,w})$ in terms of function spaces is not yet available. Thus, a direct application of the results from [121] does not lead to optimal regularity results.

To the best of our knowledge, so far there does not exist any analysis of the regularity of SPDEs in the non-linear approximation scale $(*)$ of Besov spaces—except the recent results in [22, 25, 26] by the author and collaborators, which are essential parts of this thesis.

It is worth noting that a direct application of the semigroup approach does not immediately lead to regularity results in the scale $(*)$. As already mentioned above, the semigroup framework has been used in [121] to derive regularity results in L_p-Sobolev and L_p-Besov spaces ($p \geq 2$) on sufficiently smooth domains $\mathcal{O} \subseteq \mathbb{R}^d$. The cornerstone for this theory is a generalization of Itô's stochastic integration theory to UMD Banach spaces, see Section 2.2 for details. However, for $\alpha > d(p-1)/p$, the scale $(*)$ does not consist of Banach spaces, but of *quasi*-Banach spaces. Thus, a direct application of the semigroup approach in order to obtain (sufficiently high) regularity in the scale $(*)$ requires (at least!) a fully-fledged theory of stochastic integration in proper classes of quasi-Banach spaces which is not yet available.

We also want to mention that by the same reason, we can not expect direct results from the so called *variational approach* for SPDEs initiated by E. Pradoux in [101]; we also refer to [104, Chapter 4] and the literature therein for more details. This approach has been designed particularly for the treatment of non-linear SPDEs and uses a Gelfand triple setting. In particular, the state space of the solution process needs to be a reflexive Banach space V which is continuously embedded into a Hilbert space E. It is known that any Besov space $B^\alpha_{\tau,\tau}(\mathcal{O})$ from the scale $(*)$ with $p = 2$ is continuously embedded in the Hilbert space $L_2(\mathcal{O})$. However, as already mentioned, for $\alpha > d/2$, $B^\alpha_{\tau,\tau}(\mathcal{O})$ is just a quasi-Banach space which is not reflexive. Since the reflexivity and the Banach space property are essential in this framework, we can not obtain regularity results in the non-linear approximation scale $(*)$ by a direct application of the abstract results within this approach.

However, as already mentioned in Section 1.1, the non-linear approximation scale $(*)$ has been already used for analysing the regularity of solutions to *deterministic* partial differential equations. First results on the regularity of the Dirichlet problem for harmonic functions and of the Poisson equation on general bounded Lipschitz domains in the Besov spaces from $(*)$ have been obtained by S. Dahlke and R.A. DeVore in [38]. Several extensions followed: In [34] elliptic boundary value problems with variable coefficients are analysed. The special cases of polygonal and polyhedral domains have been considered in [35] and in [36], respectively. Also, equations on smooth and polyhedral cones have been considered, see [40]. Extensions to deterministic parabolic equations have been studied in [3–5]. Simultaneously, P. Grisvard shows in [57, 58] that the Sobolev regularity of solutions to elliptic and parabolic equations on non-smooth and non-convex domains is generically limited from above. Bringing those results together shows that, in general, solutions to deterministic partial differential equations on non-smooth and non-convex domains have the behaviour described by (1.10). Thus, in this case, the decay rate of the best m-term wavelet approximation error is higher than the convergence rate of wavelet based uniform approximation methods (see Section 1.1 for details).

1.3   The thesis in a nutshell

Framework: the L_p-theory from the analytic approach

In the previous section, we explained that the abstract results from the semigroup approach and from the variational approach can not be used directly to obtain regularity results for SPDEs in the non-linear approximation scale $(*)$. Therefore, we take an indirect way to prove regularity in $(*)$ of the solutions to SPDEs of the form (1.1). Our analysis takes place in the framework of the analytic approach. We borrow (and expand) the L_p-theory for linear SPDEs from [75], which gives us the existence and uniqueness of a solution to Eq. (1.1) on general bounded Lipschitz domains $\mathcal{O} \subset \mathbb{R}^d$. Then, we analyse the spatial Besov regularity, that is topic (T1), and the Hölder regularity of the paths, that is topic (T2), of this solution. We start by proving a fundamental embedding of weighted Sobolev spaces into Besov spaces from the non-linear approximation scale $(*)$.

Embeddings of weighted Sobolev spaces into Besov spaces

The solutions to the linear SPDEs considered in [75] are elements of special classes $\mathfrak{H}_{p,\theta}^\gamma(\mathcal{O}, T)$, consisting of certain predictable p-Bochner integrable $H_{p,\theta-p}^\gamma(\mathcal{O})$-valued stochastic processes. In particular,

$$\mathfrak{H}_{p,\theta}^\gamma(\mathcal{O}, T) \hookrightarrow L_p(\Omega \times [0, T]; H_{p,\theta-p}^\gamma(\mathcal{O})). \tag{1.14}$$

('$\hookrightarrow$' means 'continuously linearly embedded'.) Hence, one way to extract regularity results in the non-linear approximation scale from this theory, is to prove an embedding of weighted Sobolev spaces into Besov spaces from $(*)$. This idea is underpinned by the fact that, in the deterministic setting, weighted Sobolev estimates have been used to establish Besov regularity in the scale $(*)$ for the solutions of elliptic boundary value problems, such as the Dirichlet problem for harmonic functions and the Poisson equation, see, e.g., [38]. This has been performed by estimating the wavelet coefficients of the unknown solution by means of weighted Sobolev (semi-)norms. Then, by using the equivalences of Besov norms and weighted sequence norms of wavelet coefficients, the desired Besov estimates were established.

Using similar techniques, we can prove that for arbitrary bounded Lipschitz domains $\mathcal{O} \subset \mathbb{R}^d$ and parameters $p \in [2, \infty)$ and $\gamma, \nu \in (0, \infty)$,

$$H_{p,d-\nu p}^\gamma(\mathcal{O}) \hookrightarrow B_{\tau,\tau}^\alpha(\mathcal{O}), \qquad \frac{1}{\tau} = \frac{\alpha}{d} + \frac{1}{p}, \qquad \text{for all} \qquad 0 < \alpha < \min\left\{\gamma, \nu\frac{d}{d-1}\right\}, \tag{1.15}$$

see Theorem 4.7. Our proof for integer $\gamma \in \mathbb{N}$ follows the line of the proof of [38, Theorem 3.2]. Additionally we use and prove the following embedding of weighted Sobolev spaces into Sobolev spaces without weights:

$$H_{p,d-\nu p}^\gamma(\mathcal{O}) \hookrightarrow \mathring{W}_p^{\gamma \wedge \nu}(\mathcal{O}), \tag{1.16}$$

which holds under the same requirements on the parameters and the shape of the domain (Proposition 4.1). By using complex interpolation we are able to prove Embedding (1.15) for arbitrary $\gamma > 0$ (Theorem 4.7). It is worth noting that this generalization has been proven in [26, Theorem 6.9] by the author and collaborators in a different more direct way without using interpolation methods.

The impact of (1.15) is obvious: Up to a certain amount, the analysis of the spatial regularity of SPDEs in the scale $(*)$ can be traced back to the analysis of the weighted Sobolev regularity of the solutions. In other words, every result on the weighted Sobolev regularity of SPDEs automatically encodes a statement about the Besov regularity in the scale $(*)$.

(T1) Spatial regularity in the non-linear approximation scale

As mentioned above, in this thesis, the solutions to SPDEs of the form (1.1) are elements of the classes $\mathfrak{H}_{p,\theta}^\gamma(\mathcal{O}, T)$ with $p \subset [2, \infty)$, $\gamma, \theta \in \mathbb{R}$. Since

$$\theta - p = d - \left(1 + \frac{d-\theta}{p}\right)p,$$

combining the embeddings (1.14) and (1.15) shows that

$$\mathfrak{H}^{\gamma}_{p,\theta}(\mathcal{O},T) \hookrightarrow L_p(\Omega \times [0,T]; B^{\alpha}_{\tau,\tau}(\mathcal{O})), \quad \frac{1}{\tau} = \frac{\alpha}{d} + \frac{1}{p}, \quad \text{for all } 0 < \alpha < \gamma \wedge \left(1 + \frac{d-\theta}{p}\right)\frac{d}{d-1}. \quad (1.17)$$

In Chapter 5 we use this embedding to prove spatial Besov regularity in the scale $(*)$ for linear and semi-linear SPDEs on general bounded Lipschitz domains $\mathcal{O} \subset \mathbb{R}^d$.

Linear equations

The L_p-theory developed in [75] provides existence and uniqueness of solutions $u \in \mathfrak{H}^{\gamma}_{p,\theta}(\mathcal{O},T)$, $p \in [2,\infty)$, $\gamma, \theta \in \mathbb{R}$, for a wide class of linear second order stochastic parabolic differential equations of the form (1.1) with vanishing L and Λ. Applying Embedding (1.17) proves that

$$u \in L_p(\Omega \times [0,T]; B^{\alpha}_{\tau,\tau}(\mathcal{O})), \quad \frac{1}{\tau} = \frac{\alpha}{d} + \frac{1}{p}, \quad \text{for all} \quad 0 < \alpha < \gamma \wedge \left(1 + \frac{d-\theta}{p}\right)\frac{d}{d-1}, \quad (1.18)$$

see Theorem 5.2. Hence, we have found an

$$\alpha^* := \min\left\{\gamma, \left(1 + \frac{d-\theta}{p}\right)\frac{d}{d-1}\right\} > 0,$$

such that for all $0 < \alpha < \alpha^*$ and $1/\tau = \alpha/d + 1/p$, the solution u to the linear SPDEs as discussed in [75] is contained in the space of (equivalence classes of) p-integrable $B^{\alpha}_{\tau,\tau}(\mathcal{O})$-valued stochastic processes. The precise conditions on the weight parameter $\theta \in \mathbb{R}$, for which (1.18) holds, can be found in the statement of our main result, Theorem 5.2. For example, in the two-dimensional case, we can choose $p = 2$, $\gamma = 2$ and $\theta = d = 2$, which yields

$$u \in L_2(\Omega \times [0,T]; B^{\alpha}_{\tau,\tau}(\mathcal{O})), \quad \frac{1}{\tau} = \frac{\alpha}{2} + \frac{1}{2}, \quad \text{for all} \quad 0 < \alpha < 2.$$

Our result together with the analysis of the maximal Sobolev regularity of SPDEs in [92] shows that, in general, on bounded Lipschitz domains $\mathcal{O} \subset \mathbb{R}^d$ which are non-convex at the singularities of $\partial\mathcal{O}$, the solutions to the linear SPDEs considered in [75] behave as described in (1.11). By our exposition in Section 1.1, this is a clear theoretical justification for the design of spatially adaptive wavelet schemes for linear SPDEs. For the detailed analysis and several examples we refer to Section 5.1.

Semi-linear equations

Many physical or chemical systems are described by equations, which are rather *non-linear*. Thus, it is an immediate question whether the results presented above can be extended to non-linear SPDEs. As a first step in this direction we consider *semi-linear* equations. That is, we consider equations of the type (1.1) with Lipschitz continuous non-linearities L and Λ.

As before, we use Embedding (1.17) to prove spatial Besov regularity in the scale $(*)$. Since there is no L_p-theory for semi-linear SPDEs on bounded Lipschitz domains, we first prove existence and uniqueness of solutions in the classes $\mathfrak{H}^{\gamma}_{p,\theta}(\mathcal{O},T)$, see Theorem 5.13. We assume that the non-linearities L and Λ in Eq. (1.1) fulfil suitable Lipschitz conditions (Assumption 5.9), such that our equation can be interpreted as a disturbed linear equation. Then, by using fixed point arguments, see Lemma 5.16, we obtain existence and uniqueness of a solution $u \in \mathfrak{H}^{\gamma}_{p,\theta}(\mathcal{O},T)$ to Eq. (1.1), which by (1.17) automatically fulfils (1.18). In this way, spatial regularity in the non-linear approximation scale $(*)$ can be established also for semi-linear SPDEs, see our main result in Theorem 5.15.

(T2) Space time regularity

After we have proven that the solutions $u \in \mathfrak{H}_{p,\theta}^{\gamma}(\mathcal{O}, T)$ to linear and semi-linear SPDEs of the form (1.1) can be considered as a $B_{\tau,\tau}^{\alpha}(\mathcal{O})$-valued stochastic processes for $0 < \alpha < \alpha^*$, where $1/\tau = \alpha/d + 1/p$, we can move on to the second main topic in this thesis: The analysis of the Hölder regularity of the paths of this process, which will be presented in Chapter 6.

The L_p-theory developed in [75] already provides Hölder estimates for elements of the classes $\mathfrak{H}_{p,\theta}^{\gamma}(\mathcal{O}, T)$, considered as stochastic processes with values in weighted Sobolev spaces. In particular, it has been shown therein that for $u \in \mathfrak{H}_{p,\theta}^{\gamma}(\mathcal{O}, T)$ and $2/p < \tilde{\beta} < \beta \le 1$,

$$\|u\|_{\mathcal{C}^{\tilde{\beta}/2-1/p}([0,T];H_{p,\theta-(1-\beta)p}^{\gamma-\beta}(\mathcal{O}))} < \infty \qquad \mathbb{P}\text{-a.s.,} \tag{1.19}$$

where for any quasi-Banach space $(E, \|\cdot\|_E)$, $(\mathcal{C}^{\kappa}([0,T];E), \|\cdot\|_{\mathcal{C}^{\kappa}([0,T];E)})$ denotes the space of κ-Hölder continuous E-valued functions on $[0,T]$, see Subsection 2.1.4 for a precise definition. An immediate idea is to use the embedding (1.15) and obtain Hölder regularity for the paths of the solutions $u \in \mathfrak{H}_{p,\theta}^{\gamma}(\mathcal{O}, T)$ considered as stochastic processes taking values in the Besov spaces from the scale (*). However, since the Hölder regularity in (1.19) depends on the summability parameter p used to measure the regularity with respect to the space variables and because of the restrictions on the weight parameters $\theta \in \mathbb{R}$ needed in [75] to establish existence of solutions in the classes $\mathfrak{H}_{p,\theta}^{\gamma}(\mathcal{O}, T)$, this does not yield satisfactory results—we refer to the introduction of Chapter 6 for more details.

We overcome these difficulties by using the following strategy. Instead of $\mathfrak{H}_{p,\theta}^{\gamma}(\mathcal{O}, T)$, we consider their counterparts $\mathfrak{H}_{p,\theta}^{\gamma,q}(\mathcal{O}, T)$, which consist of certain q-integrable $H_{p,\theta-p}^{\gamma}(\mathcal{O})$-valued stochastic processes, where the integrability parameter q in time direction (and with respect to $\omega \in \Omega$) is explicitly allowed to be greater than the summability parameter p used to measure the smoothness with respect to the space variables. We first prove that for $u \in \mathfrak{H}_{p,\theta}^{\gamma,q}(\mathcal{O}, T)$ with $2 \le p \le q < \infty$, $\gamma \in \mathbb{N}$ and $2/q < \tilde{\beta} < \beta \le 1$,

$$\|u\|_{\mathcal{C}^{\tilde{\beta}/2-1/q}([0,T];H_{p,\theta-(1-\beta)p}^{\gamma-\beta}(\mathcal{O}))} < \infty \qquad \mathbb{P}\text{-a.s.,}$$

see Theorem 6.1. In particular, the Hölder regularity of the paths does not depend on the summability parameter p with respect to the space variables. Therefore, even if the restrictions from [75] on the weight parameter θ have to be imposed, satisfactory Hölder estimates for the paths of elements $u \in \mathfrak{H}_{p,\theta}^{\gamma,q}(\mathcal{O}, T)$, considered as stochastic processes with state spaces from the scale (*), are possible (Theorem 6.2).

However, if we want to apply these results in order to obtain improved space time regularity of the solutions to SPDEs, we have to prove that—under suitable assumptions on the data of the considered equation—the solution lies in $\mathfrak{H}_{p,\theta}^{\gamma,q}(\mathcal{O}, T)$ where q and p are explicitly allowed to differ. In other words, we need to extend the L_p-theory from [75] to an $L_q(L_p)$-theory for SPDEs with $q \ne p$. In this thesis we prove a first $L_q(L_p)$-thoery result for the stochastic heat equation on general bounded Lipschitz domains (Theorem 6.11). Our proofs rely on a combination of the semigroup approach and the analytic approach. From the semigroup approach, we obtain the existence of a solution with low weighted Sobolev regularity (Proposition 6.12). Using techniques from the analytic approach we can lift this regularity, if we can increase the regularity of the free terms (Theorem 6.7). At this point, when merging results from the two different different approaches, we will need the isomorphy between the spaces $H_{p,\theta}^{\gamma}(\mathcal{O}; \ell_2)$, which are central within the analytic approach, and the corresponding spaces $\Gamma(\ell_2, H_{p,\theta}^{\gamma}(\mathcal{O}))$ of γ-radonifying operators from ℓ_2 to $H_{p,\theta}^{\gamma}(\mathcal{O})$. This will be proven in Subsection 2.3.3, see Theorem 2.54.

Finally, we can bring those results together proving Hölder regularity of the paths of the solution $u \in \mathfrak{H}_{p,\theta}^{\gamma,q}(\mathcal{O}, T)$ to the stochastic heat equation, considered as a stochastic process with values in the Besov spaces from the scale $(*)$. In particular, we prove that, under suitable assumptions on the data of the equation,

$$\|u\|_{\mathcal{C}^{\tilde{\beta}/2-1/q}([0,T];B^{\alpha}_{\tau,\tau}(\mathcal{O}))} < \infty \qquad \mathbb{P}\text{-a.s.,}$$

where

$$\frac{2}{q} < \tilde{\beta} < 1, \qquad \frac{1}{\tau} = \frac{\alpha}{d} + \frac{1}{p}, \qquad \text{and} \qquad 0 < \alpha < (1 - \tilde{\beta})\frac{d}{d-1}.$$

For the precise formulation of our main result on space time regularity, which includes also estimates of the Hölder-Besov norm of the solution by the weighted Sobolev norms of the free terms, we refer to Theorem 6.17.

1.4 Outline

This thesis starts with some preliminaries (Chapter 2). First we fix some notational and conceptual conventions in Section 2.1. Then, in Section 2.2, we give a brief inside into the theory of stochastic integration in UMD Banach spaces as developed recently in [120]. In this context, we also discuss some geometric properties of Banach spaces, like 'type' and 'UMD property', and the class of γ-radonifying operators. Afterwards, in Section 2.3, we introduce and discuss some properties of relevant function spaces, pointing out several known relationships between them. In particular, in Subsection 2.3.3, we focus on the weighted Sobolev spaces $H^{\gamma}_{p,\theta}(G)$ and their counterparts $H^{\gamma}_{p,\theta}(G;\ell_2)$ for ℓ_2-valued functions, which play an important role within the analytic approach ($G \subset \mathbb{R}^d$ is an arbitrary domain with non-empty boundary). Section 2.4 deals with semigroups of linear operators. We mainly focus on analytic semigroups and on the notion of H^{∞}-calculus, which is relevant within the semigroup approach for SPDEs. We also consider the class of variational operators.

Chapter 3 is concerned with the L_p-theory for linear SPDEs in weighted Sobolev spaces, recently developed in [75] within the analytic approach. The analysis therein takes place in the stochastic parabolic weighted Sobolev spaces $\mathfrak{H}^{\gamma}_{p,\theta}(G,T)$, $p \in [2,\infty)$, $\gamma, \theta \in \mathbb{R}$. In Section 3.1 we introduce and discuss some properties of these spaces (and of their generalizations $\mathfrak{H}^{\gamma,q}_{p,\theta}(G,T)$, $q \in [2,\infty)$). We also fix some other notation, which is common within the analytic approach. Afterwards, in Section 3.2, we present the main results from the aforementioned L_p-theory. We restrict ourselves to the case of bounded Lipschitz domains. The solution concept borrowed from [75] is introduced in Definition 3.10 and it is related to the concept of weak solutions, as it is used within the semigroup approach, in Proposition 3.18.

In Chapter 4 we leave the SPDE framework for a moment and prove Embedding (1.15) of weighted Sobolev spaces on bounded Lipschitz domains $\mathcal{O} \subset \mathbb{R}^d$ into Besov spaces from the non-linear approximation scale $(*)$, see Theorem 4.7. We also prove Embedding (1.16), see Proposition 4.1. From the latter, we can conclude that the elements of weighted Sobolev spaces are zero at the boundary in a well-defined sense, see Corollary 4.2 and Remark 4.3 for details.

Chapter 5 is devoted to the spatial regularity of SPDEs in the scale $(*)$ of Besov spaces, i.e., topic (T1). In Section 5.1, we state and prove our main result concerning linear equations, Theorem 5.2. We also present several examples and discuss the results from the point of view of approximation theory and numerical analysis. In the subsequent Section 5.2 we consider semi-linear equations. We first prove the existence of solutions in the classes $\mathfrak{H}^{\gamma}_{p,\theta}(\mathcal{O},T)$, $p \in [2,\infty)$, $\gamma, \theta \in \mathbb{R}$, under suitable assumptions on the non-linearities, see Theorem 5.13. Then, we prove our main result concerning the spatial regularity of semi-linear SPDEs in the scale $(*)$, see Theorem 5.15.

The final Chapter 6 is concerned with the space time regularity of the solution to the stochastic heat equation on bounded Lipschitz domains, i.e., with topic (T2). In Section 6.1 we analyse the Hölder regularity of the paths of elements from $\mathfrak{H}^{\gamma,q}_{p,\theta}(\mathcal{O},T)$: first, considered as stochastic processes taking values in weighted Sobolev spaces (Theorem 6.1), and, subsequently, considered as stochastic processes with state spaces from the non-linear approximation scale (Theorem 6.2). We are particularly interested in the case $q \neq p$. Afterwards, in Section 6.2 we show that the spaces $\mathfrak{H}^{\gamma,q}_{p,\theta}(\mathcal{O},T)$ with $q \geq p \geq 2$ are suitable for the analysis of SPDEs in the following sense: If we have a solution $u \in \mathfrak{H}^{\gamma,q}_{p,\theta}(\mathcal{O},T)$ with low regularity $\gamma \geq 0$, but the free terms f and g have high $L_q(L_p)$-regularity, then we can lift up the regularity of the solution (Theorem 6.7). Finally, in Section 6.3 we prove the existence and uniqueness of a solution in the class $\mathfrak{H}^{\gamma,q}_{p,\theta}(\mathcal{O},T)$ to the stochastic heat equation (Theorem 6.11). Combined with the results mentioned above, this yields our main result on the space time regularity of the stochastic heat equation, Theorem 6.17.

A short German summary of this thesis starts on page 131. A list of notation can be found starting on page 137 and an index begins on page 151.

Chapter 2

Preliminaries

In this chapter we present definitions and results needed later on for our analysis. In the first section we fix some conceptual and notational conventions from different mathematical areas. Afterwards, we give a brief inside into the theory of stochastic integration in Banach spaces developed mainly in [120] (Section 2.2). In this context we will also discuss some geometric Banach space properties and the class of γ-radonifying operators. In Section 2.3 we will introduce the function spaces appearing in this thesis and discuss and prove some of their properties which are relevant for the subsequent analysis. Finally, Section 2.4 is devoted to analytic semigroups and the concept of H^∞-calculus, and to variational operators.

2.1 Some conventions

In order to guarantee conceptual clarity, in this section we summarize the conventions made in this thesis. We give a fast overview of the notation and the basic concepts we will use later on. We start with classes of bounded operators. Then, we consider domains in $\mathbb{R}^d$ and present the definitions of different classes of domains. In particular, we substantiate the notion of a bounded Lipschitz domain, which is central in this thesis. Afterwards, we recall the basics from (quasi-) Banach space valued measure and integration theory. We continue with different aspects from probability theory and the underlying probabilistic setting. Then, we strike the subject of real and complex valued functions and distributions. At this point, we want to emphasize that in this thesis, unless explicitly stated otherwise, *functions and distributions are meant to be real-valued*. Finally, we present some miscellaneous notation. In the course of this thesis, the reader is invited to use the list of notations on page 137 and the index on page 151 and come back to this section whenever more conceptual clarity is needed.

2.1.1 Bounded operators

Let $(E, \|\cdot\|_E)$ and $(F, \|\cdot\|_F)$ be two real normed spaces. We write $\mathcal{L}(E, F)$ for the space of all linear and bounded operators from E to F, endowed with the classical norm

$$\|R\|_{\mathcal{L}(E,F)} := \sup_{x \in E, \|x\|_E \leq 1} \|Rx\|_F, \qquad R \in \mathcal{L}(E, F).$$

If $F = E$ we use the common abbreviation $\mathcal{L}(E) := \mathcal{L}(E, E)$. $E^* := \mathcal{L}(E, \mathbb{R})$ denotes the *dual space* of E. We will use the notation

$$\langle x^*, x \rangle := \langle x^*, x \rangle_{E^* \times E} := x^*(x), \qquad x^* \in E^*, x \in E,$$

for the dual pairing. The *adjoint* $R^* \in \mathcal{L}(F^*, E^*)$ of an operator $R \in \mathcal{L}(E, F)$ is uniquely determined by

$$\langle R^* y^*, x \rangle_{E^* \times E} = \langle y^*, Rx \rangle_{F^* \times F}, \qquad y^* \in F^*, x \in E.$$

If $(H, \langle \cdot, \cdot \rangle_H)$ is a real Hilbert space, we usually identify H and H^* via the Riesz isometric isomorphism $H \ni h \mapsto \langle h, \cdot \rangle_H \in H^*$ with

$$\langle h, \cdot \rangle_H : H \to \mathbb{R}$$
$$g \mapsto \langle h, g \rangle_H.$$

If H is implicitly given by the context, we write $\langle \cdot, \cdot \rangle := \langle \cdot, \cdot \rangle_H$ for short. Assume that $(U, \langle \cdot, \cdot \rangle_U)$ is a further real separable Hilbert space. Following [32] and [104] we write $\mathcal{L}_1(H, U)$ for the space of *nuclear operators* and $\mathcal{L}_2(H, U)$ for the space of *Hilbert-Schmidt operators* from H to U, see also (2.10). We will also use the common abbreviations $\mathcal{L}_1(H)$ and $\mathcal{L}_2(H)$, respectively, if $U = H$.

Recall that a *quasi-normed space* $(E, \|\cdot\|_E)$ is a vector space E endowed with a map $\|\cdot\|_E : E \to [0, \infty)$, which is positive definite and homogeneous (as a norm) but fails to fulfil the triangle inequality. Instead, there exists a constant C, which is allowed to be greater than one, such that

$$\|x + y\|_E \leq C\big(\|x\|_E + \|y\|_E\big), \qquad x, y \in E.$$

Such a map is called a *quasi-norm*. A *quasi-Banach space* is a quasi-normed space which is complete with respect to the quasi-metric $d(x, y) := \|x - y\|_E$, $x, y \in E$. We will use the notations from above also in the case of quasi-normed spaces, whenever it makes sense.

2.1.2 Domains in $\mathbb{R}^d$

Throughout this thesis, G will denote an arbitrary *domain* in $\mathbb{R}^d$, i.e., an open and connected subset of the d-dimensional Euclidian space $\mathbb{R}^d$ ($d \geq 2$). If G has a non-empty boundary, we will denote it by ∂G. In this case, we will write $\rho(x) := \rho_G(x) := \text{dist}(x, \partial G)$ for the distance of a point $x \in G$ to the boundary ∂G. Furthermore, in this thesis, $\mathcal{O}$ **will always denote a bounded Lipschitz domain in $\mathbb{R}^d$.** Let us be more precise.

Definition 2.1. We call a bounded domain $\mathcal{O} \subset \mathbb{R}^d$ a *Lipschitz domain* if, and only if, for any $x_0 = (x_0^1, x_0') \in \partial \mathcal{O}$, there exists a Lipschitz continuous function $\mu_0 : \mathbb{R}^{d-1} \to \mathbb{R}$ such that, upon relabelling and reorienting the coordinate axes if necessary, we have

(i) $\mathcal{O} \cap B_{r_0}(x_0) = \{x = (x^1, x') \in B_{r_0}(x_0) : x^1 > \mu_0(x')\}$, and

(ii) $|\mu_0(x') - \mu_0(y')| \leq K_0 |x' - y'|$, for any $x', y' \in \mathbb{R}^{d-1}$,

where r_0, K_0 are independent of x_0.

Some results will be also formulated for domains of the following class.

Definition 2.2. Let G be a domain in $\mathbb{R}^d$ with non-empty boundary ∂G. We say that G satisfies the *outer ball condition* if for each $x \in \partial G$, there exists an $r = r(x) > 0$ and a point $x_1 = x_1(x) \in \mathbb{R}^d$, such that

$$B_{r(x)}(x_1) \subset (\mathbb{R}^d \setminus G) \qquad \text{and} \qquad x \in \partial B_{r(x)}(x_1). \tag{2.1}$$

G satisfies a *uniform outer ball condition* if there exists an $R > 0$, such that for all $x \in \partial G$, $r(x) = R$ can be chosen in (2.1).

We will sometimes compare our results for SPDEs on bounded Lipschitz domains with results which can be proven for bounded domains of class $\mathcal{C}_u^1$. Equations on domains of this class have been analysed by N.V. Krylov and collaborators, see, e.g., [72,76]. We recall the definition given in [72, Assumption 2.1]. It is worth noting that the conditions imposed on the diffeomorphism Ψ therein and its inverse Ψ^{-1} are not symmetric. We fix a function $\tilde{\kappa}_0$ defined on $[0,\infty)$ such that $\tilde{\kappa}_0(\varepsilon) \downarrow 0$ for $\varepsilon \downarrow 0$. Furthermore, $\frac{\partial}{\partial x^j}\Psi^{(i)}$ denotes the classical partial derivative of the i-th coordinate of a function $\Psi : G \subseteq \mathbb{R}^d \to \mathbb{R}^d$ with respect to the j-th variable x^j, $i,j \in \{1,\dots,d\}$.

Definition 2.3. We call a domain $G \subset \mathbb{R}^d$ *of class $\mathcal{C}_u^1$* or simply a *$\mathcal{C}_u^1$-domain*, if there exist constants $r_0, K_0 > 0$ such that for any $x_0 \in \partial G$, there exists a one-to-one continuously differentiable Ψ from $B_{r_0}(x_0)$ onto a domain $J \subset \mathbb{R}^d$ such that

(i) $J_+ := \Psi(B_{r_0}(x_0) \cap G) \subset \mathbb{R}_+^d := \{y = (y^1, y') \in \mathbb{R}^d : y^1 > 0\}$ and $\Psi(x_0) = 0$;

(ii) $\Psi(B_{r_0}(x_0) \cap \partial G) = J \cap \{y \in \mathbb{R}^d : y^1 = 0\}$;

(iii) $\sup_{x \in B_{r_0}(x_0)} \left(|\Psi(x)| + \sum_{i,j=1}^d \left| \frac{\partial}{\partial x^j}(\Psi^{(i)}(x)) \right| \right) \le K_0$ and $\left| \Psi^{-1}(y_1) - \Psi^{-1}(y_2) \right| \le K_0 |y_1 - y_2|$ for any $y_1, y_2 \in J$;

(iv) $\sum_{i,j=1}^d \left| \frac{\partial}{\partial x^j}\Psi^{(i)}(x_1) - \frac{\partial}{\partial x^j}\Psi^{(i)}(x_2) \right| \le \tilde{\kappa}_0(|x_0 - x_1|)$ for any $x_1, x_2 \in B_{r_0}(x_0)$.

2.1.3 Measurable mappings and L_p-spaces

Let $(\mathfrak{M}, \mathcal{A}, \mu)$ be a σ-finite measure space and let $(E, \|\cdot\|_E)$ be a Banach space. We call a function $u : \mathfrak{M} \to E$ *$\mathcal{A}$-simple*, if it has the form $u = \sum_{k=1}^K \mathbb{1}_{A_k} x_k$ with $A_k \in \mathcal{A}$ and $x_k \in E$ for $1 \le k \le K < \infty$. A function $u : \mathfrak{M} \to E$ is called *strongly $\mathcal{A}$-measurable*, if there exists a sequence $(f_n)_{n \in \mathbb{N}}$ of $\mathcal{A}$-simple functions approximating f pointwise in $\mathfrak{M}$. It is well-known that, if E is separable, a function $u : \mathfrak{M} \to E$ is strongly $\mathcal{A}$-measurable if, and only if, it is $\mathcal{A}/\mathcal{B}(E)$-*measurable* in the classical sense, i.e., if $u^{-1}(B) \in \mathcal{A}$ for all $B \in \mathcal{B}(E)$, where $\mathcal{B}(E)$ denotes the Borel σ-field on E. In this case, we also say u is *$\mathcal{A}$-measurable* for short. Two strongly $\mathcal{A}$-measurable functions which agree μ-almost everywhere on $\mathfrak{M}$ are said to be *μ-versions* or simply *versions* of each other. For $p \in (0,\infty)$, $L_p(\mathfrak{M}, \mathcal{A}, \mu; E)$ denotes the space of all (μ-equivalence classes of) strongly $\mathcal{A}$-measurable functions $u : \mathfrak{M} \to E$ such that

$$\|u\|_{L_p(\mathfrak{M},\mathcal{A},\mu;E)} := \left(\int_{\mathfrak{M}} \|u\|_E^p \, d\mu \right)^{\frac{1}{p}} < \infty, \tag{2.2}$$

the integral being understood as a Lebesgue integral (see, e.g., [111]). As usual, we follow the convention that a μ-equivalence class $[u] \in L_p(\mathfrak{M}, \mathcal{A}, \mu; E)$ contains all functions $u : \mathfrak{M} \setminus \mathfrak{M}_0 \to E$ defined on $\mathfrak{M}$ except a μ-null set $\mathfrak{M}_0 \in \mathcal{A}$, $\mu(\mathfrak{M}_0) = 0$, such that $u\mathbb{1}_{\mathfrak{M} \setminus \mathfrak{M}_0} \in [u]$. We simply write u instead of the $[u] \in L_p(\mathfrak{M}, \mathcal{A}, \mu; E)$. We will sometimes use the common abbreviations $L_p(\mathfrak{M}; E)$ and $L_p(\mathfrak{M})$ if $E = \mathbb{R}$. $u \in L_p(\mathfrak{M}; E)$ will be called *p-Bochner integrable* or simply *p-integrable*. A function $u \in L_1(\mathfrak{M}; E)$ is also called *Bochner integrable* or simply *integrable*. In this case,

$$\int_{\mathfrak{M}} u \, d\mu = \int_{\mathfrak{M}} u(x) \, \mu(dx) = \int_{\mathfrak{M}} u(x) \, d\mu(x) \in E$$

is well-defined as a Bochner integral, see, e.g., [118, Chapter 1] for details. $L_\infty(\mathfrak{M}, \mathcal{A}, \mu; E)$ (sometimes $L_\infty(\mathfrak{M}; E)$, for short) denotes the Banach space of all (μ-equivalence classes of) strongly $\mathcal{A}$-measurable functions $u : \mathfrak{M} \to E$ for which there exists a finite number $r \ge 0$ such that $\mu(\{x \in \mathfrak{M} : \|u(x)\|_E > r\}) = 0$. It is endowed with the norm

$$\|u\|_{L_\infty(\mathfrak{M};E)} := \inf \left\{ r \ge 0 : \mu(\{x \in \mathfrak{M} : \|u(x)\|_E > r\}) = 0 \right\}, \qquad u \in L_\infty(\mathfrak{M}; E).$$

For a countable set I, we write $\ell_p(I) := L_p(I, \mathscr{P}(I), \sum_{i\in I} \delta_i; \mathbb{R})$, where $\mathscr{P}(I)$ denotes the power set of I and δ_i is the common notation for the Dirac measure at the point $i \in I$. For $\mathrm{a} \in \ell_p(I)$ we write $\mathrm{a}^i := \mathrm{a}(i)$ for the i-th coordinate. The function

$$\langle \mathrm{a}, \mathrm{b}\rangle_{\ell_2(I)} := \sum_{i \in I} \mathrm{a}^i\, \mathrm{b}^i, \qquad \mathrm{a}, \mathrm{b} \in \ell_2(I),$$

defines a scalar product on the Hilbert space $\ell_2(I)$ and $|\cdot|_{\ell_2(I)} := \sqrt{\langle \cdot, \cdot \rangle_{\ell_2(I)}}$ is the corresponding norm. If $I = \mathbb{N} := \{1, 2, 3, \ldots\}$ we write $\ell_2 := \ell_2(\mathbb{N})$, and denote by $\{\mathrm{e}_k : k \in \mathbb{N}\}$ the standard orthonormal basis of ℓ_2, i.e., for $i, k \in \mathbb{N}$, $\mathrm{e}_k^i = 0$, if $i \neq k$ and $\mathrm{e}_k^k = 1$.

By λ^d we denote the Lebesgue measure on $\mathcal{B}(\mathbb{R}^d)$ and its restriction to $B \in \mathcal{B}(\mathbb{R}^d)$. When integrating with respect to λ^d we will often write shorthand $\mathrm{d}x$ instead of $\lambda^d(\mathrm{d}x)$. If a measure μ has density g with respect to the Lebesgue measure λ^d, we write $\mu = g\lambda^d$ and $g\mathrm{d}x$ instead of $g\lambda^d(\mathrm{d}x)$. Moreover, we write L_p instead of $L_p(\mathbb{R}^d, \mathcal{B}(\mathbb{R}^d), \lambda^d; \mathbb{R})$. For $f, g : G \to \mathbb{R}$, we will use the notation

$$\langle f, g \rangle := \int_G fg\, \mathrm{d}x \tag{2.3}$$

whenever $fg \in L_1(G, \mathcal{B}(G), \lambda^d; \mathbb{R})$. We say a function $f : G \to \mathbb{R}$ is *locally integrable in G*, if it is $\mathcal{B}(G)/\mathcal{B}(\mathbb{R})$-measurable and

$$\int_K |f(x)|\, \mathrm{d}x < \infty,$$

for every compact subset K of G.

If $(E, \|\cdot\|_E)$ is just a quasi-Banach space and $p \in (0, \infty)$ we use the analogous notation $L_p(\mathfrak{M}, \mathcal{A}, \mu; E)$—and the corresponding abbreviations—to denote the set of all (μ-equivalence classes of) strongly $\mathcal{A}$-measurable E-valued functions fulfilling (2.2).

2.1.4   Probabilistic setting

Throughout this thesis $(\Omega, \mathcal{F}, \mathbb{P})$ will denote a probability space.

Random variables

A strongly $\mathcal{F}$-measurable mapping u from Ω into a quasi-Banach space E will be called E-valued *random variable*. If this function is $\mathcal{F}$-simple, we will call it an *$\mathcal{F}$-simple random variable*. If E is a Banach space and $u \in L_p(\Omega; E)$ for some $p \in [1, \infty)$, we write $\mathbb{E}[u]$ for its expectation, i.e., $\mathbb{E}[u] := \int_\Omega u\, \mathrm{d}\mathbb{P}$. If $\mathbb{E}[u] = 0$ we call the random variable *centred*. A random variable $u : \Omega \to E$ is called *Gaussian* if $\langle x^*, u \rangle$ is a real-valued Gaussian random variable for any $x^* \in E^*$. The positive definite and symmetric operator $Q \in \mathcal{L}(E^*, E)$ defined via

$$E^* \ni x^* \mapsto Qx^* := \mathbb{E}\big[\langle x^*, u - \mathbb{E}[u]\rangle(u - \mathbb{E}[u])\big] \in E$$

is called *covariance operator* of the Gaussian random variable u.

Stochastic processes

Let E be a quasi-Banach space and J an arbitrary set. A *stochastic process* $u = (u(j))_{j\in J}$ on $(\Omega, \mathcal{F}, \mathbb{P})$ with index set J is a mapping $u : \Omega \times J \to E$ such that for any $j \in J$, the mapping $u(j) = u(\cdot, j) : \Omega \to E$ is strongly $\mathcal{F}$-measurable. We will sometimes use the notation $(u_j)_{j\in J}$ instead of $(u(j))_{j\in J}$. For any $\omega \in \Omega$, the map $J \ni j \mapsto u(\omega, \cdot) \in E$ is called *path* or *trajectory* of the process u. Two stochastic processes $(u(j))_{j\in J}$ and $(v(j))_{j\in J}$ on a common probability space

$(\Omega, \mathcal{F}, \mathbb{P})$ are *modifications* of each other, if $\mathbb{P}(\{\omega \in \Omega : u(\omega, j) = v(\omega, j)\}) = 1$ for each $j \in J$. If $\mathbb{P}(\{\omega \in \Omega : u(\omega, j) = v(\omega, j) \text{ for all } j \in J\}) = 1$, the two processes are called *indistinguishable*. Assume that the index set $(J, \leq)$ is partially ordered (e.g., $J = [0, T]$ with $T > 0$ or J is a separable subset of $\mathbb{R}$). A *filtration* $(\mathscr{F}_j)_{j \in J}$ on $(\Omega, \mathcal{F}, \mathbb{P})$ is an increasing family of sub-σ-fields of $\mathcal{F}$. An E-valued stochastic process $u = (u(j))_{j \in J}$ is called *adapted* to the filtration $(\mathscr{F}_j)_{j \in J}$ $((\mathscr{F}_j)_{j \in J}$-*adapted*, for short), if for any $j \in J$ the random variable $u(j)$ is strongly $\mathscr{F}_j$-measurable. Let E be a Banach space and $(J, \leq)$ be partially ordered. An $(\mathscr{F}_j)_{j \in J}$-adapted stochastic process $u : \Omega \times [0, T] \to E$ is called *martingale* with respect to $(\mathscr{F}_j)_{j \in J}$ if $u_j \in L_1(\Omega; E)$ for any $j \in J$, and for any $i, j \in J$ with $i \leq j$,

$$\mathbb{E}(u_j | \mathscr{F}_i) = u_i \qquad (\mathbb{P}\text{-a.s.}),$$

where $\mathbb{E}(u_j | \mathscr{F}_i)$ denotes the conditional expectation of u_j with respect to $\mathscr{F}_i$. If, furthermore, for some $p \in [1, \infty)$, $u_j \in L_p(\Omega; E)$ for all $j \in J$, the martingale $(u_j)_{j \in J}$ is called an L_p-*martingale*.

Regularity of paths

Let $(u(t))_{t \in [0,T]}$ be a stochastic process with index set $J := [0, T]$ taking values in a quasi-Banach space E. We will measure the smoothness of the paths of u by means of their Hölder regularity. For $\kappa \in (0, 1)$ and a quasi-Banach space $(E, \|\cdot\|_E)$ we denote by $\mathcal{C}^\kappa([0, T]; E)$ the Hölder space of continuous E-valued functions on $[0, T]$ with finite norm $\|\cdot\|_{\mathcal{C}^\kappa([0,T];E)}$ defined by

$$[u]_{\mathcal{C}^\kappa([0,T];E)} := \sup_{\substack{s,t \in [0,T] \\ s \neq t}} \frac{\|u(t) - u(s)\|_E}{|t - s|^\kappa},$$

$$\|u\|_{\mathcal{C}([0,T];E)} := \sup_{t \in [0,T]} \|u(t)\|_E,$$

$$\|u\|_{\mathcal{C}^\kappa([0,T];E)} := \|u\|_{\mathcal{C}([0,T];E)} + [u]_{\mathcal{C}^\kappa([0,T];E)}.$$

Operator valued stochastic processes

Let E_1, E_2 be two Banach spaces. An operator valued function $\Phi : \Omega \times [0, T] \to \mathcal{L}(E_1, E_2)$ is called an E_1-*strongly measurable stochastic process* if for any $x \in E_1$, the E_2 valued stochastic process $\Phi x : \Omega \times [0, T] \to E_2$, $(\omega, t) \mapsto \Phi x(\omega, t) := \Phi(\omega, t)x$ is strongly $\mathcal{F} \otimes \mathcal{B}([0, T])$-measurable. An E_1-strongly measurable stochastic process is called *adapted* to a filtration $(\mathcal{F}_t)_{t \in [0,T]}$ on $(\Omega, \mathcal{F}, \mathbb{P})$ (or, simply $(\mathcal{F}_t)_{t \in [0,T]}$-*adapted*) if for any $x_1 \in E_1$, the process Φx_1 is adapted to $(\mathcal{F}_t)_{t \in [0,T]}$.

Let $(H, \langle \cdot, \cdot \rangle_H)$ be a Hilbert space and let $(E, \|\cdot\|_E)$ be a Banach space. Let $\Phi : \Omega \times [0, T] \to \mathcal{L}(H, E)$ be an H-strongly measurable stochastic process. We write

$$\Phi^* : \Omega \times [0, T] \to \mathcal{L}(E^*, H)$$
$$(\omega, t) \mapsto \Phi^*(\omega, t) := \Phi(\omega, t)^*,$$

identifying H and its dual H^* via the Riesz isomorphism. Φ is said to *belong to* $L_2([0, T]; H)$ *scalarly almost surely* if for all $x^* \in E^*$,

$$\Phi^*(\omega, \cdot)x^* \in L_2([0, T]; H) \qquad \text{for } \mathbb{P}\text{-almost all } \omega \in \Omega.$$

Note that the exceptional set may depend on x^*. For $p \in [2, \infty)$, Φ is said to *belong to* $L_p(\Omega; L_2([0, T]; H))$ *scalarly* if for all $x^* \in E^*$,

$$\Phi^* x^* \in L_p(\Omega; L_2([0, T]; H)).$$

A stochastic process $\Phi : \Omega \times [0,T] \to \mathcal{L}(H,E)$ which belongs to $L_2([0,T];H)$ scalarly almost surely is said to *represent* a random variable $X : \Omega \to \mathcal{L}(L_2([0,T];H),E)$, if for all $f \in L_2([0,T];H)$ and $x^* \in E^*$ we have

$$\langle x^*, X(\omega)f \rangle_{E^* \times E} = \int_0^T \langle \Phi^*(\omega,t)x^*, f(t) \rangle_H \, dt \qquad \text{for } \mathbb{P}\text{-almost all } \omega \in \Omega.$$

Stochastic integration in Hilbert spaces

We assume that the reader is familiar with the issue of stochastic integration (in the sense of Itô) with respect to (cylindrical) Q-Wiener processes in the Hilbert space setting, as described, e.g., in [32] or [104]. Let $(H, \langle \cdot, \cdot \rangle_H)$ and $(U, \langle \cdot, \cdot \rangle_U)$ be two real Hilbert spaces. Furthermore, assume that $(W^Q(t))_{t \in [0,T]}$ is an H-valued Q-Wiener process with $Q \in \mathcal{L}_1(H)$. We write

$$\left(\int_0^t u(s) \, dW^Q(s) \right)_{t \in [0,T]}$$

for the stochastic Itô integral of a process $u : \Omega_T \to \mathcal{L}_2(H_0,U)$ which is stochastically integrable with respect to W^Q. Here, $(H_0, \langle \cdot, \cdot \rangle_{H_0}) := (Q^{1/2}H, \langle Q^{-1/2} \cdot, Q^{-1/2} \cdot \rangle_H)$ is the reproducing kernel Hilbert space, $Q^{-1/2}$ being the pseudo-inverse of $Q^{1/2}$. In particular, if a real valued stochastic process $(g_t)_{t \in [0,T]}$ is stochastically integrable with respect to a real valued Brownian motion $(w_t)_{t \in [0,T]}$, we write

$$\left(\int_0^t g_s \, dw_s \right)_{t \in [0,T]}$$

for the stochastic integral process. A brief overview of the extension of this theory to certain classes of Banach spaces will be given later on in Section 2.2.

Miscellaneous conventions on the probabilistic setting

In this thesis, $T > 0$ will always denote a finite time horizon and $(w_t^k)_{t \in [0,T]}$, $k \in \mathbb{N}$, will be a sequence of independent real-valued standard Brownian motions with respect to a normal filtration $(\mathcal{F}_t)_{t \in [0,T]}$ on a complete probability space $(\Omega, \mathcal{F}, \mathbb{P})$. By *normal* we mean that $(\mathcal{F}_t)_{t \in [0,T]}$ is right continuous and that $\mathcal{F}_0$ contains all $\mathbb{P}$-null sets. We will use the common abbreviation $\Omega_T := \Omega \times [0,T]$ as well as

$$\mathcal{P}_T := \sigma\Big(\big\{ F_s \times (s,t] : 0 \le s < t \le T, F_s \in \mathcal{F}_s \big\} \cup \big\{ F_0 \times \{0\} : F_0 \in \mathcal{F}_0 \big\} \Big) \subseteq \mathcal{F} \otimes \mathcal{B}([0,T])$$

for the $(\mathcal{F}_t)_{t \in [0,T]}$-*predictable σ-field*. Furthermore, we will write $\mathbb{P}_T$ for the product measure $\mathbb{P} \otimes dt$ on $\mathcal{F} \otimes \mathcal{B}([0,T])$ and for its restriction to $\mathcal{P}_T$. The abbreviation

$$L_p(\Omega_T; E) := L_p(\Omega_T, \mathcal{P}_T, \mathbb{P}_T; E), \qquad p \in (0,\infty],$$

will often be used in this thesis to denote the set of predictable p-integrable stochastic processes with values in a (quasi-)Banach space E.

2.1.5   Functions, distributions and the Fourier transform

In this thesis, unless explicitly stated otherwise, *functions and distributions are meant to be real valued*. For a domain $G \subseteq \mathbb{R}^d$ and $r \in \mathbb{N}$, $\mathcal{C}^r(G)$ denotes the space of all r-times continuously differentiable functions, whereas $\mathcal{C}(G)$ is the space of continuous functions. We will write $\mathcal{C}_0^\infty(G)$ for the set of test functions, i.e., the collection of all infinitely differentiable functions with compact support in G. For bounded domains $G \subset \mathbb{R}^d$, the notation $\mathcal{C}^r(\overline{G})$ is used for the set of all functions which are continuous on the closure $\overline{G}$ of G and possess derivatives up to and including the order $r \in \mathbb{N}$, which are continuous on G and can be extended by continuity to $\overline{G}$. $\mathcal{C}(\overline{G})$ stands for the space of continuous functions on $\overline{G}$. For a multi-index $\alpha = (\alpha^1, \ldots, \alpha^d) \in \mathbb{N}_0^d$ with $|\alpha| := \alpha^1 + \ldots + \alpha^d = r$ and an r-times differentiable function $u : G \to \mathbb{R}$, we write

$$D^{(\alpha)}u = \frac{\partial^{|\alpha|}}{\partial(x^1)^{\alpha^1} \cdots \partial(x^d)^{\alpha^d}}u$$

for the corresponding classical partial derivative. $\mathcal{S}(\mathbb{R}^d)$ denotes the Schwartz space of rapidly decreasing functions, see e.g. [108, Section 7.3]. The set of distributions on G will be denoted by $\mathcal{D}'(G)$, whereas $\mathcal{S}'(\mathbb{R}^d)$ denotes the set of tempered distributions on $\mathbb{R}^d$. The terms *distribution* and *generalized function* will be used synonymously. For the application of a distribution $u \in \mathcal{D}'(G)$ to a test function $\varphi \in \mathcal{C}_0^\infty(G)$ we write (u, φ). The same notation will be used if $u \in \mathcal{S}'(\mathbb{R}^d)$ and $\varphi \in \mathcal{S}(\mathbb{R}^d)$. Let $u \in \mathcal{D}'(G)$. If there exists a locally integrable function $f : G \to \mathbb{R}$ such that

$$(u, \varphi) = \langle f, \varphi \rangle = \int_G f(x)\varphi(x)\,dx, \quad \text{for all} \quad \varphi \in \mathcal{C}_0^\infty(G),$$

we say that the distribution u is *regular*. Since such an f is uniquely determined, we do not distinguish between u and f. For $u \in \mathcal{D}'(G)$ and a multi-index $\alpha = (\alpha^1, \ldots, \alpha^d) \in \mathbb{N}_0^d$, we write $D^\alpha u$ for the α-th *generalized*, *weak* or *distributional derivative* of u with respect to $x = (x^1, \ldots, x^d) \in G$, i.e., $D^\alpha u$ is a distribution on G, uniquely determined by the formula

$$(D^\alpha u, \varphi) := (-1)^{|\alpha|}(u, D^{(\alpha)}\varphi), \qquad \varphi \in \mathcal{C}_0^\infty(G),$$

see e.g. [108, Section 6.12]; $D^0 := \mathrm{Id}$. By making slight abuse of notation, for $m \in \mathbb{N}_0$, we write $D^m u$ for any (generalized) m-th order derivative of u and for the vector of all m-th order derivatives of u. E.g. if we write $D^m u \in E$, where E is a function space on G, we mean $D^\alpha u \in E$ for all $\alpha \in \mathbb{N}_0^d$ with $|\alpha| = m$. We also use the notation $u_{x^i} := D^{e_i}u$ and $u_{x^i x^j} := D^{e_i}D^{e_j}u$, where for $i \in \{1, \ldots, d\}$, e_i denotes the i-th unit vector in $\mathbb{R}^d$, i.e., $e_i^i = 1$ and $e_i^k = 0$ for $i \neq k$. The notation u_x (respectively u_{xx}) is used synonymously for $Du := D^1 u$ (respectively for $D^2 u$), whereas $\|u_x\|_E := \sum_i \|u_{x^i}\|_E$ (respectively $\|u_{xx}\|_E := \sum_{i,j} \|u_{x^i x^j}\|_E$). We write

$$\Delta u := \sum_{i=1}^d u_{x^i x^i}$$

whenever it makes sense. If we consider spaces of complex-valued functions and distributions we will indicate this explicitly by writing e.g. $\mathcal{S}'(\mathbb{R}^d; \mathbb{C})$ or $\mathcal{D}'(G; \mathbb{C})$. The notation $(\cdot, \cdot)$ is generalized to

$$(u, \varphi) := \int_{\mathbb{R}^d} u(x)\overline{\varphi(x)}\,dx$$

on $\mathcal{S}'(\mathbb{R}^d; \mathbb{C}) \times \mathcal{S}(\mathbb{R}^d; \mathbb{C})$. Here, $\overline{\varphi(x)}$ denotes the complex conjugate of $\varphi(x)$, $x \in \mathbb{R}^d$. The analogous meaning is given to $(\cdot, \cdot)$ on $\mathcal{D}'(G; \mathbb{C}) \times \mathcal{D}(G; \mathbb{C})$. In thesis, we denote by

$$\mathfrak{F} : \mathcal{S}'(\mathbb{R}^d; \mathbb{C}) \to \mathcal{S}'(\mathbb{R}^d; \mathbb{C})$$

the *Fourier transform* on the space of complex valued tempered distributions $\mathcal{S}'(\mathbb{R}^d; \mathbb{C})$. For $u \in \mathcal{S}'(\mathbb{R}^d; \mathbb{C})$, it is defined by

$$(\mathfrak{F}u, \varphi) := (u, \mathfrak{F}\varphi), \qquad \varphi \in \mathcal{S}(\mathbb{R}^d; \mathbb{C}),$$

where

$$\mathfrak{F}\varphi := \frac{1}{(2\pi)^{d/2}} \int_{\mathbb{R}^d} \varphi(x)e^{-i\langle x,\xi\rangle}\,\mathrm{d}\xi, \qquad \varphi \in \mathcal{S}(\mathbb{R}^d;\mathbb{C}).$$

Its inverses is denoted by $\mathfrak{F}^{-1}$. For more details on the Fourier transform of tempered distributions we refer to [108, Chapter 7].

2.1.6 Miscellaneous notation

For two quasi-normed spaces $(E_i, \|\cdot\|_{E_i})$, $i = 1, 2$, $E_1 \hookrightarrow E_2$ means that E_1 is continuously linearly embedded in E_2, i.e., there exists a linear continuous embedding $j : E_1 \to E_2$. If we want to specify the embedding, we write $E_1 \xhookrightarrow{j} E_2$. If E_1 and E_2 are normed and there exists a (topological) *isomorphism* between E_1 and E_2, i.e., a bijective linear continuous mapping from E_1 to E_2 with bounded inverse, we write $E_1 \simeq E_2$. The two spaces are then called *isomorphic*. The notation $E_1 \cong E_2$ is used if there exists an *isometric isomorphism* between E_1 and E_2, i.e., if there exists a norm preserving isomorphism from E_1 to E_2. The space $E_1 \times E_2 := \{(x_1, x_2) : x_1 \in E_1, x_2 \in E_2\}$, is endowed with the norm $\|(x_1, x_2)\|_{E_1 \times E_2} := \|x_1\|_{E_1} + \|x_2\|_{E_2}$, $(x_1, x_2) \in E_1 \times E_2$. For a compatible couple (E_1, E_2) of Banach spaces, $[E_1, E_2]_\eta$ denotes the interpolation space of exponent $\eta \in (0, 1)$ arising from the complex interpolation method, see, e.g., [13, Chapter 4]. Furthermore, the intersection $E_1 \cap E_2$ is endowed with the norm

$$\|x\|_{E_1 \cap E_2} := \max\big\{\|x\|_{E_1}, \|x\|_{E_2}\big\}, \qquad x \in E_1 \cap E_2.$$

Throughout this thesis, C denotes a positive and finite constant which may change its value with every new appearance. If we have two terms depending on a parameter u, which might be a distribution or a distribution valued process or something else, and write

$$f_1(u) \le C\,f_2(u)$$

we mean: There exists a constant $C \in (0, \infty)$, which does *not* depend on u, such that, if $f_2(u)$ makes sense and is finite, so does $f_1(u)$, and the inequality holds. We will also write

$$f_1(u) \asymp f_2(u)$$

if simultaneously

$$f_1(u) \le C\,f_2(u) \quad \text{and} \quad f_2(u) \le C\,f_1(u).$$

2.2 Stochastic integration in UMD Banach spaces

The predominant part of this thesis is based on the generalization of Itô's stochastic integration theory for operator valued stochastic processes with respect to (cylindrical) Wiener processes, where the operators are mappings from a real separable Hilbert space into another. Even more, in the most parts we use just results from finite-dimensional stochastic analysis. However, in Chapter 6, we will use regularity results in L_p-Sobolev spaces ($p \ge 2$) as derived by van Neerven, Veraar and Weis [121]. The cornerstone for this theory is a generalization of Itô's stochastic integration theory to certain classes of Banach spaces, see [120] as well as [123] and [17]. It is the aim of this section to give a brief overview of this theory and to present basic results which we will use later on.

2.2.1 Geometric properties of Banach spaces

The generalization of Itô's stochastic integration theory does not work for arbitrary Banach spaces. What we need are certain geometric properties of these spaces, which we present now. Throughout this subsection, $(\Omega, \mathcal{F}, \mathbb{P})$ denotes a probability space and $(E, \|\cdot\|_E)$ is an arbitrary Banach space. For further notation commonly used in probability theory, we refer to Subsection 2.1.4.

The first geometric property we introduce is the so-called 'UMD' property. Before we define this notion, we need to recall what is said to be a martingale difference sequence. Our definitions are taken from [18].

Definition 2.4. A sequence $(d_k)_{k\in\mathbb{N}}$ of E-valued Bochner integrable random variables is a *martingale difference sequence* relative to a filtration $(\mathcal{F}_k)_{k\in\mathbb{N}}$ if for every $k \in \mathbb{N}$, the random variable $d_k : \Omega \to \mathbb{R}$ is $\mathcal{F}_k$-measurable and

$$\mathbb{E}(d_{k+1}|\mathcal{F}_k) = 0. \tag{2.4}$$

If, additionally, $(d_k)_{k\in\mathbb{N}} \subseteq L_p(\Omega; E)$ for some $p \in [1, \infty)$, then we call $(d_k)_{k\in\mathbb{N}}$ an L_p-*martingale difference sequence*.

Definition 2.5. A Banach space $(E, \|\cdot\|_E)$ is said to be a UMD *space* (or to satisfy the UMD *property*) if for some (equivalently, for all) $p \in (1, \infty)$ there exists a constant C, depending only on p and E, such that for any L_p-martingale difference sequence $(d_k)_{k\in\mathbb{N}}$ and any finite $\{-1, 1\}$-valued sequence $(\varepsilon_k)_{k=1}^K$, one has:

$$\left(\mathbb{E}\left[\left\| \sum_{k=1}^K \varepsilon_k d_k \right\|_E^p \right] \right)^{\frac{1}{p}} \leq C \left(\mathbb{E}\left[\left\| \sum_{k=1}^K d_k \right\|_E^p \right] \right)^{\frac{1}{p}}.$$

The abbreviation 'UMD' stands for 'unconditional martingale differences'. The following well-known facts about UMD spaces will be useful later on. Their proofs, as well as a proof of the p-independence of the UMD property, can be found in [118, Chapter 12], see also the references therein.

Lemma 2.6. (i) *Banach spaces isomorphic to closed subspaces of* UMD *spaces satisfy the* UMD *property.*

(ii) *Every Hilbert space is a* UMD *space.*

(iii) *Let E be a* UMD *space and $(\mathfrak{M}, \mathcal{A}, \mu)$ be a σ-finite measure space. Then $L_p(\mathfrak{M}, \mathcal{A}, \mu; E)$ is a* UMD *space for any $p \in (1, \infty)$.*

(iv) *E is a* UMD *space if, and only if, the dual E^* is a* UMD *space.*

The second geometric property of Banach spaces required later on is the so-called 'type' of a Banach space. Before we present its definition, we need to recall what is said to be a Rademacher sequence.

Definition 2.7. A *Rademacher sequence* is a sequence $\{r_k : k \in \mathbb{N}\}$ of independent and identically distributed $\{-1, 1\}$-valued random variables with

$$\mathbb{P}(r_1 = 1) = \mathbb{P}(r_1 = -1) = 1/2.$$

In what follows, $\{r_k : k \in \mathbb{N}\}$ will always denote a Rademacher sequence.

Definition 2.8. Fix $p \in [1, 2]$. A Banach space $(E, \|\cdot\|_E)$ is said to have *type p*, if there exists a constant C such that for all finite sequences $(x_k)_{k=1}^K \subseteq E$ the following estimate holds:

$$\left(\mathbb{E} \left[\left\| \sum_{k=1}^K r_k \, x_k \right\|_E^p \right] \right)^{\frac{1}{p}} \leq C \left(\sum_{k=1}^K \|x_k\|_E^p \right)^{\frac{1}{p}}. \tag{2.5}$$

Obviously, every Banach space E has type 1. If the type of E is strictly greater than one, we say E has *non-trivial type*. In the following lemma we collect some useful and well-known facts regarding the type of Banach spaces, see, e.g., [103] and the references therein.

Lemma 2.9. (i) *If a Banach space E has type p' for some $p' \in [1, 2]$, then E has type p for all $p \in [1, p']$.*

(ii) *Every Hilbert space has type 2.*

(iii) *Let $(\mathfrak{M}, \mathcal{A}, \mu)$ be a σ-finite measure space and let $p \in [1, \infty)$. Then, $L_p(\mathfrak{M}, \mathcal{A}, \mu; \mathbb{R})$ has type $r := \min\{2, p\}$.*

(iv) *Let E_1 and E_2 be isomorphic Banach spaces and let $p \in [1, 2]$. Then, E_1 has type p if, and only if, E_2 has type p.*

(v) *Let E be a* UMD *Banach space. Then, the following assertions are equivalent:*

    (1) *E has type 2.*

    (2) *E has M-type 2, i.e., there exists a constant C such that for every E-valued L_2-martingale $(M_k)_{k \in \mathbb{N}}$ the following inequality holds:*

$$\sup_{k \in \mathbb{N}} \mathbb{E} \left[\|M_k\|_E^2 \right] \leq C \sum_{k=1}^{\infty} \mathbb{E} \left[\|M_k - M_{k-1}\|_E^2 \right],$$

with the usual convention $M_0 := 0$.

The term on the left hand side of (2.5) depends on p only up to a constant. This is due to the *Kahane-Khintchine inequality*, which is the content of the next lemma. A proof based on Lévy's inequality can be found e.g. in [118, Theorem 3.11].

Lemma 2.10. *For all $p, q \in [1, \infty)$, there exists a constant $C_{p,q}$, depending only on p and q, such that for all finite sequences $(x_k)_{k=1}^K \subseteq E$ we have*

$$\left(\mathbb{E} \left[\left\| \sum_{k=1}^K r_k \, x_k \right\|_E^p \right] \right)^{\frac{1}{p}} \leq C_{p,q} \left(\mathbb{E} \left[\left\| \sum_{k=1}^K r_k \, x_k \right\|_E^q \right] \right)^{\frac{1}{q}}.$$

This result can be extended to the case where the Rademacher sequence is replaced by a Gaussian sequence. Let us first recall what we mean by that.

Definition 2.11. A *Gaussian sequence* is a sequence $\{\gamma_k : k \in \mathbb{N}\}$ of independent real valued random variables, each of which is standard Gaussian.

In what follows, $\{\gamma_k : k \in \mathbb{N}\}$ will always denote a Gaussian sequence. A proof of the following generalization of Lemma 2.10, which is based on the central limit theorem, can be found in [118, Theorem 3.12]. We will refer to it as the *Kahane-Khintchine inequality for Gaussian sums*.

Lemma 2.12. *For all $p, q \in [1, \infty)$, there exists a constant $C_{p,q}$, depending only on p and q, such that for all finite sequences $(x_k)_{k=1}^K \subseteq E$ we have*

$$\left(\mathbb{E} \left[\left\| \sum_{k=1}^K \gamma_k \, x_k \right\|_E^p \right] \right)^{\frac{1}{p}} \leq C_{p,q} \left(\mathbb{E} \left[\left\| \sum_{k=1}^K \gamma_k \, x_k \right\|_E^q \right] \right)^{\frac{1}{q}}.$$

2.2.2 γ-radonifying operators

In this subsection we discuss the notion of γ-radonifying operators from a real Hilbert space $(H, \langle \cdot, \cdot \rangle_H)$ into a Banach space $(E, \|\cdot\|_E)$. These operators usually appear in the context of Banach space valued Gaussian random variables. It is a well-known result that an operator $Q \in \mathcal{L}(E^*, E)$ is the covariance operator of a centred E-valued Gaussian random variable if, and only if, $Q = RR^*$ with R being a γ-radonifying operator from a Hilbert space H into E, see e.g. [118, Theorem 5.12]. This class of operators is also central in the development of the notion of stochastic integration in UMD Banach spaces with respect to H-cylindrical Brownian motions of van Neerven, Veraar and Weis [120], which we will discuss in the next subsection. Our exposition follows the lines of [118, Chapter 5], see also the survey [119]. The notation used in the subsection before is still valid. Remember that $\{\gamma_k : k \in \mathbb{N}\}$ is a *Gaussian sequence*, see Definition 2.11.

Definition 2.13. A linear operator $R : H \to E$ is called γ-*summing* if

$$\|R\|_{\Gamma^\infty(H,E)} := \sup \left(\mathbb{E}\left[\left\| \sum_{k=1}^{K} \gamma_k R h_k \right\|_E^2 \right] \right)^{\frac{1}{2}} < \infty, \tag{2.6}$$

where the supremum is taken over all finite orthonormal systems $\{h_1, \ldots, h_K\} \subseteq H$. The space of γ-summing operators from H to E will be denoted by $\Gamma^\infty(H, E)$ and endowed with the norm $\|\cdot\|_{\Gamma^\infty(H,E)}$ introduced in (2.6).

Remark 2.14. **(i)** $\Gamma^\infty(H, E) \hookrightarrow \mathcal{L}(H, E)$, since for any $R \in \Gamma^\infty(H, E)$,

$$\|R\|_{\mathcal{L}(H,E)} = \sup_{\|h\|_H=1} \|Rh\|_E = \sup_{\|h\|_H=1} \left(\mathbb{E}\left[\|\gamma_1 Rh\|_E^2 \right] \right)^{\frac{1}{2}} \leq \|R\|_{\Gamma^\infty(H,E)}.$$

(ii) Let $p \in [1, \infty)$. Let $\Gamma_p^\infty(H, E)$ be the set of all linear operators $R : H \to E$ such that

$$\|R\|_{\Gamma_p^\infty(H,E)} := \sup \left(\mathbb{E}\left[\left\| \sum_{k=1}^{K} \gamma_k R h_k \right\|_E^p \right] \right)^{\frac{1}{p}} < \infty,$$

where the supremum is taken over all finite orthonormal systems $\{h_1, \ldots, h_K\} \subseteq H$. Then, by the Kahane-Khintchine inequality for Gaussian sums, see Lemma 2.12,

$$\Gamma_p^\infty(H, E) = \Gamma^\infty(H, E).$$

Moreover, the norms $\|\cdot\|_{\Gamma_p^\infty(H,E)}$ and $\|\cdot\|_{\Gamma^\infty(H,E)}$ are equivalent.

(iii) $(\Gamma^\infty(H, E), \|\cdot\|_{\Gamma^\infty(H,E)})$ is a Banach space, see, e.g., [118, Theorem 5.2].

For $h \in H$ and $x \in E$, we sometimes write $h \otimes x := \langle h, \cdot \rangle_H x \in \mathcal{L}(H, E)$ for the *rank one operator*

$$H \ni \tilde{h} \mapsto \langle h, \tilde{h} \rangle_H x \in E. \tag{2.7}$$

Furthermore, we will use the notation

$$\mathcal{L}_f(H, E) := \bigcup_{J \in \mathbb{N}} \left\{ \sum_{j=1}^{J} h_j \otimes x_j : h_j \in H, x_j \in E \text{ for } j = 1, \ldots, J \right\} \subseteq \mathcal{L}(H, E)$$

for the subspace of linear and bounded *finite rank operators*. Obviously, $\mathcal{L}_f(H, E) \subseteq \Gamma^\infty(H, E)$.

Definition 2.15. The *space* $\Gamma(H, E)$ *of γ-radonifying operators* is defined as the closure of the space of finite rank operators $\mathcal{L}_f(H, E)$ in $\Gamma^\infty(H, E)$. I.e.,

$$\Gamma(H, E) := \overline{\mathcal{L}_f(H, E)}^{\|\cdot\|_{\Gamma^\infty(H,E)}} \subseteq \Gamma^\infty(H, E).$$

Remark 2.16.   **(i)** By definition, $\Gamma(H, E)$, endowed with the norm $\|\cdot\|_{\Gamma^\infty(H,E)}$ inherited from $\Gamma^\infty(H, E)$ is a Banach space. We will use the abbreviations

$$\|R\|_{\Gamma(H,E)} := \|R\|_{\Gamma^\infty(H,E)}, \qquad R \in \Gamma(H, E),$$

and, for $p \in [1, \infty)$,

$$\|R\|_{\Gamma_p(H,E)} := \|R\|_{\Gamma_p^\infty(H,E)}, \qquad R \in \Gamma(H, E).$$

Note that for any $p \in [1, \infty)$, the norm equivalence

$$\|R\|_{\Gamma(H,E)} \asymp \|R\|_{\Gamma_p(H,E)}, \qquad R \in \Gamma(H, E) \tag{2.8}$$

holds, see also Remark 2.14(ii).

(ii) In general, $\Gamma(H, E) \subsetneq \Gamma^\infty(H, E)$. An example for a γ-summable operator which is not γ-radonifying can be found in [90]. However, if the Banach space E does not contain a closed subspace isomorphic to the space c_0 of sequences converging to zero endowed with the supremum norm, then $\Gamma(H, E) = \Gamma^\infty(H, E)$. This can be proven by using the results of Hoffmann-Jørgensen and Kwapień concerning sums of independent symmetric Banach space valued random variables [65, 88], see [119, Theorem 4.3].

Now we collect some properties and useful characterizations of $\Gamma(H, E)$. The proofs can be found in [118, Chapter 5]. We start with the ideal property of γ-radonifying operators.

Theorem 2.17. *Let $R \in \Gamma(H, E)$. Let H' be another real Hilbert space and E' be another Banach space. Then for all $U \in \mathcal{L}(E, E')$ and $S \in \mathcal{L}(H', H)$ we have $URS \in \Gamma(H', E')$ and for all $p \in [1, \infty)$,*

$$\|URS\|_{\Gamma_p(H',E')} \leq \|U\|_{\mathcal{L}(E,E')} \|R\|_{\Gamma_p(H,E)} \|S\|_{\mathcal{L}(H',H)}.$$

Often, the Hilbert space H is assumed to be separable and the following characterization of γ-radonifying operators is taken as a definition.

Theorem 2.18. *Let H be a separable real Hilbert space. Then for an operator $R \in \mathcal{L}(H, E)$ the following assertions are equivalent:*

(i) *$R \in \Gamma(H, E)$.*

(ii) *For all orthonormal bases $\{h_k : k \in \mathbb{N}\}$ of H and all $p \in [1, \infty)$ the series $\sum_{k=1}^\infty \gamma_k R h_k$ converges in $L_p(\Omega; E)$.*

(iii) *For some orthonormal basis $\{h_k : k \in \mathbb{N}\}$ of H and some $p \in [1, \infty)$ the series $\sum_{k=1}^\infty \gamma_k R h_k$ converges in $L_p(\Omega; E)$.*

In this situation, the sums in (ii) and (iii) converge almost surely and for all orthonormal bases $\{h_k : k \in \mathbb{N}\}$ of H and $p \in [1, \infty)$,

$$\|R\|_{\Gamma_p(H,E)}^p = \mathbb{E}\left[\left\|\sum_{k=1}^\infty \gamma_k R h_k\right\|_E^p\right]. \tag{2.9}$$

Remark 2.19. Let $(H, \langle \cdot, \cdot \rangle_H)$ be a separable real Hilbert space and assume that $(E, \langle \cdot, \cdot \rangle_E)$ is also a separable real Hilbert space. Furthermore, let $\{h_k : k \in \mathbb{N}\}$ be an orthonormal basis of H. Recall that

$$\mathcal{L}_2(H, E) := \left\{ T \in \mathcal{L}(H, E) : \|T\|^2_{\mathcal{L}_2(H,E)} := \sum_{k=1}^{\infty} \|Th_k\|^2_E < \infty \right\} \tag{2.10}$$

is the space of *Hilbert-Schmidt operators* from H to E. Then, $\mathcal{L}_2(H, E) = \Gamma(H, E)$ and for any $R \in \Gamma(H, E)$ we have

$$\|R\|_{\Gamma(H,E)} = \|R\|_{\mathcal{L}_2(H,E)}.$$

This is an immediate consequence of Theorem 2.18 above and Pythagoras' theorem.

If E is an L_p-space on a σ-finite measure space, $\Gamma(H, E)$ can be also characterized as follows. The proof relies on the Kahane-Khintchine inequality for Gaussian sums, see Lemma 2.12.

Theorem 2.20. *Let H be a separable real Hilbert space. Furthermore, let $(\mathfrak{M}, \mathcal{A}, \mu)$ be a σ-finite measure space and $p \in [1, \infty)$. For an operator $R \in \mathcal{L}(H, L_p(\mathfrak{M}))$ the following assertions are equivalent:*

(i) *$R \in \Gamma(H, L_p(\mathfrak{M}))$.*

(ii) *For all orthonormal bases $\{h_k : k \in \mathbb{N}\}$ of H the function $\left(\sum_{k=1}^{\infty} |Rh_k|^2 \right)^{\frac{1}{2}}$ belongs to $L_p(\mathfrak{M})$.*

(iii) *For some orthonormal basis $\{h_k : k \in \mathbb{N}\}$ of H the function $\left(\sum_{k=1}^{\infty} |Rh_k|^2 \right)^{\frac{1}{2}}$ belongs to $L_p(\mathfrak{M})$.*

In this case, there exists a constant $C = C(p)$ independent of $R \in \Gamma(H, L_p(\mathfrak{M}))$ such that

$$\frac{1}{C} \|R\|_{\Gamma(H,L_p(\mathfrak{M}))} \leq \left\| \left(\sum_{k=1}^{\infty} |Rh_k|^2 \right)^{\frac{1}{2}} \right\|_{L_p(\mathfrak{M})} \leq C \|R\|_{\Gamma(H,L_p(\mathfrak{M}))}. \tag{2.11}$$

2.2.3 Stochastic integration for cylindrical Brownian motions

Using the notions introduced in the previous subsections, we are now able to give a brief introduction to the theory of stochastic integration in Banach spaces as developed in [120, 123]. The integrator is an H-cylindrical Brownian motion. We collect the relevant definitions and results without proofs. We follow [118] and [31] in our exposition.

Throughout this subsection, $(H, \langle \cdot, \cdot \rangle_H)$ denotes a separable real Hilbert space and $(\mathcal{F}_t)_{t \in [0,T]}$ is a normal filtration on a complete probability space $(\Omega, \mathcal{F}, \mathbb{P})$. $(E, \|\cdot\|_E)$ denotes an arbitrary real Banach space.

Definition 2.21. An *H-cylindrical Brownian motion* with respect to $(\mathcal{F}_t)_{t \in [0,T]}$ is a family $W = (W_H(t))_{t \in [0,T]}$ of linear operators from H to $L_2(\Omega)$ with the following properties:

[W1] For every $h \in H$, $(W_H(t)h)_{t \in [0,T]}$ is a real-valued Brownian motion with respect to $(\mathcal{F}_t)_{t \in [0,T]}$.

[W2] For every $t_1, t_2 \in [0, T]$ and $h_1, h_2 \in H$, we have

$$\mathbb{E}\left[W_H(t_1)h_1 \, W_H(t_2)h_2\right] = \min\{t_1, t_2\} \langle h_1, h_2 \rangle_H.$$

From now on, $W_H = (W_H(t))_{t \in [0,T]}$ denotes an H-cylindrical Brownian motion with respect to $(\mathcal{F}_t)_{t \in [0,T]}$. We do not specify $(\mathcal{F}_t)_{t \in [0,T]}$ if it is clear from the context which filtration is meant.

Example 2.22. Let H be a separable real Hilbert space with orthonormal basis $\{h_k : k \in \mathbb{N}\}$. Furthermore, let $\{(w_t^k)_{t \in [0,T]} : k \in \mathbb{N}\}$ be a collection of independent real-valued standard Brownian motions with respect to $(\mathcal{F}_t)_{t \in [0,T]}$ on $(\Omega, \mathcal{F}, \mathbb{P})$. Then,

$$H \ni h \mapsto W_H(t)h := \sum_{k=1}^{\infty} w_t^k \langle h_k, h \rangle_H \in L_2(\Omega), \qquad t \in [0,T], \tag{2.12}$$

defines an H-cylindrical Brownian motion with respect to $(\mathcal{F}_t)_{t \in [0,T]}$.

Stochastic integration of functions: the Wiener integral

We first define the stochastic integral of operator valued functions having the following simple structure.

Definition 2.23. An operator valued function $\Phi : [0,T] \to \mathcal{L}(H,E)$ of the form

$$\Phi(t) := \sum_{j=1}^{J} \mathbb{1}_{(t_{j-1}, t_j]}(t) \sum_{k=1}^{K} \langle h_k, \cdot \rangle_H \, x_{j,k}, \qquad t \in [0,T], \tag{2.13}$$

with $0 = t_0 < t_1 < \ldots < t_J = T$, $\{h_1, \ldots, h_K\} \subseteq H$ orthonormal, and $x_{j,k} \in E$, $1 \leq j \leq J$, $1 \leq k \leq K$, for some finite $J, K \in \mathbb{N}$, is called *finite rank step function*.

The stochastic integral of a finite rank step function is defined in the following natural way.

Definition 2.24. Let Φ be a finite rank step function of the form (2.13). The *stochastic integral* of Φ on $[0,T]$ with respect to W_H is the E-valued random variable

$$\int_0^T \Phi(t) \, dW_H(t) := \sum_{j=1}^{J} \sum_{k=1}^{K} (W_H(t_j)h_k - W_H(t_{j-1})h_k) \, x_{j,k}.$$

Note that the stochastic integral of a finite rank step function is a centred E-valued Gaussian random variable. The class of stochastically integrable functions is introduced as follows.

Definition 2.25. A function $\Phi : [0,T] \to \mathcal{L}(H,E)$ is said to be *stochastically integrable* with respect to W_H if there exists a sequence $(\Phi_n)_{n \in \mathbb{N}}$ of $\mathcal{L}(H,E)$-valued finite rank step functions on $[0,T]$ such that:

(i) for all $h \in H$ we have $\lim_{n \to \infty} \Phi_n h = \Phi h$ in measure on $[0,T]$, i.e., for any $\varepsilon > 0$,

$$\lim_{n \to \infty} \lambda\left(\left\{t \in [0,T] : \|\Phi_n(t)h - \Phi(t)h\|_E > \varepsilon\right\}\right) = 0 \, ;$$

(ii) there exists an E-valued random variable X, such that $\lim_{n \to \infty} \int_0^T \Phi_n(t) \, dW_H(t) = X$ in probability.

In this situation, the *stochastic integral* of Φ with respect to W_H is defined as the limit in probability

$$\int_0^T \Phi(t) \, dW_H(t) := \lim_{n \to \infty} \int_0^T \Phi_n(t) \, dW_H(t). \tag{2.14}$$

Since, as mentioned above, $(\int_0^T \Phi_n(t) \, dW_H(t))_{n \in \mathbb{N}}$ in (2.14) is a sequence of centred Gaussian random variables, it converges also in $L_p(\Omega; E)$ for any $p \in [1, \infty)$, and the stochastic integral $\int_0^T \Phi(t) \, dW_H(t)$ is again centred Gaussian, see, e.g., [118, Theorem 4.15].

In [123], the following analogue of the Itô isometry has been proven for finite rank step functions $\Phi : [0, T] \to \mathcal{L}(H, E)$:

$$\left(\mathbb{E} \left[\left\| \int_0^T \Phi(t)\, dW_H(t) \right\|_E^2 \right] \right)^{\frac{1}{2}} = \| R_\Phi \|_{\Gamma(L_2([0,T];H), E)}, \qquad (2.15)$$

where $R_\Phi : L_2([0, T]; H) \to E$ is the operator represented by Φ, i.e.,

$$R_\Phi f := \int_0^T \Phi(t) f(t)\, dt, \qquad f \in L_2([0, T]; H). \qquad (2.16)$$

Furthermore, the following alternative characterization of the set of stochastically integrable functions holds: An H-strongly measurable function $\Phi : [0, T] \to \mathcal{L}(H, E)$ is stochastically integrable on $[0, T]$ with respect to W_H if, and only if, $(t \mapsto \Phi^* x^*(t) := \Phi(t)^* x^*) \in L_2([0, T]; H)$ for all $x^* \in E^*$ and there exists an operator $R_\Phi \in \Gamma(L_2([0, T]; H), E)$, such that

$$R_\Phi^* x^* = \Phi^* x^* \quad \text{in } L_2([0, T]; H) \text{ for all } x^* \in E^*.$$

The isometry (2.15) extends to this situation. Remember that, unless explicitly stated otherwise, we identify H with its dual space H^* via the Riesz isometric isomorphism $h \mapsto \langle h, \cdot \rangle_H$.

Stochastic integration of stochastic processes

The integral defined above has been extended to $\mathcal{L}(H, E)$-valued *stochastic processes* for UMD Banach spaces E in [120]. The construction starts with the definition of the stochastic integral for so-called finite rank step processes.

Definition 2.26. A *finite rank $(\mathcal{F}_t)_{t \in [0,T]}$-adapted step process* is an $\mathcal{L}(H, E)$-valued stochastic process $\Phi = (\Phi(t))_{t \in [0,T]}$ of the form

$$\Phi(\omega, t) := \sum_{j=0}^J \mathbb{1}_{(t_{j-1}, t_j]}(t) \sum_{m=1}^M \mathbb{1}_{A_{j,m}}(\omega) \sum_{k=1}^K \langle h_k, \cdot \rangle_H\, x_{j,m,k}, \qquad (\omega, t) \in \Omega_T, \qquad (2.17)$$

where $0 = t_0 < t_1 < \ldots < t_J = T$, and the sets $\{A_{j,1}, \ldots, A_{j,M}\} \subseteq \mathcal{F}_{t_{j-1}}$ are disjoint for each $1 \le j \le J$ (with the convention $(t_{-1}, t_0] = \{0\}$ and $\mathcal{F}_{-1} = \mathcal{F}_0$), the vectors $\{h_1, \ldots, h_K\} \subseteq H$ are orthonormal, and $x_{j,m,k} \in E$, $1 \le j \le J$, $1 \le m \le M$, $1 \le k \le K$, for some finite $J, M, K \in \mathbb{N}$. (In [120] such processes are called *elementary adapted* to $(\mathcal{F}_t)_{t \in [0,T]}$.)

Definition 2.27. Let Φ be a finite rank $(\mathcal{F}_t)_{t \in [0,T]}$-adapted step process of the form (2.17). The *stochastic integral* of Φ with respect to W_H is defined as the E-valued random variable

$$\int_0^T \Phi(t)\, dW_H(t) := \sum_{j=1}^J \sum_{m=1}^M \mathbb{1}_{A_{j,m}} \sum_{k=1}^K (W_H(t_j) h_k - W_H(t_{j-1}) h_k)\, x_{j,m,k}.$$

Note that the stochastic integral of a finite rank step process is centred and p-integrable for any $p \in [1, \infty)$. The class of stochastically integrable processes is defined as follows.

Definition 2.28. Let E be a Banach space and fix $p \in (1, \infty)$. Let W_H be an H-cylindrical Brownian motion with respect to $(\mathcal{F}_t)_{t \in [0,T]}$. An H-strongly measurable process $\Phi : \Omega \times [0, T] \to \mathcal{L}(H, E)$ is called L_p-*stochastically integrable with respect to W_H* if there exists a sequence of finite rank $(\mathcal{F}_t)_{t \in [0,T]}$-adapted step processes $\Phi_n : \Omega \times [0, T] \to \mathcal{L}(H, E)$, $n \in \mathbb{N}$, such that:

(i) for all $h \in H$ we have $\lim_{n \to \infty} \Phi_n h = \Phi h$ in measure on Ω_T, i.e., for any $\varepsilon > 0$,

$$\lim_{n \to \infty} \mathbb{P}_T \left(\{ (\omega, t) \in \Omega_T : \| \Phi_n(\omega, t) h - \Phi(\omega, t) h \|_E > \varepsilon \} \right) = 0 ;$$

(ii) there exists a random variable $X \in L_p(\Omega; E)$ such that

$$\lim_{n \to \infty} \int_0^T \Phi_n(t) \, dW_H(t) = X \qquad \text{in } L_p(\Omega; E).$$

In this situation, the *stochastic integral* of Φ with respect to W_H is defined as the $L_p(\Omega; E)$-limit

$$\int_0^T \Phi(t) \, dW_H(t) := \lim_{n \to \infty} \int_0^T \Phi_n(t) \, dW_H(t).$$

It is easy to see that, if $\Phi : \Omega \times [0, T] \to \mathcal{L}(H, E)$ is L_p-stochastically integrable with respect to W_H and $S \in \mathcal{L}(E, F)$, F being another Banach space, then, $S\Phi : \Omega \times [0, T] \to \mathcal{L}(H, F)$ is L_p-stochastically integrable with respect to W_H. Furthermore,

$$S \int_0^T \Phi(t) \, dW_H(t) = \int_0^T S\Phi(t) \, dW_H(t) \qquad \text{in } L_p(\Omega; F). \tag{2.18}$$

The paths of finite rank step processes are finite rank step functions. Thus, for any $\omega \in \Omega$, the path $t \mapsto \Phi_\omega(t) := \Phi(\omega, t)$ defines an operator $R_{\Phi_\omega} \in \Gamma(L_2([0, T]; H), E)$ by (2.16). This leads to an $\mathcal{F}$-simple random variable $R_\Phi : \Omega \to \Gamma(L_2([0, T]; H), E)$. If E is a UMD Banach space, one can use 'decoupling' and prove that for finite rank adapted step processes Φ,

$$\left(\mathbb{E}\left[\left\| \int_0^T \Phi(t) \, dW_H(t) \right\|_E^p \right] \right)^{\frac{1}{p}} \eqsim \left(\mathbb{E}\left[\|R_\Phi\|_{\Gamma(L_2([0,T];H),E)}^p \right] \right)^{\frac{1}{p}},$$

where the constants involved do not depend on Φ. This extension of the analogue (2.15) of Itô's isometry can be generalized in the following way (see [120, Theorem 3.6 and Remark 3.7]).

Theorem 2.29. *Let E be a UMD Banach space and fix $p \in (1, \infty)$. Furthermore, let $\Phi : \Omega \times [0, T] \to \mathcal{L}(H, E)$ be an H-strongly measurable $(\mathcal{F}_t)_{t \in [0,T]}$-adapted process. Assume that for any $x^* \in E^*$, the stochastic process*

$$\Phi^* x^* : \Omega \times [0, T] \to H$$
$$(\omega, t) \mapsto \Phi(\omega, t)^* x^*$$

belongs to $L_p(\Omega; L_2([0, T]; H))$, i.e., Φ is belongs to $L_p(\Omega; L_2([0, T]; H))$ scalarly. Then, the following assertions are equivalent:

[S1] *Φ is L_p-stochastically integrable with respect to W_H.*

[S2] *There exists a strongly measurable random variable $X \in L_p(\Omega; E)$ such that for all $x^* \in E^*$ we have*

$$\langle x^*, X \rangle = \int_0^T \Phi^* x^*(t) \, dW_H(t) \qquad \text{in } L_p(\Omega). \tag{2.19}$$

[S3] *There exists $R_\Phi \in L_p(\Omega; \Gamma(L_2([0, T]; H), E))$ such that for all $x^* \in E^*$ we have*

$$R_\Phi^* x^* = \Phi^* x^* \qquad \text{in } L_p(\Omega; L_2([0, T]; H)).$$

In this situation, $X = \int_0^T \Phi(t) \, dW_H(t)$ and R_Φ in [S3] above is uniquely determined. Moreover,

$$\left(\mathbb{E}\left[\left\| \int_0^T \Phi(t) \, dW_H(t) \right\|_E^p \right] \right)^{\frac{1}{p}} \eqsim \left(\mathbb{E}\left[\|R_\Phi\|_{\Gamma(L_2([0,T];H),E)}^p \right] \right)^{\frac{1}{p}},$$

where the constants involved do not depend on Φ and T.

Remark 2.30. In the setting of Theorem 2.29, [S3] means that Φ represents an element $R_\Phi \in L_p(\Omega; \Gamma(L_2([0,T]; H), E))$. By [120, Propositions 2.11 and 2.12], R_Φ belongs to the closure of the finite rank step processes in $L_p(\Omega; \Gamma(L_2([0,T]; H), E))$, which, in analogy to the notation used in [120], will be denoted by $L_p^{\mathbb{F}}(\Omega; \Gamma(L_2([0,T]; H), E))$.

Later on, we will need the following series expansion of the stochastic integral which can be found in [120, Corollary 3.9].

Theorem 2.31. *Let E be a* UMD *Banach space and fix $p \in (1, \infty)$. Assume that the H-strongly measurable and $(\mathcal{F}_t)_{t \in [0,T]}$-adapted process $\Phi : \Omega \times [0, T] \to \mathcal{L}(H, E)$ is L_p-stochastically integrable with respect to W_H. Then, for all $h \in H$ the process $\Phi h : \Omega \times [0, T] \to E$ is L_p-stochastically integrable with respect to $W_H h$. Moreover, if $\{h_k : k \in \mathbb{N}\}$ is an orthonormal basis of H, then*

$$\int_0^T \Phi(t) \, dW_H(t) = \sum_{k=1}^\infty \int_0^T \Phi(t) h_k \, dW_H(t) h_k \qquad (\text{convergence in } L_p(\Omega; E)).$$

We have already mentioned in Lemma 2.9(v), that a UMD Banach space has type 2 if, and only if, it has M-type 2. Stochastic integration of processes in M-type 2 Banach spaces has been studied by several authors, see the literature overview in [120, p. 1460] for a short list. As mentioned therein, according to [106, 124], if E has type 2, then

$$L_2([0,T]; \Gamma(H, E)) \overset{R_.}{\hookrightarrow} \Gamma(L_2([0,T]; H), E), \tag{2.20}$$

with $R_. : \Phi \mapsto R_\Phi$ given by (2.16). Consequently, the following result holds, see [120, Corollary 3.10].

Theorem 2.32. *Let E be a* UMD *Banach space and fix $p \in (1, \infty)$. If E has type 2, then every H-strongly measurable and adapted process $\Phi \in L_p(\Omega; L_2([0,T]; \Gamma(H, E)))$ belongs to $L_p(\Omega; L_2([0,T]; H))$ scalarly, is L_p-stochastically integrable with respect to W_H and we have*

$$\left(\mathbb{E}\left[\left\| \int_0^T \Phi(t) \, dW_H(t) \right\|_E^p \right] \right)^{\frac{1}{p}} \le C \left(\mathbb{E}\left[\|\Phi\|_{L_2([0,T]; \Gamma(H,E))}^p \right] \right)^{\frac{1}{p}},$$

where the constant C does not depend on Φ.

2.3 Function spaces

In this section we introduce some function spaces, which will be used later on for the analysis of the regularity of SPDEs. We will also collect and prove some useful properties of these spaces, especially of the weighted Sobolev spaces in Subsection 2.3.3.

2.3.1 Sobolev spaces

The Sobolev spaces on an arbitrary domain $G \subseteq \mathbb{R}^d$ are defined as follows.

Definition 2.33. Let $p \in (1, \infty)$.

(i) For $m \in \mathbb{N}_0$,

$$W_p^m(G) := \left\{ u \in L_p(G) : D^\alpha u \in L_p(G) \text{ for all } \alpha \in \mathbb{N}_0^d \text{ with } |\alpha| \le m \right\}$$

is called the *Sobolev space of (smoothness) order m with summability parameter p*. It is endowed with the norm

$$\|u\|_{W_p^m(G)} := \left(\sum_{|\alpha| \le m} \|D^\alpha u\|_{L_p(G)}^p \right)^{1/p}, \qquad u \in W_p^m(G).$$

(ii) For $s = m + \sigma$ with $m \in \mathbb{N}_0$ and $\sigma \in (0,1)$, we denote by

$$W_p^s(G) := \left\{ u \in W_p^m(G) \,:\, |u|_{W_p^s(G)}^p := \sum_{|\alpha|=k} \int_G \int_G \frac{|D^\alpha u(x) - D^\alpha u(y)|^p}{|x-y|^{\sigma p + d}} \, dx dy < \infty \right\}$$

the *Sobolev space of fractional (smoothness) order* $s \in (0,\infty) \setminus \mathbb{N}$ *with summability parameter p*. It is endowed with the norm

$$\|u\|_{W_p^s(G)} := \left(\|u\|_{W_p^m(G)}^p + |u|_{W_p^s(G)}^p \right)^{1/p}, \qquad u \in W_p^s(G).$$

(iii) For $s \ge 0$ we denote by $\mathring{W}_p^s(G)$ the closure of the test functions $C_0^\infty(G)$ in $W_p^s(G)$, endowed with the norm $\|\cdot\|_{W_p^s(G)}$.

(iv) For $s < 0$, we denote by $W_p^s(G)$ the dual space of $\mathring{W}_{p'}^{-s}(G)$, $1/p + 1/p' = 1$, endowed with the canonical dual norm.

Remark 2.34. **(i)** Remember that in this thesis, $D^\alpha u$ denotes the α-th generalized derivative of a distribution u (see Subsection 2.1.5 for details). Thus, for $m \in \mathbb{N}$ and $p \in (1,\infty)$, $u \in W_p^m(G)$ means that $u \in L_p(G)$ and that for all $\alpha \in \mathbb{N}_0^d$ with $|\alpha| \le m$, the α-th generalized derivative is regular and is (interpreted as a function) an element of $L_p(G)$.

(ii) For arbitrary domains $G \subseteq \mathbb{R}^d$ and arbitrary $p \in (1,\infty)$ and $s \ge 0$, $W_p^s(G)$ is a Banach space, see, e.g., [2, Theorem 3.3] for the case $s \in \mathbb{N}_0$ and [51, Theorem 6.3.3] for fractional $s \in (0,\infty) \setminus \mathbb{N}$. Also, $\mathring{W}_p^s(G)$ is complete, since it is a closed subspace of $W_p^s(G)$. The corresponding duals $W_p^s(G)$, $s < 0$, are consequently also Banach spaces.

(iii) For fractional $s \in (0,\infty) \setminus \mathbb{N}$, the space $W_p^s(G)$ is sometimes called *Slobodeckij* space and the scale $W_p^s(G)$, $s \ge 0$, is referred to as the scale of *Sobolev-Slobodeckij* spaces. Even if this terminology would be historically more correct, in this thesis we will call the space $W_p^s(G)$ for any $s \in \mathbb{R}$ a Sobolev space. For further details regarding the historical background, we refer to [115].

(iv) Let $G \subset \mathbb{R}^d$ be a bounded domain and $p \in (1,\infty)$. For $m \in \mathbb{N}$, the expression

$$[u]_{W_p^m(G)} := \left(\sum_{|\alpha|=m} \|D^\alpha u\|_{L_p(G)}^p \right)^{1/p}, \qquad u \in \mathring{W}_p^m(G),$$

is an equivalent norm on $\mathring{W}_p^m(G)$. This is a consequence of Poincaré's inequality, see, e.g. [53, Theorem 5.6/3].

2.3.2   Spaces of Bessel potentials

In this subsection we introduce the spaces $H_p^\gamma(\mathbb{R}^d)$ of Bessel potentials and discuss some of their properties which we will use in this thesis. Furthermore, we present the definition of the spaces $H_p^\gamma(\mathbb{R}^d; \ell_2)$. They have been used in [80] by N.V. Krylov for the development of the analytic approach to SPDEs on the whole space $\mathbb{R}^d$. We prove that the spaces $H_p^\gamma(\mathbb{R}^d; \ell_2)$ are isomorphic to the corresponding spaces $\Gamma(\ell_2, H_p^\gamma(\mathbb{R}^d))$ of γ-radonifying operators from ℓ_2 to $H_p^\gamma(\mathbb{R}^d)$. Due to this fact, under suitable assumptions, the SPDEs from [80] can be rewritten as Banach space valued stochastic differential equations in the sense of van Neerven, Veraar and Weis [121].

For $\gamma \in \mathbb{R}$, we denote by $(1 - \Delta)^{\gamma/2}$ the pseudo-differential operator with symbol

$$\mathbb{R}^d \ni \xi \mapsto (1 + |\xi|^2)^{\gamma/2} \in \mathbb{R}_+.$$

That is,

$$(1 - \Delta)^{\gamma/2} : \mathcal{S}'(\mathbb{R}^d; \mathbb{C}) \to \mathcal{S}'(\mathbb{R}^d; \mathbb{C})$$
$$u \mapsto (1 - \Delta)^{\gamma/2} u := \mathfrak{F}^{-1}\big(\xi \mapsto (1 + |\xi|^2)^{\gamma/2} \mathfrak{F}(u)(\xi)\big),$$

where $\mathfrak{F}$ denotes the Fourier transform on the space of complex valued tempered distributions $\mathcal{S}'(\mathbb{R}^d; \mathbb{C})$, see Subsection 2.1.5 for details.

Definition 2.35. Let $p \in (1, \infty)$ and $\gamma \in \mathbb{R}$. Then

$$H_p^\gamma := H_p^\gamma(\mathbb{R}^d) := (1 - \Delta)^{-\gamma/2} L_p(\mathbb{R}^d) = \Big\{(1 - \Delta)^{-\gamma/2} f : f \in L_p(\mathbb{R}^d)\Big\}$$

is the *space of Bessel potentials of order γ with summability parameter p*. It is endowed with the norm

$$\|u\|_{H_p^\gamma(\mathbb{R}^d)} := \big\|(1 - \Delta)^{\gamma/2} u\big\|_{L_p(\mathbb{R}^d)}, \qquad u \in H_p^\gamma(\mathbb{R}^d).$$

Remark 2.36.   **(i)** Recall that in Section 2.1 we have postulated that in this thesis, unless explicitly stated otherwise, functions and distributions are meant to be real valued. In particular, $L_p(\mathbb{R}^d)$ stands for the space of real valued p-Bochner integrable functions on $\mathbb{R}^d$. What about the (generalized) functions in the spaces $H_p^\gamma(\mathbb{R}^d)$ defined above? At first view, even if f is real valued, we can not guarantee that $(1 - \Delta)^{-\gamma/2} f$ is real valued, since the pseudo-differential operator maps into the space of complex valued tempered distributions $\mathcal{S}'(\mathbb{R}^d; \mathbb{C})$. However, the following arguments show that $(1 - \Delta)^{-\gamma/2} f$ is indeed a *real valued* tempered distribution if $f \in L_p(\mathbb{R}^d)$, and, therefore, that $H_p^\gamma(\mathbb{R}^d)$ as defined above, consists of real valued (generalized) functions.

Fix $\gamma \in \mathbb{R}$ and $f \in L_p(\mathbb{R}^d)$ for some $p \in (1, \infty)$. Then, $(1 - \Delta)^{-\gamma/2} f \in \mathcal{S}'(\mathbb{R}^d; \mathbb{C})$ is given by the formula

$$\big((1 - \Delta)^{-\gamma/2} f, \varphi\big) = \int_{\mathbb{R}^d} f \,\overline{(1 - \Delta)^{-\gamma/2} \varphi}\, dx, \qquad \varphi \in \mathcal{S}(\mathbb{R}^d; \mathbb{C}). \tag{2.21}$$

Assume that $\varphi \in \mathcal{C}_0^\infty(\mathbb{R}^d)$. Then, $(1 - \Delta)^{-\gamma/2} \varphi$ can be expressed as

$$(1 - \Delta)^{-\gamma/2} \varphi(x) = \int_{\mathbb{R}^d} \mathcal{G}(y) \varphi(x - y)\, dy, \qquad x \in \mathbb{R}^d, \tag{2.22}$$

where $\mathcal{G}$ denotes the Green function of $(1 - \Delta)^{\gamma/2}$ and is given by

$$\mathcal{G}(y) = \frac{1}{(2\pi)^d} \lim_{R \to \infty} \int_{\{|\xi| \leq R\}} \frac{1}{(1 + |\xi|^2)^{\gamma/2}} e^{i\langle x, \xi \rangle}\, d\xi, \qquad y \in \mathbb{R}^d,$$

see [84, Chapter 12] for details. Due to the symmetry properties of the function

$$\mathbb{R}^d \ni \xi \mapsto \frac{1}{(1+|\xi|^2)^{\gamma/2}} \in \mathbb{R}_+,$$

$\mathcal{G}$ is real valued. Therefore and by (2.22), $(1-\Delta)^{-\gamma/2}\varphi$ is real valued too. Inserting this into (2.21), we obtain

$$\left((1-\Delta)^{-\gamma/2}f, \varphi\right) = \int_{\mathbb{R}^d} f (1-\Delta)^{-\gamma/2}\varphi \, dx \in \mathbb{R}, \qquad \varphi \in \mathcal{C}_0^\infty(\mathbb{R}^d).$$

Hence, $(1-\Delta)^{-\gamma/2}f$ is a real valued (generalized) function. Consequently, the tempered distribution $(1-\Delta)^{-\gamma/2}f \in \mathcal{S}'(\mathbb{R}^d; \mathbb{C})$ is real valued, since for any $\varphi \in \mathcal{S}(\mathbb{R}^d)$, there exists a sequence $(\varphi_n)_{n\in\mathbb{N}} \subseteq \mathcal{C}_0^\infty(\mathbb{R}^d)$ converging to φ in $\mathcal{S}(\mathbb{R}^d)$. All in all, we have shown that

$$H_p^\gamma(\mathbb{R}^d) = \left\{ u \in \mathcal{S}'(\mathbb{R}^d) : u = (1-\Delta)^{-\gamma/2}f \text{ for some } f \in L_p(\mathbb{R}^d) \right\}.$$

(ii) For $p \in (1, \infty)$ and $\gamma \in \mathbb{R}$, the space $H_p^\gamma(\mathbb{R}^d)$ of Bessel potentials is a Banach space. For a proof see, e.g., [84, Theorem 13.3.3(i)] and use part (i) of this remark.

The following result shows that the spaces of Bessel potentials are generalizations of the Sobolev spaces $W_p^m(\mathbb{R}^d)$, $m \in \mathbb{N}_0$. A proof can be found e.g. in [84, Theorem 13.3.12].

Lemma 2.37. *Let $p \in (1, \infty)$ and $\gamma = m \in \mathbb{N}_0$. Then,*

$$H_p^m(\mathbb{R}^d) = W_p^m(\mathbb{R}^d) \qquad (\text{equivalent norms}).$$

It is well-known that for $p, p' \in (1, \infty)$ with $1/p + 1/p' = 1$, the mapping

$$\Psi : L_{p'}(\mathbb{R}^d) \to \left(L_p(\mathbb{R}^d)\right)^*, \qquad (\Psi f)(g) := \int_{\mathbb{R}^d} f(x)g(x) \, dx,$$

is an isometric isomorphism. In particular, $L_{p'}(\mathbb{R}^d) \cong \left(L_p(\mathbb{R}^d)\right)^*$. This duality relation can be extended to the scale of Bessel potential spaces, as the following result shows. The proof is left to the reader.

Lemma 2.38. *Let $\gamma \in \mathbb{R}$ and $p, p' \in (1, \infty)$ with $1/p + 1/p' = 1$. Then, for any $u \in H_{p'}^{-\gamma}$ and $\varphi \in \mathcal{C}_0^\infty(\mathbb{R}^d)$,*

$$\left|(u, \varphi)\right| \leq \|u\|_{H_{p'}^{-\gamma}} \|\varphi\|_{H_p^\gamma}.$$

Thus, for any $u \in H_{p'}^{-\gamma}$, there exists a unique functional $\Psi_u \in (H_p^\gamma)^*$ such that its restriction to the test functions $\mathcal{C}_0^\infty(\mathbb{R}^d)$ coincides with $(u, \cdot)$, i.e.,

$$\Psi_u|_{\mathcal{C}_0^\infty(\mathbb{R}^d)} = (u, \cdot). \tag{2.23}$$

Moreover, the expression

$$\Psi : H_{p'}^{-\gamma} \to H_p^\gamma$$
$$u \mapsto \Psi(u) := \Psi_u,$$

with Ψ_u from above, is an isometric isomorphism. In particular, $H_{p'}^{-\gamma} \cong (H_p^\gamma)^*$.

Recall that the notation (u, φ) is used to denote the application of a distribution u to a test function φ. In what follows we extend the meaning of $(\cdot, \cdot)$ and write

$$(u, \varphi) := \Psi_u(\varphi), \quad \text{for all} \quad u \in H_{p'}^{-\gamma} \text{ and } \varphi \in H_p^\gamma, \tag{2.24}$$

with Ψ_u as in Lemma 2.38. This is justified by (2.23).

The following can be said about the geometric properties of the spaces of Bessel potentials.

Remark 2.39. For any $p \in (1, \infty)$ and $\gamma \in \mathbb{R}$, $H_p^\gamma(\mathbb{R}^d)$ is a UMD Banach space, since $(1 - \Delta)^{\gamma/2} : H_p^\gamma(\mathbb{R}^d) \to L_p(\mathbb{R}^d)$ is an isometric isomorphism and $L_p(\mathbb{R}^d)$ is UMD, see Lemma 2.6. The same argument, together with Lemma 2.9(iii) and (iv), shows that $H_p^\gamma(\mathbb{R}^d)$ has type $r := \min\{2, p\}$.

Now we recall the definition of the spaces $H_p^\gamma(\mathbb{R}^d; \ell_2)$ given in [80]. As already mentioned, these spaces are used therein for the formulation of the SPDEs under consideration.

Definition 2.40. For $p \in (1, \infty)$ and $\gamma \in \mathbb{R}$,

$$H_p^\gamma(\ell_2) := H_p^\gamma(\mathbb{R}^d; \ell_2) := \Big\{ g = (g^k)_{k\in\mathbb{N}} \in \big(H_p^\gamma(\mathbb{R}^d)\big)^\mathbb{N} :$$

$$\|g\|_{H_p^\gamma(\ell_2)} := \Big\| \big\| \big((1 - \Delta)^{\gamma/2} g^k\big)_{k\in\mathbb{N}} \big\|_{\ell_2} \Big\|_{L_p} < \infty \Big\}.$$

Remark 2.41. Let $p \in (1, \infty)$. Remember that in this thesis we write $\{e_k : k \in \mathbb{N}\}$ for the standard orthonormal basis of $\ell_2 = \ell_2(\mathbb{N})$.

(i) The mapping $\Phi : H_p^0(\mathbb{R}^d; \ell_2) \to L_p(\mathbb{R}^d, \mathcal{B}(\mathbb{R}^d), \lambda^d; \ell_2)$ assigning the function (equivalence class)

$$\Phi(g) : \mathbb{R}^d \to \ell_2$$

$$x \mapsto \sum_{k=1}^\infty g^k(x)\, e_k \quad \text{(convergence in } \ell_2)$$

to each $g \in H_p^0(\mathbb{R}^d; \ell_2)$ is an isometric isomorphism. In particular, we have

$$H_p^0(\mathbb{R}^d; \ell_2) \cong L_p(\mathbb{R}^d, \mathcal{B}(\mathbb{R}^d), \lambda^d; \ell_2).$$

Consequently, since $L_p(\mathbb{R}^d; \ell_2)$ is complete, $H_p^0(\mathbb{R}^d; \ell_2)$ is a Banach space.

(ii) Let $\gamma \in \mathbb{R}$. Then,

$$\Phi_\gamma : H_p^\gamma(\mathbb{R}^d; \ell_2) \to H_p^0(\mathbb{R}^d; \ell_2)$$

$$g = (g^k)_{k\in\mathbb{N}} \mapsto \big((1 - \Delta)^{\gamma/2} g^k\big)_{k\in\mathbb{N}}$$

is an isometric mapping. Furthermore,

$$\widetilde{\Phi}_\gamma : H_p^0(\mathbb{R}^d; \ell_2) \to H_p^\gamma(\mathbb{R}^d; \ell_2)$$

$$g = (g^k)_{k\in\mathbb{N}} \mapsto \big((1 - \Delta)^{-\gamma/2} g^k\big)_{k\in\mathbb{N}}$$

defines a right inverse for Φ_γ. Thus, Φ_γ is a surjective isometric mapping, and therefore an isometric isomorphism. Consequently, $H_p^\gamma(\mathbb{R}^d; \ell_2)$ is a Banach space, since $H_p^0(\mathbb{R}^d; \ell_2)$ is complete by part (i) of this remark.

Finally, we prove that the space $H_p^\gamma(\mathbb{R}^d; \ell_2)$ is isomorphic to $\Gamma(\ell_2, H_p^\gamma(\mathbb{R}^d))$.

Theorem 2.42. *Let* $\gamma \in \mathbb{R}$ *and* $p \in [2, \infty)$. *Then the operator*

$$\Phi : H_p^\gamma(\mathbb{R}^d; \ell_2) \to \Gamma(\ell_2, H_p^\gamma(\mathbb{R}^d))$$

$$(g^k)_{k \in \mathbb{N}} \mapsto \sum_{k=1}^\infty \langle \mathrm{e}_k, \cdot \rangle_{\ell_2} g^k \qquad (\text{convergence in } \Gamma(\ell_2, H_p^\gamma(\mathbb{R}^d)))$$

is an isomorphism, and, therefore,

$$H_p^\gamma(\mathbb{R}^d; \ell_2) \simeq \Gamma(\ell_2, H_p^\gamma(\mathbb{R}^d)).$$

Proof. Fix $\gamma \in \mathbb{R}$. First of all we have to show that the operator Φ is well defined. To this end, since $\Gamma(\ell_2, H_p^\gamma(\mathbb{R}^d))$ is complete, it is enough to prove that for any fixed $g \in H_p^\gamma(\mathbb{R}^d; \ell_2)$, the sequence

$$(R_n)_{n \in \mathbb{N}} := \left(\sum_{k=1}^n \langle \mathrm{e}_k, \cdot \rangle_{\ell_2} g^k \right)_{n \in \mathbb{N}} \subseteq \mathcal{L}_f(\ell_2, H_p^\gamma(\mathbb{R}^d))$$

is a Cauchy sequence in $\Gamma(\ell_2, H_p^\gamma(\mathbb{R}^d))$. Fix $g \in H_p^\gamma(\mathbb{R}^d; \ell_2)$. Using the fact that $(1 - \Delta)^{-\gamma/2} : H_p^0(\mathbb{R}^d) \to H_p^\gamma(\mathbb{R}^d)$ is an isometric isomorphism together with the ideal property of γ-radonifying operators, see Theorem 2.17 above, we obtain that for arbitrary $m, n \in \mathbb{N}$,

$$\|R_n - R_m\|_{\Gamma(\ell_2, H_p^\gamma(\mathbb{R}^d))} = \left\| \sum_{k=m+1}^n \langle \mathrm{e}_k, \cdot \rangle_{\ell_2} g^k \right\|_{\Gamma(\ell_2, H_p^\gamma(\mathbb{R}^d))}$$

$$= \left\| (1 - \Delta)^{-\gamma/2} \sum_{k=m+1}^n \langle \mathrm{e}_k, \cdot \rangle_{\ell_2} (1 - \Delta)^{\gamma/2} g^k \right\|_{\Gamma(\ell_2, H_p^\gamma(\mathbb{R}^d))}$$

$$\leq \left\| (1 - \Delta)^{-\gamma/2} \right\|_{\mathcal{L}(H_p^0(\mathbb{R}^d), H_p^\gamma(\mathbb{R}^d))} \left\| \sum_{k=m+1}^n \langle \mathrm{e}_k, \cdot \rangle_{\ell_2} (1 - \Delta)^{\gamma/2} g^k \right\|_{\Gamma(\ell_2, H_p^0(\mathbb{R}^d))}$$

$$= \left\| \sum_{k=m+1}^n \langle \mathrm{e}_k, \cdot \rangle_{\ell_2} (1 - \Delta)^{\gamma/2} g^k \right\|_{\Gamma(\ell_2, H_p^0(\mathbb{R}^d))}.$$

Thus, since $H_p^0(\mathbb{R}^d) = L_p(\mathbb{R}^d)$, we obtain from (2.11),

$$\|R_n - R_m\|_{\Gamma(\ell_2, H_p^\gamma(\mathbb{R}^d))} \leq C \left\| \left(\sum_{k=m}^n |(1 - \Delta)^{\gamma/2} g^k|^2 \right)^{\frac{1}{2}} \right\|_{L_p(\mathbb{R}^d)}.$$

Therefore, $(R_n)_{n \in \mathbb{N}}$ converges in $\Gamma(\ell_2, H_p^\gamma(\mathbb{R}^d))$ and the series $\sum_{k=1}^\infty \langle \mathrm{e}_k, \cdot \rangle_{\ell_2} g^k$ is well-defined. Moreover, using the same arguments, it follows that

$$\|\Phi(g)\|_{\Gamma(\ell_2, H_p^\gamma(\mathbb{R}^d))} = \left\| \sum_{k=1}^\infty \langle \mathrm{e}_k, \cdot \rangle_{\ell_2} g^k \right\|_{\Gamma(\ell_2, H_p^\gamma(\mathbb{R}^d))}$$

$$\leq \left\| \sum_{k=1}^\infty \langle \mathrm{e}_k, \cdot \rangle_{\ell_2} (1 - \Delta)^{\gamma/2} g^k \right\|_{\Gamma(\ell_2, H_p^0(\mathbb{R}^d))}$$

$$\leq C \left\| \left(\sum_{k=1}^\infty |(1 - \Delta)^{\gamma/2} g^k|^2 \right)^{\frac{1}{2}} \right\|_{L_p(\mathbb{R}^d)} = C \|g\|_{H_p^\gamma(\mathbb{R}^d; \ell_2)}.$$

Consequently, the obviously linear operator Φ is bounded. Simultaneously, one easily checks that

$$\widetilde{\Phi} : \Gamma(\ell_2, H_p^\gamma(\mathbb{R}^d)) \to H_p^\gamma(\mathbb{R}^d; \ell_2)$$

$$R \mapsto (R\mathrm{e}_k)_{k \in \mathbb{N}}$$

is the inverse of Φ, which is well-defined, linear and bounded by Theorem 2.20. $\qquad\square$

2.3.3 Weighted Sobolev spaces

In this subsection we recall the definition and some basic properties of the weighted Sobolev spaces $H_{p,\theta}^\gamma(G)$ as introduced e.g. in [93]. These spaces serve as state spaces for the solution processes $u = (u(t))_{t\in[0,T]}$ of SPDEs on domains in the L_p-theory of N.V. Krylov and collaborators (see, e.g., [72, 73, 75, 76, 85, 86]). Although in this thesis we consider only SPDEs on bounded Lipschitz domains, we introduce and analyse the weighted Sobolev spaces on arbitrary domains G in $\mathbb{R}^d$ with non-empty boundary ∂G. Among others, we will prove that $H_{p,\theta}^\gamma(G)$ possesses the UMD property and has type $r := \min\{p,2\}$, see Lemma 2.50. We will also consider the spaces $H_{p,\theta}^\gamma(G;\ell_2)$ used in the mentioned L_p-theory and prove that they are isomorphic to $\Gamma(\ell_2, H_{p,\theta}^\gamma(G))$, see Theorem 2.54, which is a generalization of the just proven Theorem 2.42. This will allow us to apply the stochastic integration theory from Section 2.2 to stochastic processes $\Phi : \Omega \times [0,T] \to H_{p,\theta}^\gamma(G;\ell_2)$ in Chapter 6.

Let $G \subset \mathbb{R}^d$ be an arbitrary domain with non-empty boundary ∂G. Remember that $\rho(x) = \rho_G(x)$ denotes the distance of a point $x \in G$ to the boundary ∂G. We fix an infinitely differentiable function ψ defined on G such that for all $x \in G$,

$$\rho(x) \leq C\psi(x), \quad \rho(x)^{m-1}|D^m\psi(x)| \leq C(m) < \infty \text{ for all } m \in \mathbb{N}_0, \tag{2.25}$$

where C and $C(m)$ do not depend on $x \in G$. For a detailed construction of such a function see, e.g., [113, Chapter VI, Section 2.1]. Let $\zeta \in \mathcal{C}_0^\infty(\mathbb{R}_+)$ be a non-negative function satisfying

$$\sum_{n\in\mathbb{Z}} \zeta(e^{n+t}) > c > 0 \text{ for all } t \in \mathbb{R}. \tag{2.26}$$

Note that any non-negative smooth function $\zeta \in \mathcal{C}_0^\infty(\mathbb{R}_+)$ with $\zeta > 0$ on $[e^{-1}, e]$ satisfies (2.26). Without loss of generality, wee assume that ζ takes values in the interval $[0,1]$. For $x \in G$ and $n \in \mathbb{Z}$, define

$$\zeta_n(x) := \zeta(e^n\psi(x)). \tag{2.27}$$

Then, there exists $k_0 > 0$ such that, for all $n \in \mathbb{Z}$,

$$\operatorname{supp}\zeta_n \subset G_n := \{x \in G : e^{-n-k_0} < \rho(x) < e^{-n+k_0}\},$$

i.e., $\zeta_n \in \mathcal{C}_0^\infty(G_n)$. Moreover, $|D^m\zeta_n(x)| \leq C(\zeta,m)e^{mn}$ for all $x \in G$ and $m \in \mathbb{N}_0$, and $\sum_{n\in\mathbb{Z}} \zeta_n(x) \geq \delta > 0$ for all $x \in G$. Using this localisation sequence the weighted Sobolev spaces $H_{p,\theta}^\gamma(G)$ can be introduced as follows.

Definition 2.43. Let G be an arbitrary domain in $\mathbb{R}^d$ with non-empty boundary. Furthermore, let ζ_n, $n \in \mathbb{Z}$, be as above, $p \in (1,\infty)$, and $\gamma,\theta \in \mathbb{R}$. Then

$$H_{p,\theta}^\gamma(G) := \left\{ u \in \mathcal{D}'(G) : \|u\|_{H_{p,\theta}^\gamma(G)} := \left(\sum_{n\in\mathbb{Z}} e^{n\theta} \|\zeta_{-n}(e^n\cdot)u(e^n\cdot)\|_{H_p^\gamma}^p \right)^{\frac{1}{p}} < \infty \right\}.$$

It is called *weighted Sobolev space of (smoothness) order γ with summability parameter p and weight parameter θ*.

Remark 2.44. The reason why $H_{p,\theta}^\gamma(G)$ is called weighted Sobolev space is the fact that

$$L_{p,\theta}(G) := H_{p,\theta}^0(G) = L_p(G, \mathcal{B}(G), \rho^{\theta-d}\lambda^d; \mathbb{R}),$$

with equivalent norms, and that, if $\gamma = m \in \mathbb{N}$ is a positive integer,

$$H_{p,\theta}^m(G) = \left\{ u \in L_{p,\theta}(G) : \rho^{|\alpha|}D^\alpha u \in L_{p,\theta}(G) \text{ for all } \alpha \in \mathbb{N}_0^d \text{ with } |\alpha| \leq m \right\},$$

$$\|u\|^p_{H^m_{p,\theta}(G)} \asymp \sum_{k=0}^{m} |u|^p_{H^k_{p,\theta}(G)} \qquad (2.28)$$

where

$$|u|^p_{H^k_{p,\theta}(G)} := \sum_{\substack{\alpha \in \mathbb{N}_0^d \\ |\alpha|=k}} \int_G |\rho(x)^{|\alpha|} D^\alpha u(x)|^p \rho(x)^{\theta-d} \, dx, \qquad (2.29)$$

for $k \in \{0, \dots, m\}$; see, e.g., [93, Proposition 2.2].

Now we present some useful properties of the space $H^\gamma_{p,\theta}(G)$ taken from [93], see also [80,81].

Lemma 2.45. *Let $G \subset \mathbb{R}^d$ be a domain with non-empty boundary ∂G, $\gamma, \theta \in \mathbb{R}$, and $p \in (1, \infty)$.*

(i) *$H^\gamma_{p,\theta}(G)$ is a separable and reflexive Banach space.*

(ii) *The space $C_0^\infty(G)$ is dense in $H^\gamma_{p,\theta}(G)$.*

(iii) *$u \in H^\gamma_{p,\theta}(G)$ if, and only if, $u, \psi u_x \in H^{\gamma-1}_{p,\theta}(G)$ and*

$$\|u\|_{H^\gamma_{p,\theta}(G)} \leq C\|\psi u_x\|_{H^{\gamma-1}_{p,\theta}(G)} + C\|u\|_{H^{\gamma-1}_{p,\theta}(G)} \leq C\|u\|_{H^\gamma_{p,\theta}(G)}.$$

Also, $u \in H^\gamma_{p,\theta}(G)$ if, and only if, $u, (\psi u)_x \in H^{\gamma-1}_{p,\theta}(G)$ and

$$\|u\|_{H^\gamma_{p,\theta}(G)} \leq C\|(\psi u)_x\|_{H^{\gamma-1}_{p,\theta}(G)} + C\|u\|_{H^{\gamma-1}_{p,\theta}(G)} \leq C\|u\|_{H^\gamma_{p,\theta}(G)}.$$

(iv) *For any $\nu, \gamma \in \mathbb{R}$, $\psi^\nu H^\gamma_{p,\theta}(G) = H^\gamma_{p,\theta-\nu p}(G)$ and*

$$\|u\|_{H^\gamma_{p,\theta-\nu p}(G)} \leq C\|\psi^{-\nu} u\|_{H^\gamma_{p,\theta}(G)} \leq C\|u\|_{H^\gamma_{p,\theta-\nu p}(G)}.$$

(v) *If $0 < \eta < 1$, $\gamma = (1-\eta)\nu_0 + \eta\nu_1$, $1/p = (1-\eta)/p_0 + \eta/p_1$ and $\theta = (1-\eta)\theta_0 + \eta\theta_1$ with $\nu_0, \nu_1, \theta_0, \theta_1 \in \mathbb{R}$ and $p_0, p_1 \in (1, \infty)$, then*

$$H^\gamma_{p,\theta}(G) = \left[H^{\nu_0}_{p_0,\theta_0}(G), H^{\nu_1}_{p_1,\theta_1}(G)\right]_\eta \qquad \text{(equivalent norms)}.$$

Consequently, if $\gamma \in (\nu_0, \nu_1)$ then, for any $\varepsilon > 0$, there exists a constant C, depending on $\nu_0, \nu_1, \theta, p,$ and ε, such that

$$\|u\|_{H^\gamma_{p,\theta}(G)} \leq \varepsilon\|u\|_{H^{\nu_1}_{p,\theta}(G)} + C(\nu_0, \nu_1, \theta, p, \varepsilon)\|u\|_{H^{\nu_0}_{p,\theta}(G)}.$$

Also, if $\theta \in (\theta_0, \theta_1)$ then, for any $\varepsilon > 0$, there exists a constant C, depending on $\theta_0, \theta_1, \gamma, p,$ and ε, such that

$$\|u\|_{H^\gamma_{p,\theta}(G)} \leq \varepsilon\|u\|_{H^\gamma_{p,\theta_0}(G)} + C(\theta_0, \theta_1, \gamma, p, \varepsilon)\|u\|_{H^\gamma_{p,\theta_1}(G)}.$$

(vi) *There exists a constant $c_0 > 0$ depending on p, θ, γ and the function ψ such that, for all $c \geq c_0$, the operator $\psi^2\Delta - c$ is an isomorphism from $H^{\gamma+1}_{p,\theta}(G)$ to $H^{\gamma-1}_{p,\theta}(G)$.*

(vii) *If G is bounded, then $H^\gamma_{p,\theta_1}(G) \hookrightarrow H^\gamma_{p,\theta_2}(G)$ for $\theta_1 < \theta_2$.*

(viii) *The dual of $H_{p,\theta}^{\gamma}(G)$ and the weighted Sobolev space $H_{p',\theta'}^{-\gamma}(G)$ with $1/p + 1/p' = 1$ and $\theta/p + \theta'/p' = d$ are isomorphic. That is,*

$$\left(H_{p,\theta}^{\gamma}(G)\right)^{*} \simeq H_{p',\theta'}^{-\gamma}(G) \quad \text{where} \quad \frac{1}{p} + \frac{1}{p'} = 1 \quad \text{and} \quad \frac{\theta}{p} + \frac{\theta'}{p'} = d. \tag{2.30}$$

Remark 2.46. Assertions (iv) and (vi) in Lemma 2.45 imply the following: If $u \in H_{p,\theta-p}^{\gamma}(G)$ and $\Delta u \in H_{p,\theta+p}^{\gamma}(G)$, then $u \in H_{p,\theta-p}^{\gamma+2}(G)$ and there exists a constant C, which does not depend on u, such that

$$\|u\|_{H_{p,\theta-p}^{\gamma+2}(G)} \leq C\|\Delta u\|_{H_{p,\theta+p}^{\gamma}(G)} + C\|u\|_{H_{p,\theta-p}^{\gamma}(G)}.$$

A proof of the following equivalent characterization of the weighted Sobolev spaces $H_{p,\theta}^{\gamma}(G)$ can be found in [93, Proposition 2.2].

Lemma 2.47. *Let $\{\xi_n : n \in \mathbb{Z}\} \subseteq C_0^{\infty}(G)$ be such that for all $n \in \mathbb{Z}$ and $m \in \mathbb{N}_0$,*

$$|D^m \xi_n| \leq C(m)\, c^{nm} \quad \text{and} \quad \operatorname{supp}\xi_n \subseteq \{x \in G : c^{-n-k_0} < \rho(x) < c^{-n+k_0}\} \tag{2.31}$$

for some $c > 1$ and $k_0 > 0$, where the constant $C(m)$ does not depend on $n \in \mathbb{Z}$ and $x \in G$. Then, for any $u \in H_{p,\theta}^{\gamma}(G)$,

$$\sum_{n \in \mathbb{Z}} c^{n\theta} \|\xi_{-n}(c^n \cdot)u(c^n \cdot)\|_{H_p^{\gamma}}^p \leq C\,\|u\|_{H_{p,\theta}^{\gamma}(G)}^p.$$

If in addition

$$\sum_{n \in \mathbb{Z}} \xi_n(x) \geq \delta > 0 \text{ for all } x \in G, \tag{2.32}$$

then the converse inequality also holds.

The following sequences $\{\xi_n : n \in \mathbb{Z}\}$ will be useful, when we apply Lemma 2.47 in the proofs of Lemma 3.5(i), Lemma 4.9 and Theorem 6.7, respectively.

Remark 2.48. **(i)** It can be shown that for any $i, j \in \{1, \ldots, d\}$, both

$$\left\{\xi_n^{(1)} := e^{-n}(\zeta_n)_{x^i} : n \in \mathbb{Z}\right\} \quad \text{and} \quad \left\{\xi_n^{(2)} := e^{-2n}(\zeta_n)_{x^i x^j} : n \in \mathbb{Z}\right\}$$

satisfy (2.31) with $c := e$. Thus, for any $p \in (1, \infty)$ and $\theta \in \mathbb{R}$,

$$\sum_{n \in \mathbb{Z}} c^{n\theta}\left(\|c^n(\zeta_{-n})_{x^i}(e^n \cdot)u(e^n \cdot)\|_{H_p^{\gamma}}^p + \|e^{2n}(\zeta_{-n})_{x^i x^j}(e^n \cdot)u(e^n \cdot)\|_{H_p^{\gamma}}^p\right) \leq C\|u\|_{H_{p,\theta}^{\gamma}(G)}^p.$$

(ii) Let $c_0 > 1$ and $k_1 > 0$. Fix a non-negative function $\tilde{\zeta} \in C_0^{\infty}(\mathbb{R}_+)$ with

$$\tilde{\zeta}(t) = 1 \quad \text{for all} \quad t \in \left[\frac{1}{C}c_0^{-k_1}, C(0)c_0^{k_1}\right],$$

where C and $C(0)$ are as in (2.25). Then, the sequence $\{\xi_n : n \in \mathbb{Z}\} \subseteq C_0^{\infty}(G)$ defined by

$$\xi_n := \tilde{\zeta}(c_0^n \psi(\cdot)), \qquad n \in \mathbb{Z},$$

fulfils the conditions (2.31) and (2.32) from Lemma 2.47 with $c = c_0$ and a suitable $k_0 > 0$. Furthermore,

$$\xi_n(x) = 1 \quad \text{on} \quad \{x \in G : c_0^{-n-k_1} \leq \rho(x) \leq c_0^{-n+k_1}\}.$$

(iii) Let $\{\xi_n : n \in \mathbb{Z}\} \subseteq C_0^\infty(G)$ fulfil the conditions (2.31) and (2.32) from Lemma 2.47 for some fixed constants $c > 1$ and $k_0 > 0$. Consider the sequence $\{\xi_n^* : n \in \mathbb{Z}\} \subseteq C_0^\infty(G)$ given by

$$\xi_n^* := \frac{\xi_n}{\sum_{j \in \mathbb{Z}} \xi_j}, \quad n \in \mathbb{Z}.$$

Obviously,

$$\sum_{n \in \mathbb{Z}} \xi_n^*(x) = 1 \quad \text{for all} \quad x \in G. \tag{2.33}$$

By standard calculations one can check that the sequence $\{\xi_n^* : n \in \mathbb{Z}\} \subseteq C_0^\infty(G)$ also fulfils the condition (2.31) from Lemma 2.47. The following fact might be useful: Any $x \in G$ is contained in at most finitely many stripes

$$G_n(c, k_0) := \{x \in G : c^{-n-k_0} < \rho(x) < c^{-n+k_0}\}, \quad n \in \mathbb{Z}. \tag{2.34}$$

Even more, there exists a finite constant $C = C(c, k_0)$ which does not depend on $x \in G$ such that

$$\left|\{n \in \mathbb{Z} : x \in G_n(c, k_0)\}\right| \leq C. \tag{2.35}$$

Let us also be a little bit more precise on the duality statement from Lemma 2.45(viii).

Remark 2.49. Fix $\gamma \in \mathbb{R}$, $p \in (1, \infty)$, $\theta \in \mathbb{R}$ and let p' and θ' be as in (2.30). Fix $\{\xi_n : n \in \mathbb{Z}\} \subseteq C_0^\infty(G)$ with $\sum_{n \in \mathbb{Z}} \xi_n = 1$ on G satisfying (2.31) from Lemma 2.47 for some fixed $c > 1$ and $k_0 > 0$. Simultaneously, assume that we have a sequence $\{\tilde{\xi}_n : n \in \mathbb{Z}\} \subseteq C_0^\infty(G)$ such that for every $n \in \mathbb{Z}$, $\tilde{\xi}_n$ equals one on the support of ξ_n, i.e.,

$$\tilde{\xi}_n\Big|_{\text{supp}\xi_n} = 1, \tag{2.36}$$

and satisfying (2.31)—with the same $c > 1$ but possibly different $k_0 > 0$—and (2.32) from Lemma 2.47. By Remark 2.48(ii) and (iii), it is clear that we can construct such sequences. The assertion of Lemma 2.45(viii) has been proven in [93, Proposition 2.4] by showing that the mapping

$$\Psi : H_{p', \theta'}^{-\gamma}(G) \to \left(H_{p, \theta}^{\gamma}(G)\right)^*$$
$$u \mapsto \Psi(u) := [u, \cdot]$$

with

$$[\cdot, \cdot] : H_{p', \theta'}^{-\gamma}(G) \times H_{p, \theta}^{\gamma}(G) \to \mathbb{R}, \quad [u, v] := \sum_{n \in \mathbb{Z}} c^{nd}\big(\tilde{\xi}_{-n}(c^n \cdot)u(c^n \cdot), \xi_{-n}(c^n \cdot)v(c^n \cdot)\big) \tag{2.37}$$

is an isomorphism; see (2.24) for the meaning of $(\cdot, \cdot)$ on $H_{p'}^{-\gamma}(\mathbb{R}^d) \times H_p^{\gamma}(\mathbb{R}^d)$. From now on we will use this notation also on $H_{p', \theta'}^{-\gamma}(G) \times H_{p, \theta}^{\gamma}(G)$ and define

$$(\cdot, \cdot) := [\cdot, \cdot] \quad \text{on} \quad H_{p', \theta'}^{-\gamma}(G) \times H_{p, \theta}^{\gamma}(G), \tag{2.38}$$

with $[\cdot, \cdot]$ as in (2.37). This is justified by the following calculation: Let $u \in H_{p', \theta'}^{-\gamma}(G) \subseteq \mathcal{D}'(G)$ and $\varphi \in C_0^\infty(G)$. Then, since $\sum_{n \in \mathbb{Z}} \xi_n = 1$ and $\tilde{\xi}_n$ is constructed in such a way that (2.36) holds, we obtain

$$(u, \varphi) = \Big(u, \sum_{n \in \mathbb{Z}} \xi_{-n}\varphi\Big) = \sum_{n \in \mathbb{Z}} \big(u, \xi_{-n}\varphi\big) = \sum_{n \in \mathbb{Z}} \big(\tilde{\xi}_{-n}u, \xi_{-n}\varphi\big) = [u, \varphi].$$

After presenting and discussing these fundamental properties of weighted Sobolev spaces, we prove now that they satisfy the following geometric Banach space properties.

Lemma 2.50. *Let G be an arbitrary domain in $\mathbb{R}^d$ with non-empty boundary. Let $\gamma, \theta \in \mathbb{R}$ and $p \in (1, \infty)$. Then $H_{p,\theta}^\gamma(G)$ is a UMD space with type $r := \min\{2, p\}$.*

Proof. First we prove the UMD property. Obviously, the linear operator

$$S : H_{p,\theta}^\gamma(G) \to L_p\Big(\mathbb{Z}, \mathscr{P}(\mathbb{Z}), \sum_{n \in \mathbb{Z}} e^{n\theta} \delta_n; H_p^\gamma(\mathbb{R}^d)\Big)$$

$$u \mapsto \Big(n \mapsto \zeta_{-n}(e^n \cdot) u(e^n \cdot)\Big).$$

is isometric. Therefore, and since $H_{p,\theta}^\gamma(G)$ is complete, the range of S is a closed subspace of $L_p\big(\mathbb{Z}, \mathscr{P}(\mathbb{Z}), \sum_{n \in \mathbb{Z}} e^{n\theta} \delta_n; H_p^\gamma(\mathbb{R}^d)\big)$, which satisfies the UMD property by Lemma 2.6(iii). Thus, $H_{p,\theta}^\gamma(G)$ is isomorphic to a closed subspace of a UMD Banach spaces. Hence, due to Lemma 2.6(i) and Remark 2.39, the UMD property of $H_{p,\theta}^\gamma(G)$ follows.

In order to prove that $H_{p,\theta}^\gamma(G)$ has type $r = \min\{2, p\}$ we argue as follows: Fix an arbitrary Rademacher sequence $(r_k)_{k=1}^\infty$ and a finite set $\{u_1, \ldots, u_K\} \subseteq H_{p,\theta}^\gamma(G)$. Then, by the Kahane-Khintchine inequality, see Lemma 2.10, we have

$$\left(\mathbb{E}\left[\Big\| \sum_{k=1}^K r_k u_k \Big\|_{H_{p,\theta}^\gamma(G)}^r \right] \right)^{\frac{1}{r}} \leq C \left(\mathbb{E}\left[\Big\| \sum_{k=1}^K r_k u_k \Big\|_{H_{p,\theta}^\gamma(G)}^p \right] \right)^{\frac{1}{p}}.$$

By the definition of the weighted Sobolev norm, this yields

$$\left(\mathbb{E}\left[\Big\| \sum_{k=1}^K r_k u_k \Big\|_{H_{p,\theta}^\gamma(G)}^r \right] \right)^{\frac{1}{r}} \leq C \left(\mathbb{E}\left[\sum_{n \in \mathbb{Z}} e^{n\theta} \Big\| \zeta_{-n}(e^n \cdot)\Big(\sum_{k=1}^K r_k u_k \Big)(e^n \cdot) \Big\|_{H_p^\gamma(\mathbb{R}^d)}^p \right] \right)^{\frac{1}{p}}.$$

Using Fubini's theorem and the fact that $\zeta_{-n}(e^n \cdot)\big(\sum_{k=1}^K r_k u_k \big)(e^n \cdot) = \sum_{k=1}^K r_k \zeta_{-n}(e^n \cdot) u_k(e^n \cdot)$, we obtain the estimate

$$\left(\mathbb{E}\left[\Big\| \sum_{k=1}^K r_k u_k \Big\|_{H_{p,\theta}^\gamma(G)}^r \right] \right)^{\frac{1}{r}} \leq C \left(\sum_{n \in \mathbb{Z}} e^{n\theta} \mathbb{E}\left[\Big\| \sum_{k=1}^K r_k \zeta_{-n}(e^n \cdot) u_k(e^n \cdot) \Big\|_{H_p^\gamma(\mathbb{R}^d)}^p \right] \right)^{\frac{1}{p}}.$$

Thus, using again the Kahane-Khintchine inequality, we have

$$\left(\mathbb{E}\left[\Big\| \sum_{k=1}^K r_k u_k \Big\|_{H_{p,\theta}^\gamma(G)}^r \right] \right)^{\frac{1}{r}} \leq C \left(\sum_{n \in \mathbb{Z}} e^{n\theta} \left(\mathbb{E}\left[\Big\| \sum_{k=1}^K r_k \zeta_{-n}(e^n \cdot) u_k(e^n \cdot) \Big\|_{H_p^\gamma(\mathbb{R}^d)}^r \right] \right)^{\frac{p}{r}} \right)^{\frac{1}{p}}.$$

Since $H_p^\gamma(\mathbb{R}^d)$ has type r, see Remark 2.39, this leads to

$$\left(\mathbb{E}\left[\Big\| \sum_{k=1}^K r_k u_k \Big\|_{H_{p,\theta}^\gamma(G)}^r \right] \right)^{\frac{1}{r}} \leq C \left(\sum_{n \in \mathbb{Z}} e^{n\theta} \left(\sum_{k=1}^K \big\| \zeta_{-n}(e^n \cdot) u_k(e^n \cdot) \big\|_{H_p^\gamma(\mathbb{R}^d)}^r \right)^{\frac{p}{r}} \right)^{\frac{1}{p}}.$$

Therefore, applying the triangle inequality in $L_{p/r}(\mathbb{Z}, \mathscr{P}(\mathbb{Z}), \sum_n e^{n\theta}\delta_n; \mathbb{R})$ yields

$$
\left(\mathbb{E}\left[\left\| \sum_{k=1}^{K} r_k u_k \right\|_{H^\gamma_{p,\theta}(G)}^r \right] \right)^{\frac{1}{r}} \leq C \left\| \sum_{k=1}^{K} \left(n \mapsto \left\| \zeta_{-n}(e^n \cdot) u_k(e^n \cdot) \right\|_{H^\gamma_p(\mathbb{R}^d)}^r \right) \right\|_{L_{p/r}\left(\mathbb{Z}, \mathscr{P}(\mathbb{Z}), \sum_n e^{n\theta}\delta_n; \mathbb{R}\right)}^{\frac{1}{r}}
$$

$$
\leq C \left(\sum_{k=1}^{K} \left\| n \mapsto \left\| \zeta_{-n}(e^n \cdot) u_k(e^n \cdot) \right\|_{H^\gamma_p(\mathbb{R}^d)}^r \right\|_{L_{p/r}\left(\mathbb{Z}, \mathscr{P}(\mathbb{Z}), \sum_n e^{n\theta}\delta_n; \mathbb{R}\right)} \right)^{\frac{1}{r}}
$$

$$
= C \left(\sum_{k=1}^{K} \left(\sum_{n\in\mathbb{Z}} e^{n\theta} \left\| \zeta_{-n}(e^n \cdot) u_k(e^n \cdot) \right\|_{H^\gamma_p(\mathbb{R}^d)}^p \right)^{\frac{r}{p}} \right)^{\frac{1}{r}}
$$

$$
= C \left(\sum_{k=1}^{K} \| u_k \|_{H^\gamma_{p,\theta}(G)}^r \right)^{\frac{1}{r}}.
$$

The assertion follows. □

Remember that in this thesis we are mainly concerned with SPDEs on bounded Lipschitz domains $\mathcal{O} \subset \mathbb{R}^d$. As already mentioned, the weighted Sobolev spaces introduced above will serve as state spaces for the solutions processes $u = (u(t))_{t\in[0,T]}$ of the SPDEs under consideration. Therefore, since we are interested in solutions fulfilling a zero Dirichlet boundary condition, we need to check whether the elements of the the weighted Sobolev spaces introduced above 'vanish at the boundary' in an adequate way. In order to answer this question for the relevant range of parameters (this will be done in Remark 4.3) we will need the following lemma. It is an immediate consequence of [87, Theorem 9.7]. Let us mention that this result holds for a wider class of domains. E.g. it stays true for bounded domains with Hölder continuous boundary, see [87, Remark 9.8(ii)] for details. However, in the course of this thesis, we will not need these generalizations.

Lemma 2.51. *For a bounded Lipschitz domain $\mathcal{O} \subset \mathbb{R}^d$ and $k \in \mathbb{N}_0$,*

$$
\mathring{W}^k_p(\mathcal{O}) = H^k_{p,d-kp}(\mathcal{O})
$$

with equivalent norms.

In order to formulate the stochastic equations under consideration, we will use the following spaces $H^\gamma_{p,\theta}(G; \ell_2)$. They are counterparts of the spaces $H^\gamma_p(\mathbb{R}^d; \ell_2)$ introduced in the previous subsection. We define and discuss them for the general case of arbitrary domains with non-empty boundary, although later on we are mainly interested in the case of bounded Lipschitz domains.

Definition 2.52. Let G be an arbitrary domain in $\mathbb{R}^d$ with non-empty boundary. For $\gamma \in \mathbb{R}$, $p \in (1, \infty)$ and $\theta \in \mathbb{R}$, we define

$$
H^\gamma_{p,\theta}(G; \ell_2) := \left\{ g = (g^k)_{k\in\mathbb{N}} \in \left(H^\gamma_{p,\theta}(G) \right)^{\mathbb{N}} : \right.
$$

$$
\left. \| g \|_{H^\gamma_p(G; \ell_2)}^p := \sum_{n\in\mathbb{Z}} e^{n\theta} \left\| \left(\zeta_{-n}(e^n \cdot) g^k(e^n \cdot) \right)_{k\in\mathbb{N}} \right\|_{H^\gamma_p(\ell_2)}^p < \infty \right\},
$$

with ζ_n, $n \in \mathbb{Z}$, from above, cf. (2.27).

Remark 2.53. For $p \in (1, \infty)$ and $\gamma, \theta \in \mathbb{R}$, $H^\gamma_{p,\theta}(G; \ell_2)$ is a Banach space. This can be proven by following the lines of the proof of the completeness of $H^\gamma_{p,\theta}(G)$ presented in [93, Proposition 2.4.1]. The details are left to the reader.

In Chapter 6 we will need the fact that $H^\gamma_{p,\theta}(G)$ is isomorphic to the corresponding class of γ-radonifying operators from $\ell_2(\mathbb{N})$ to $H^\gamma_{p,\theta}(G)$. We prove this now. As in Subsection 2.2.2, from now on $\{\gamma_k : k \in \mathbb{N}\}$ denotes a Gaussian sequence, see Definition 2.11. Remember that for $h \in \ell_2$ and $u \in E$, where E is a Banach space, we use to write $h \otimes u$ for the rank one operator $\langle h, \cdot\rangle_{\ell_2} u$, see also (2.7).

Theorem 2.54. *Let G be an arbitrary domain in $\mathbb{R}^d$ with non-empty boundary. Furthermore, let $\gamma, \theta \in \mathbb{R}$ and $p \in [2, \infty)$. Then, the operator*

$$\Phi : H^\gamma_{p,\theta}(G; \ell_2) \to \Gamma(\ell_2, H^\gamma_{p,\theta}(G))$$

$$(g^k)_{k\in\mathbb{N}} \mapsto \sum_{k=1}^{\infty} e_k \otimes g^k \qquad (\text{convergence in } \Gamma(\ell_2, H^\gamma_{p,\theta}(G)))$$

is an isomorphism, and therefore,

$$H^\gamma_{p,\theta}(G; \ell_2) \simeq \Gamma(\ell_2, H^\gamma_{p,\theta}(G)).$$

Proof. First of all we show that Φ is well-defined and bounded. Fix $g \in H^\gamma_{p,\theta}(G; \ell_2)$. Then, using the equality (2.9) from Theorem 2.18 together with the norm equivalence (2.8), for any $m_1, m_2 \in \mathbb{N}$, we can estimate the norm of the finite rank operator $\sum_{k=m_1}^{m_2} e_k \otimes g^k$ as follows

$$\left\| \sum_{k=m_1}^{m_2} e_k \otimes g^k \right\|^p_{\Gamma(\ell_2, H^\gamma_{p,\theta}(G))} \leq C\, \mathbb{E}\left[\left\| \sum_{k=m_1}^{m_2} \gamma_k\, g^k \right\|^p_{H^\gamma_{p,\theta}(G)} \right].$$

Since for every $\omega \in \Omega$,

$$\left\| \sum_{k=m_1}^{m_2} \gamma_k(\omega) g^k \right\|^p_{H^\gamma_{p,\theta}(G)} = \sum_{n\in\mathbb{Z}} e^{n\theta} \left\| \sum_{k=m_1}^{m_2} \gamma_k(\omega)\zeta_{-n}(e^n\cdot)g^k(e^n\cdot) \right\|^p_{H^\gamma_p(\mathbb{R}^d)},$$

with $\{\zeta_n : n \in \mathbb{Z}\}$ as defined in (2.27), an application of Beppo-Levi's theorem yields

$$\left\| \sum_{k=m_1}^{m_2} e_k \otimes g^k \right\|^p_{\Gamma(\ell_2, H^\gamma_{p,\theta}(G))} \leq C \sum_{n\in\mathbb{Z}} e^{n\theta} \mathbb{E}\left[\left\| \sum_{k=m_1}^{m_2} \gamma_k\, \zeta_{-n}(e^n\cdot)g^k(e^n\cdot) \right\|^p_{H^\gamma_p(\mathbb{R}^d)} \right]. \qquad (2.39)$$

For every $n \in \mathbb{Z}$, we can apply Equality (2.9) from Theorem 2.18 to the finite rank operator

$$\sum_{k=m_1}^{m_2} e_k \otimes \left(\zeta_{-n}(e^n\cdot)g^k(e^n\cdot)\right) \in \mathcal{L}_f(\ell_2, H^\gamma_p(\mathbb{R}^d)) \subseteq \Gamma(\ell_2, H^\gamma_p(\mathbb{R}^d)),$$

followed by the norm equivalence (2.8), and obtain

$$\mathbb{E}\left[\left\| \sum_{k=m_1}^{m_2} \gamma_k\, \zeta_{-n}(e^n\cdot)g^k(e^n\cdot) \right\|^p_{H^\gamma_p(\mathbb{R}^d)} \right] = C \left\| \sum_{k=m_1}^{m_2} e_k \otimes \left(\zeta_{-n}(e^n\cdot)g^k(e^n\cdot)\right) \right\|^p_{\Gamma_p(\ell_2, H^\gamma_p(\mathbb{R}^d))}$$

$$\leq C \left\| \sum_{k=m_1}^{m_2} e_k \otimes \left(\zeta_{-n}(e^n\cdot)g^k(e^n\cdot)\right) \right\|^p_{\Gamma(\ell_2, H^\gamma_p(\mathbb{R}^d))}.$$

Thus, if we set

$$\tilde{g}^k_n(m_1, m_2) := \left\{ \begin{array}{ll} \zeta_{-n}(e^n\cdot)g^k(e^n\cdot), & \text{if } k \in \{m_1, \ldots, m_2\} \\ 0 & , \text{ else} \end{array} \right\},$$

obviously, $\left(\tilde{g}_n^k(m_1, m_2)\right)_{k \in \mathbb{N}} \in H_p^\gamma(\mathbb{R}^d; \ell_2)$, and an application of Theorem 2.42 leads to

$$\mathbb{E}\left[\left\|\sum_{k=m_1}^{m_2} \gamma_k \, \zeta_{-n}(e^n \cdot) g^k(e^n \cdot)\right\|_{H_p^\gamma(\mathbb{R}^d)}^p\right] \leq C \left\|\left(\tilde{g}_n^k(m_1, m_2)\right)_{k \in \mathbb{N}}\right\|_{H_p^\gamma(\mathbb{R}^d; \ell_2)}^p,$$

the constant C being independent of $n \in \mathbb{Z}$ and g. Inserting this into the estimate (2.39), we obtain

$$\left\|\sum_{k=m_1}^{m_2} e_k \otimes g^k\right\|_{\Gamma(\ell_2, H_{p,\theta}^\gamma(G))}^p \leq C \sum_{n \in \mathbb{Z}} e^{n\theta} \left\|\left(\tilde{g}_n^k(m_1, m_2)\right)_{k \in \mathbb{N}}\right\|_{H_p^\gamma(\mathbb{R}^d; \ell_2)}^p.$$

Since $g \in H_{p,\theta}^\gamma(G; \ell_2)$, the right hand side converges to zero for $m_1, m_2 \to \infty$ by Lebesgue's dominated convergence theorem. Thus, the sequence

$$(R_m)_{m \in \mathbb{N}} := \left(\sum_{k=1}^m e_k \otimes g^k\right)_{m \in \mathbb{N}} \subseteq \mathcal{L}_f(\ell_2, H_{p,\theta}^\gamma(G))$$

converges in $\Gamma(\ell_2, H_{p,\theta}^\gamma(G))$ and its limit $\sum_{k=1}^\infty e_k \otimes g^k$ is well-defined. The boundedness of Φ can now be proven by repeating the calculations above with $m_1 = 1$ and $m_2 = \infty$.

By the open mapping theorem, showing that

$$\widetilde{\Phi} : \Gamma(\ell_2, H_{p,\theta}^\gamma(G)) \to H_{p,\theta}^\gamma(G; \ell_2)$$
$$R \mapsto (Re_k)_{k \in \mathbb{N}}$$

is the inverse of Φ, would finish the proof. Let us check whether this operator is well-defined. If so, then the fact that it is the inverse of Φ follows by simple calculations. Fix $R \in \Gamma(\ell_2, H_{p,\theta}^\gamma(G))$. Since for every $n \in \mathbb{Z}$, the operator

$$S_n : H_{p,\theta}^\gamma(G) \to H_p^\gamma(\mathbb{R}^d)$$
$$u \mapsto \zeta_{-n}(e^n \cdot) u(e^n \cdot)$$

is obviously bounded, the composition $S_n R$ is γ-radonifying, i.e., $S_n R \in \Gamma(\ell_2, H_p^\gamma(\mathbb{R}^d))$, see Theorem 2.17. Furthermore, by Theorem 2.42,

$$\left\|(S_n Re_k)_{k \in \mathbb{N}}\right\|_{H_p^\gamma(\mathbb{R}^d; \ell_2)} = \left\|(\zeta_{-n}(e^n \cdot) Re_k(e^n \cdot))_{k \in \mathbb{N}}\right\|_{H_p^\gamma(\mathbb{R}^d; \ell_2)} \leq C \left\|S_n R\right\|_{\Gamma(\ell_2, H_p^\gamma(\mathbb{R}^d))}$$

with a constant C independent of $n \in \mathbb{Z}$ and R. Using this together with Equality (2.9) from Theorem 2.18 together with the norm equivalence (2.8), yields

$$\sum_{n \in \mathbb{Z}} e^{n\theta} \left\|(\zeta_{-n}(e^n \cdot) Re_k(e^n \cdot))_{k \in \mathbb{N}}\right\|_{H_p^\gamma(\mathbb{R}^d; \ell_2)}^p \leq C \sum_{n \in \mathbb{Z}} e^{n\theta} \left\|S_n R\right\|_{\Gamma(\ell_2, H_p^\gamma(\mathbb{R}^d))}^p$$

$$\leq C \sum_{n \in \mathbb{Z}} e^{n\theta} \mathbb{E}\left[\left\|\sum_{k=1}^\infty \gamma_k S_n Re_k\right\|_{H_p^\gamma(\mathbb{R}^d)}^p\right].$$

Applying Beppo-Levi's theorem and using the definitions of the norms we obtain

$$\sum_{n \in \mathbb{Z}} e^{n\theta} \left\|(\zeta_{-n}(e^n \cdot) Re_k(e^n \cdot))_{k \in \mathbb{N}}\right\|_{H_p^\gamma(\mathbb{R}^d; \ell_2)}^p \leq C \mathbb{E}\left[\sum_{n \in \mathbb{Z}} e^{n\theta} \left\|\sum_{k=1}^\infty \gamma_k S_n Re_k\right\|_{H_p^\gamma(\mathbb{R}^d)}^p\right]$$

$$= C \mathbb{E}\left[\sum_{n \in \mathbb{Z}} e^{n\theta} \left\|\sum_{k=1}^\infty \gamma_k \, \zeta_{-n}(e^n \cdot) Re_k(e^n \cdot)\right\|_{H_p^\gamma(\mathbb{R}^d)}^p\right]$$

$$= C \mathbb{E}\left[\left\|\sum_{k=1}^\infty \gamma_k Re_k\right\|_{H_{p,\theta}^\gamma(G)}^p\right].$$

Therefore, another application of Equality (2.9) from Theorem 2.18 followed by the norm equivalence (2.8), leads to

$$\sum_{n\in\mathbb{Z}} e^{n\theta} \left\| \left(\zeta_{-n}(e^n \cdot) Re_k(e^n \cdot) \right)_{k\in\mathbb{N}} \right\|_{H_p^\gamma(\mathbb{R}^d; \ell_2)}^p \le C \|R\|_{\Gamma(\ell_2, H_{p,\theta}^\gamma(G))}^p.$$

Thus, $(Re_k)_{k\in\mathbb{N}} \in H_{p,\theta}^\gamma(G; \ell_2)$. $\square$

We occasionally use the following properties of the spaces $H_{p,\theta}^\gamma(G; \ell_2)$ in this thesis. In several publications like [73, 75], these properties are stated and used without proof. Since we did not find any proof in the literature, we sketch a proof based on the isomorphy from Theorem 2.54 above. The details are left to the reader.

Lemma 2.55. *Let G be an arbitrary domain in $\mathbb{R}^d$ with non-empty boundary, $p \in (1, \infty)$ and $\gamma, \theta \in \mathbb{R}$.*

(i) *$g = (g^k)_{k\in\mathbb{N}} \in H_{p,\theta}^\gamma(G; \ell_2)$ if, and only if, $g, (\psi g_x^k)_{k\in\mathbb{N}} \in H_{p,\theta}^{\gamma-1}(G; \ell_2)$ and*

$$\|g\|_{H_{p,\theta}^\gamma(G;\ell_2)} \le C \left\| (\psi g_x^k)_{k\in\mathbb{N}} \right\|_{H_{p,\theta}^{\gamma-1}(G;\ell_2)} + C\|g\|_{H_{p,\theta}^{\gamma-1}(G;\ell_2)} \le C\|g\|_{H_{p,\theta}^\gamma(G;\ell_2)}.$$

Also, $g = (g^k)_{k\in\mathbb{N}} \in H_{p,\theta}^\gamma(G; \ell_2)$ if, and only if, $g, ((\psi g^k)_x)_{k\in\mathbb{N}} \in H_{p,\theta}^{\gamma-1}(G; \ell_2)$ and

$$\|g\|_{H_{p,\theta}^\gamma(G;\ell_2)} \le C \left\| ((\psi g^k)_x)_{k\in\mathbb{N}} \right\|_{H_{p,\theta}^{\gamma-1}(G;\ell_2)} + C\|g\|_{H_{p,\theta}^{\gamma-1}(G;\ell_2)} \le C\|g\|_{H_{p,\theta}^\gamma(G;\ell_2)}.$$

(ii) *For any $\nu, \gamma \in \mathbb{R}$, $\psi^\nu H_{p,\theta}^\gamma(G; \ell_2) = H_{p,\theta-\nu p}^\gamma(G; \ell_2)$ and*

$$\|g\|_{H_{p,\theta-\nu p}^\gamma(G;\ell_2)} \le C \left\| (\psi^{-\nu} g^k)_{k\in\mathbb{N}} \right\|_{H_{p,\theta}^\gamma(G;\ell_2)} \le C\|g\|_{H_{p,\theta-\nu p}^\gamma(G;\ell_2)}.$$

(iii) *There exists a constant $c_0 > 0$ depending on p, θ, γ and the function ψ such that, for all $c \ge c_0$, the operator*

$$\psi^2 \Delta - c : H_{p,\theta}^{\gamma+1}(G; \ell_2) \to H_{p,\theta}^{\gamma-1}(G; \ell_2)$$
$$g = (g^k)_{k\in\mathbb{N}} \mapsto (\psi^2 \Delta - c)g := \left((\psi^2 \Delta - c)g^k \right)_{k\in\mathbb{N}}$$

is an isomorphism.

(iv) *If G is bounded, then $H_{p,\theta_1}^\gamma(G; \ell_2) \hookrightarrow H_{p,\theta_2}^\gamma(G; \ell_2)$ for $\theta_1 < \theta_2$.*

(v) *If $0 < \eta < 1$, $\gamma = (1-\eta)\nu_0 + \eta\nu_1$, $1/p = (1-\eta)/p_0 + \eta/p_1$ and $\theta = (1-\eta)\theta_0 + \eta\theta_1$ with $\nu_0, \nu_1, \theta_0, \theta_1 \in \mathbb{R}$ and $p_0, p_1 \in (1, \infty)$, then*

$$H_{p,\theta}^\gamma(G; \ell_2) = \left[H_{p_0,\theta_0}^{\nu_0}(G; \ell_2), H_{p_1,\theta_1}^{\nu_1}(G; \ell_2) \right]_\eta \qquad \text{(equivalent norms)}.$$

Sketch of proof. The assertions can be proven by using the isomorphism from Theorem 2.54 together with the ideal property of γ-radonifying operators (Theorem 2.17) and the corresponding properties of the spaces $H_{p,\theta}^\gamma(G)$, $\gamma, \theta \in \mathbb{R}$, $p \in (1, \infty)$, from Lemma 2.45. In order to prove the interpolation statement (v) one additionally needs the fact that

$$\left[\Gamma(\ell_2, H_{p_0,\theta_0}^{\nu_0}(G)), \Gamma(\ell_2, H_{p_1,\theta_1}^{\nu_1}(G)) \right]_\eta = \Gamma(\ell_2, H_{p,\theta}^\gamma(G)).$$

This follows from [114, Theorem 2.1] with $H_0 = H_1 = \ell_2$, $X_0 = H_{p_0,\theta_0}^{\nu_0}(G)$ and $X_1 = H_{p_1,\theta_1}^{\nu_1}(G)$. (Note that, by a result of G. Pisier, the B-convexity of X_0 and X_1 assumed in [114, Theorem 2.1] is equivalent to the fact that the Banach spaces have non-trivial type, see, e.g., [49, Theorem 13.10] for a proof.) $\square$

2.3.4 Besov spaces

One of the main goals of this thesis is to analyse the spatial regularity of the solutions to SPDEs on bounded Lipschitz domains in the particular scale $(*)$ of Besov spaces. In this subsection we recall the definition of Besov spaces by means of the Fourier transform and present an alternative (intrinsic) characterization via differences. In certain publications, such as [27], [46] and [48], the latter is used as a definition of Besov spaces for $0 < p, q < \infty$ and $s > 0$.

Denote by $\varphi_0 \in C_0^\infty(\mathbb{R}^d)$ a compactly supported, infinitely differentiable function having the properties

$$\varphi_0(x) = 1 \text{ if } |x| \leq 1 \quad \text{and} \quad \varphi_0(x) = 0 \text{ if } |x| \geq 3/2. \tag{2.40}$$

For $k \in \mathbb{N}$ define

$$\varphi_k(x) := \varphi_0(2^{-k}x) - \varphi_0(2^{-k+1}x) \quad \text{for} \quad x \in \mathbb{R}^d \tag{2.41}$$

to obtain a smooth dyadic resolution of unity on $\mathbb{R}^d$, i.e., $\varphi_k \in C_0^\infty(\mathbb{R}^d)$ for all $k \in \mathbb{N}$, and

$$\sum_{k=0}^\infty \varphi_k(x) = 1 \quad \text{for all} \quad x \in \mathbb{R}^d. \tag{2.42}$$

Definition 2.56. Let $\{\varphi_k\}_{k \in \mathbb{N}_0} \subseteq C_0^\infty(\mathbb{R}^d)$ be a resolution of unity according to (2.40)–(2.42).

(i) Let $0 < p, q < \infty$, $s \in \mathbb{R}$, and

$$\|u\|_{B_{p,q}^s(\mathbb{R}^d)} := \left(\sum_{k=0}^\infty 2^{ksq} \|\mathfrak{F}^{-1}[\varphi_k \mathfrak{F} u]\|_{L_p(\mathbb{R}^d)}^q \right)^{1/q}.$$

Then

$$B_{p,q}^s(\mathbb{R}^d) := \left\{ u \in \mathcal{S}'(\mathbb{R}^d) \, : \, \|u\|_{B_{p,q}^s(\mathbb{R}^d)} < \infty \right\}$$

is the *Besov space of (smoothness) order s with summability parameter p and fine tuning parameter q*.

(ii) Let $G \subseteq \mathbb{R}^d$ be an arbitrary domain. Then, for $0 < p, q < \infty$ and $s \in \mathbb{R}$, the *Besov space $B_{p,q}^s(G)$ of (smoothness) order s with summability parameter p and fine tuning parameter q on G* is defined as follows:

$$B_{p,q}^s(G) := \left\{ u \in \mathcal{D}'(G) \, : \, \text{there exists } g \in B_{p,q}^s(\mathbb{R}^d) \, : \, g|_G = u \right\}.$$

It is endowed with the norm

$$\|u\|_{B_{p,q}^s(G)} := \inf_{\substack{g \in B_{p,q}^s(\mathbb{R}^d) \\ g|_G = u}} \|g\|_{B_{p,q}^s(\mathbb{R}^d)}, \qquad u \in B_{p,q}^s(G). \tag{2.43}$$

Remark 2.57. For $0 < p, q < \infty$ and $s \in \mathbb{R}$, the Besov space $B_{p,q}^s(G)$, endowed with the quasi-norm $\|\cdot\|_{B_{p,q}^s(G)}$, is a quasi-Banach space. If $1 \leq p, q < \infty$, then $\|\cdot\|_{B_{p,q}^s(G)}$ is a norm and therefore $(B_{p,q}^s(G), \|\cdot\|_{B_{p,q}^s(G)})$ is a Banach space. A proof can be found in [115, Theorem 2.3.3(i)] for the case $G = \mathbb{R}^d$. For general domains we refer to the proof of [115, Proposition 3.2.3(i)]. Note that the assumed smoothness property for the boundary of the underlying domain therein does not have any relevance in the proof of the completeness of the Besov spaces.

As already mentioned in the introduction to this subsection, besides the definition given above, Besov spaces are frequently defined by means of differences. In general, it is not immediately clear whether the two definitions yield the same spaces. However, if $G = \mathbb{R}^d$, for $p, q \in (0, \infty)$ and $s > \max\{0, d(1/p - 1)\}$, the two ways of defining Besov spaces match. This follows from [115, Theorem 2.5.12] and the integral transformation formula for rotationally symmetric functions, see, e.g. [111, Corollary 15.14]. In this thesis, we will also need this statement for the case that G is a bounded Lipschitz domain in $\mathbb{R}^d$. It has been proven in [50, Theorem 3.18]. In the next theorem, we present these results in detail. We use the following notation. Let G be an arbitrary domain in $\mathbb{R}^d$. For a function $u : G \to \mathbb{R}$ and a natural number $n \in \mathbb{N}$ let

$$\Delta_h^n[u](x) := \Delta_h^n u(x) := \prod_{i=0}^{n} \mathbb{1}_G(x + ih) \cdot \sum_{j=0}^{n} \binom{n}{j} (-1)^{n-j} \, u(x + jh)$$

be the *n-th difference* of u with step $h \in \mathbb{R}^d$. For $p \in (0, \infty)$, the *n*-th order L_p-modulus of smoothness of u is given by

$$\omega^n(t, u, G)_p := \omega^n(t, u)_p := \sup_{|h| < t} \left\| \Delta_h^n u \right\|_{L_p(G)}, \qquad t > 0.$$

Theorem 2.58. *Let G be either $\mathbb{R}^d$ or a bounded Lipschitz domain in $\mathbb{R}^d$. Let $p, q \in (0, \infty)$, $s > \max\{0, d(1/p - 1)\}$ and $n \in \mathbb{N}$ with $n > s$. Then $B_{p,q}^s(G)$ is the collection of all functions $u \in L_p(G)$ such that*

$$|u|_{B_{p,q}^s(G)} := \left(\int_0^\infty \left(t^{-s} \, \omega^n(t, u, G)_p \right)^q \frac{dt}{t} \right)^{1/q} < \infty. \tag{2.44}$$

The function

$$B_{p,q}^s(G) \ni u \mapsto \|u\|_{L_p(G)} + |u|_{B_{p,q}^s(G)} \tag{2.45}$$

is an equivalent (quasi-)norm for $\|\cdot\|_{B_{p,q}^s(G)}$ on $B_{p,q}^s(G)$.

Remark 2.59. In this thesis we will be mainly concerned with Besov spaces $B_{p,p}^s(G)$ with $p \geq 2$ and $s > 0$, and on the non-linear approximation scale $B_{\tau,\tau}^\alpha(G)$, $1/\tau = \alpha/d + 1/p$, $\alpha > 0$, with $p \geq 2$, where either $G = \mathbb{R}^d$ or G is a bounded Lipschitz domain in $\mathbb{R}^d$. In both cases, the parameters fulfil the assumptions from Theorem 2.58. Therefore, the definition based on the Fourier transform and the intrinsic characterization of Besov spaces via differences are equivalent.

In the next theorem we collect some parameter constellations, for which Besov and Sobolev spaces coincide. As in the theorem before, we restrict ourselves to the cases which are relevant for this thesis and assume that the underlying domain is either the whole space or a bounded Lipschitz domain in $\mathbb{R}^d$. For $G = \mathbb{R}^d$ the statements are taken from [116]: The first assertion can be found in [116, Section 2.5.1, especially Remark 4] and the second follows from [116, Theorem 2.3.2(d) together with Theorem 2.3.3(b)]. For the case that G is a bounded Lipschitz domain, the first assertion can be found in [116, Remark 4.4.2/2], whereas for the second statement we additionally need [116, Proposition 4.2.4 together with Definition 4.2.1/1 and Theorem 4.6.1(b)]. It is worth noting that in the just mentioned references, the statements are proven for more general bounded domains. That is, the statements hold for bounded domains of cone-type in the sense of [116, Definition 4.2.3]. However, bounded Lipschitz domains are of cone-type, see e.g. [2, Sections 4.8, 4.9 and 4.11].

Theorem 2.60. *Let G be either $\mathbb{R}^d$ or a bounded Lipschitz domain in $\mathbb{R}^d$.*

(i) *For $p \in (1, \infty)$ and $s \in (0, \infty) \setminus \mathbb{N}$ the following equality holds (equivalent norms):*

$$W_p^s(G) = B_{p,p}^s(G).$$

(ii) *For $s \in (0, \infty)$ the following equality holds (equivalent norms):*

$$W_2^s(G) = B_{2,2}^s(G).$$

We present now three embeddings of Besov spaces, which we will frequently use in this thesis.

Theorem 2.61. **(i)** *Let G be an arbitrary domain in $\mathbb{R}^d$. Then, for any $p \in (0, \infty)$ and $s_0, s_1 \in \mathbb{R}$ with $s_1 > s_0$ the following embedding holds:*

$$B_{p,p}^{s_1}(G) \hookrightarrow B_{p,p}^{s_0}(G). \tag{2.46}$$

(ii) *Let G be either $\mathbb{R}^d$ or a bounded Lipschitz domain in $\mathbb{R}^d$ and fix $p \in (1, \infty)$. Furthermore, assume that $\alpha_2 > \alpha_1 > 0$ and let $\tau_1, \tau_2 > 0$ fulfil*

$$\frac{1}{\tau_i} = \frac{\alpha_i}{d} + \frac{1}{p}, \qquad i = 1, 2.$$

Then the following embeddings hold:

$$B_{\tau_2,\tau_2}^{\alpha_2}(G) \hookrightarrow B_{\tau_1,\tau_1}^{\alpha_1}(G) \hookrightarrow L_p(G). \tag{2.47}$$

(iii) *Let $\mathcal{O}$ be a bounded Lipschitz domain in $\mathbb{R}^d$ and let $0 < p < q < \infty$. Then, for*

$$s > s - \varepsilon > \max\left\{0, d\left(\frac{1}{p} - 1\right)\right\}, \tag{2.48}$$

the following embedding holds:

$$B_{q,q}^s(\mathcal{O}) \hookrightarrow B_{p,p}^{s-\varepsilon}(\mathcal{O}). \tag{2.49}$$

Proof. **(i)** This is an immediate consequence of the definition of Besov spaces given above.

(ii) For $G = \mathbb{R}^d$, the first embedding in (2.47) is proved in [27, Corollary 3.7.1], whereas the second embedding can be found in [117, Theorem 1.73(i)]. If G is a bounded Lipschitz domain, the assertion follows from the case $G = \mathbb{R}^d$ by using the existence and boundedness of the extension operator introduced in [110] for bounded Lipschitz domains.

(iii) Note that since (2.48) holds, we are in the setting of Theorem 2.58. Thus, using the equivalent characterisation of Besov spaces via differences, we immediately obtain

$$B_{q,q}^s(\mathcal{O}) \hookrightarrow B_{p,q}^s(\mathcal{O}),$$

by an application of Hölder's inequality to the moduli of smoothness (remember that $\mathcal{O}$ is bounded). In order to prove that simultaneously

$$B_{p,q}^s(\mathcal{O}) \hookrightarrow B_{p,p}^{s-\varepsilon}(\mathcal{O})$$

holds, we can argue as follows. First one can check the equivalence of the (quasi-)semi-norm (2.44) and

$$\left\| \left(2^{js} \omega^n (2^{-j}, u, \mathcal{O})_p\right)_{j \in \mathbb{N}} \right\|_{\ell_q},$$

see also [27, Remark 3.2.1]. Then, the arguments from the proof of [115, Proposition 2.3.2/2(ii)] yield the asserted embedding. □

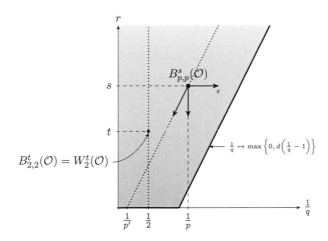

Figure 2.1:   Visualisation of Besov spaces on bounded Lipschitz
domains $\mathcal{O} \subset \mathbb{R}^d$ in a DeVore/Triebel diagram.

Remark 2.62. **(i)** In order to prove the embeddings (2.47) for bounded Lipschitz domains, we only need the existence of a linear and bounded extension operator as presented in [110]. Thus, the statement of Theorem 2.61(ii) stays true for any domain $G \subseteq \mathbb{R}^d$, where a linear and bounded extension operator from the proper Besov spaces on G to the corresponding Besov spaces on $\mathbb{R}^d$ (and from $L_p(G)$ to $L_p(\mathbb{R}^d)$) exist.

(ii) The short proof of Theorem 2.61(iii) presented above reveals that, if the Besov spaces are defined via differences, then the embedding (2.49) holds for any for any $s > s - \varepsilon > 0$ with $\mathcal{O}$ replaced by an arbitrary bounded domain $G \subset \mathbb{R}^d$.

Remark 2.63. In Figure 2.1 we use a so-called DeVore/Triebel diagram for a visualisation of the results presented above for bounded Lipschitz domains $\mathcal{O} \subset \mathbb{R}^d$. In this $(1/q, r)$-diagram, a point $(1/p, s)$ in the first quadrant $(0, \infty)^2$ stands for the Besov space $B^s_{p,p}(\mathcal{O})$ as introduced in Definition 2.56. The shaded area delimited by the coordinate axes and the ray with slope d starting at the point $(1, 0)$ represents the range of parameters $(1/q, r) \in (0, \infty)^2$ fulfilling

$$r > \max\left\{0, d\left(\frac{1}{q} - 1\right)\right\}.$$

In particular, for any $(1/q, r)$ in this area, the alternative characterization of the corresponding Besov space $B^r_{q,q}(\mathcal{O})$ via differences from Theorem 2.58 holds. As we have seen in Theorem 2.60(ii), for $p = 2$, the Besov spaces coincide with the Sobolev spaces introduced in Subsection 2.3.1. In our diagram these spaces are represented by the points above $(1/2, 0)$. The three arrows starting at $(1/p, s)$ stand for the three embeddings from Theorem 2.61. In clockwise orientation: Firstly, the arrow pointing to the right stands for (2.49); the ε at the arrowhead indicates that the smoothness decreases by an arbitrarily small $\varepsilon > 0$ in this case. Secondly, the trivial embedding (2.46) is represented by the arrow pointing straight down. Finally, the third arrow starting at $(1/p, s)$ and pointing to the south-west with a slope d stands for the first embedding in (2.47). This embedding is a generalization of the well-known Sobolev embedding. Therefore, a ray contained in the shaded area, starting at a point $(1/p', 0)$ with slope d is usually

called a *Sobolev embedding line*. Note that for $p \in (1, \infty)$, the non-linear approximation scale
(∗) of Besov spaces is represented by a Sobolev embedding line.

2.3.5 Triebel-Lizorkin spaces

In this subsection we present a Fourier analytic definition of the Triebel-Lizorkin spaces. They
will be used when analysing the relationship between Sobolev spaces with and without weights,
respectively, in Chapter 4.

Definition 2.64. Let $\{\varphi_k\}_{k \in \mathbb{N}_0} \subseteq \mathcal{C}_0^\infty(\mathbb{R}^d)$ be a resolution of unity according to (2.40)–(2.42).

(i) Let $0 < p, q < \infty$, $s \in \mathbb{R}$, and

$$\|u\|_{F_{p,q}^s(\mathbb{R}^d)} := \left\| \left(\sum_{k=0}^\infty 2^{ksq} \left| \mathfrak{F}^{-1} \left[\varphi_k \mathfrak{F} u \right] \right|^q \right)^{1/q} \right\|_{L_p(\mathbb{R}^d)}.$$

Then

$$F_{p,q}^s(\mathbb{R}^d) := \left\{ u \in \mathcal{S}'(\mathbb{R}^d) \ : \ \|u\|_{F_{p,q}^s(\mathbb{R}^d)} < \infty \right\}$$

is the Triebel-Lizorkin space of (smoothness) order s with summability parameters p, q.

(ii) Let $G \subseteq \mathbb{R}^d$ be an arbitrary domain. Then, for $0 < p, q < \infty$ and $s \in \mathbb{R}$, the Triebel-Lizorkin
space $F_{p,q}^s(G)$ of order s on G is defined as follows:

$$F_{p,q}^s(G) := \left\{ u \in \mathcal{D}'(G) \ : \ \text{there exists } g \in F_{p,q}^s(\mathbb{R}^d) \ : \ g|_G = u \right\}.$$

It is endowed with the norm

$$\|u\|_{F_{p,q}^s(G)} := \inf_{\substack{g \in F_{p,q}^s(\mathbb{R}^d) \\ g|_G = u}} \|g\|_{F_{p,q}^s(\mathbb{R}^d)}, \qquad u \in F_{p,q}^s(G).$$

Remark 2.65. For $0 < p, q < \infty$ and $s \in \mathbb{R}$, the Triebel-Lizorkin space $F_{p,q}^s(G)$, endowed with
the quasi-norm $\|\cdot\|_{F_{p,q}^s(G)}$, is a quasi-Banach space. If $1 \leq p, q < \infty$, then $\|\cdot\|_{F_{p,q}^s(G)}$ is a norm and
therefore $(F_{p,q}^s(G), \|\cdot\|_{F_{p,q}^s(G)})$ is a Banach space. A proof can be found in [115, Theorem 2.3.3(ii)]
for the case $G = \mathbb{R}^d$. For general domains we refer to the proof of [115, Proposition 3.2.3(iii)].
Note that the assumed smoothness property for the boundary of the underlying domain therein
does not have any relevance in the proof of the completeness of the Triebel-Lizorkin spaces.

The following relationships of Triebel-Lizorkin spaces to Besov and Sobolev spaces respectively will be used in this thesis. A proof of assertion (i) can be found in [116, Theorem 4.6.1(b)].
The second part of the theorem is taken from [117, Proposition 1.122(i)]—at least the case $m \in \mathbb{N}$.
A proof for the more general case of bounded domains of cone-type in the sense of [116, Definition 4.2.3] can be found in [116, Theorem 4.2.4]. For $m = 0$, (2.50) holds for arbitrary domains
$G \subseteq \mathbb{R}^d$ instead of $\mathcal{O}$. This follows from the fact that $F_{p,2}^0(\mathbb{R}^d) = L_p(\mathbb{R}^d)$. A proof of the latter
can be found in [115, Proposition 2.5.6].

Theorem 2.66. (i) *Let G be an arbitrary domain in $\mathbb{R}^d$. Then, for $p \geq 2$ and $s \in \mathbb{R}$,*

$$F_{p,2}^s(G) \hookrightarrow B_{p,p}^s(G).$$

(ii) *Let $\mathcal{O}$ be a bounded Lipschitz domain in $\mathbb{R}^d$. Then, for $p \in (1, \infty)$ and $m \in \mathbb{N}_0$,*

$$F_{p,2}^m(\mathcal{O}) = W_p^m(\mathcal{O}) \qquad \text{(equivalent norms)}. \tag{2.50}$$

As a consequence, one obtains the following relationship between Sobolev spaces and Besov spaces on bounded Lipschitz domains, in the case that the smoothness parameter is a non-negative integer.

Corollary 2.67. *Let $\mathcal{O}$ be a bounded Lipschitz domain in $\mathbb{R}^d$. Then, for $p \geq 2$ and $m \in \mathbb{N}_0$,*

$$W_p^m(\mathcal{O}) \hookrightarrow B_{p,p}^m(\mathcal{O}).$$

2.4 Semigroups of linear operators

In the *semigroup approach* to SPDEs, the equation is rewritten as a vector-valued ordinary stochastic differential equation of the form

$$\left.\begin{aligned} \mathrm{d}U(t) + AU(t)\,\mathrm{d}t &= F(U(t),t)\,\mathrm{d}t + \Sigma(U(t),t)\,\mathrm{d}W_H(t), \qquad t \in [0,T] \\ U(0) &= u_0. \end{aligned}\right\}$$

The leading operator A is usually unbounded and $-A$ generates a strongly continuous semigroup on a suitable Banach space. In this section we recall the terminology from the semigroup theory, focusing first on analytic semigroups. Afterwards, we give a definition of what is sometimes called variational operators and collect some properties needed later on.

Let E be a Banach space (real or complex). In general, we call a linear mapping $B : D(B) \subseteq E \to E$, defined on a linear subspace $D(B)$ of E, a linear operator with *domain* $D(B)$. $(B, D(B))$ is said to be *closed*, if its graph $\{(x, Bx) : x \in D(B)\}$ is a closed subset of $E \times E$. It is *densely defined*, if $D(B)$ is dense in E, i.e., if

$$\overline{D(B)}^{\|\cdot\|_E} = E.$$

The *resolvent set* of B is the set $\rho(B)$ consisting of all $\lambda \in \mathbb{C}$ for which there exists a bounded inverse $(\lambda - B)^{-1} : (E, \|\cdot\|_E) \to (D(B), \|\cdot\|_E)$ of $(\lambda - B) := (\lambda \mathrm{Id} - B)$. The *spectrum* of B is its complement $\sigma(B) := \mathbb{C} \setminus \rho(B)$. If B is an operator on a real Banach space we put $\rho(B) = \rho(B_{\mathbb{C}})$ and $\sigma(B) := \sigma(B_{\mathbb{C}})$, where $B_{\mathbb{C}}$ is a complexification of B, see, e.g., [98] or [8, p. 4ff.] for details.

Recall that a family $S = \{S(t)\}_{t \geq 0} \subseteq \mathcal{L}(E)$ of bounded linear operators is called a $\mathcal{C}_0$-*semigroup* (or, alternatively, a *strongly continuous semigroup*), if $S(0) = \mathrm{Id}$, $S(t)S(s) = S(t+s)$ for any $t, s \geq 0$, and $\lim_{t \downarrow 0} \|S(t)x - x\|_E = 0$ for every $x \in E$. $S = \{S(t)\}_{t \geq 0}$ is called a *contraction semigroup*, if additionally $\|S(t)\|_{\mathcal{L}(E)} \leq 1$ for all $t \geq 0$. The *infinitesimal generator*, or briefly the *generator*, of a $\mathcal{C}_0$-semigroup $S = \{S(t)\}_{t \geq 0}$ is the (unbounded) linear operator $B : D(B) \subseteq E \to E$ defined by

$$D(B) := \left\{ x \in E : \lim_{t \downarrow 0} \frac{S(t)x - x}{t} \text{ exists in } E \right\},$$

$$Bx := \lim_{t \downarrow 0} \frac{S(t)x - x}{t}, \qquad x \in D(B).$$

By [102, Corollary 2.5], any generator $(B, D(B))$ of a $\mathcal{C}_0$-semigroup is densely defined and closed. Therefore, the domain $D(B)$ endowed with the graph norm

$$\|x\|_{D(B)} := \|x\|_E + \|Bx\|_E, \qquad x \in D(B),$$

becomes a Banach space. Obviously, if $0 \in \rho(B)$, the graph norm is equivalent to $\|B \cdot\|_E$. A contraction semigroup $S = \{S(t)\}_{t \geq 0}$ is said to be *of negative type*, if there exists an $\omega < 0$ such that

$$\|S(t)\|_{\mathcal{L}(E)} \leq e^{\omega t}, \qquad t \geq 0.$$

From [102, Theorem 5.3] one can deduce that, if $(B, D(B))$ is the generator of a semigroup S of negative type, then $(A, D(A)) := (-B, D(B))$ is *positive* in the sense of [115, Definition 1.14.1], i.e., $(-\infty, 0] \subseteq \rho(A)$ and there exists a constant $C \in (0, \infty)$ such that

$$\|(A - \lambda)^{-1}\|_{\mathcal{L}(E)} \leq \frac{C}{1 + |\lambda|}, \qquad \lambda \in (-\infty, 0].$$

A $\mathcal{C}_0$-semigroup $S = \{S(t)\}_{t \geq 0}$ on the Banach space $E := L_p(G)$ with $p \in [1, \infty)$ is called *positive* if for each $t \geq 0$,

$$f \in L_p(G), \quad f \geq 0 \quad \text{a.e. on } G \quad \text{implies} \quad S(t)f \geq 0 \quad \text{a.e. on } G,$$

see [52, p. 353].

Analytic semigroups

Now we collect some definitions and results from the theory of analytic semigroups. We restrict ourselves to the topics we will need in this thesis and refer to the monographs [102], [52], or [94] for an in-depth treatment of the theory. For $\sigma \in (0, \pi)$, we write $\Sigma_\sigma := \{z \in \mathbb{C} \setminus \{0\} : |\arg(z)| < \sigma\}$.

Definition 2.68. Let $\sigma \in (0, \pi/2)$. A $\mathcal{C}_0$-semigroup $S = \{S(t)\}_{t \geq 0} \subseteq \mathcal{L}(E)$ acting on a Banach space E is called *analytic on Σ_σ* if

A1. S extends to an analytic function $S : \Sigma_\sigma \to \mathcal{L}(E)$, $z \mapsto S(z)$;

A2. $\lim_{z \to 0, z \in \Sigma_\sigma} S(z)x = x$ for every $x \in E$;

A3. $S(z_1)S(z_2) = S(z_1 + z_2)$ for $z_1, z_2 \in \Sigma_\sigma$.

We say that a $\mathcal{C}_0$-semigroup S is *analytic*, if it is analytic on Σ_σ for some $\sigma \in (0, \pi/2)$. If, in addition,

A4. $z \mapsto \|S(z)\|_{\mathcal{L}(E)}$ is bounded in $\Sigma_{\sigma'}$ for every $0 < \sigma' < \sigma$,

we call S a *bounded* analytic semigroup.

Next, we introduce the notion 'H^∞-calculus' of a sectorial operator. Originally developed by McIntosh and collaborators [6, 11, 95], it has found various applications in the context of (stochastic) partial differential equations. Our definition is taken from [122]. Let $(-A, D(-A))$ be the generator of a bounded analytic semigroup on a Banach space E. Then, see, e.g., [8, Proposition I.1.4.1],

$$\sigma(A) \subseteq \overline{\Sigma}_{\sigma_0} \text{ for some } \sigma_0 \in (0, \pi/2),$$

and for all $\sigma \in (\sigma_0, \pi)$,

$$\sup_{z \in \mathbb{C} \setminus \Sigma_\sigma} \|z(z - A)^{-1}\|_{\mathcal{L}(E)} < \infty,$$

i.e., in the terminology used e.g. in [60, Chapter 2], A is a *sectorial* operator. Let $H^\infty(\Sigma_\sigma)$ denote the Banach space of all bounded analytic functions $f : \Sigma_\sigma \to \mathbb{C}$ endowed with the supremum norm. Furthermore, $H_0^\infty(\Sigma_\sigma)$ denotes the subspace of $H^\infty(\Sigma_\sigma)$, consisting of all functions satisfying

$$|f(z)| \leq C \frac{|z|^\varepsilon}{(1 + |z|^2)^\varepsilon}, \qquad z \in \Sigma_\sigma, \tag{2.51}$$

for some $\varepsilon > 0$. For $f \in H_0^\infty(\Sigma_\sigma)$ and $\sigma' \in (\sigma_0, \sigma)$, due to (2.51), the $\mathcal{L}(E)$-valued Bochner integral

$$f(A) := \frac{1}{2\pi i} \int_{\partial \Sigma_{\sigma'}} f(z)(z - A)^{-1} \, \mathrm{d}z$$

converges absolutely. Furthermore, it is independent of σ'. We say that the operator $(A, D(A))$ admits a *bounded $H^\infty(\Sigma_\sigma)$-calculus* if there exists a constant $C \in (0, \infty)$ such that

$$\|f(A)\|_{\mathcal{L}(E)} \le C \, \|f\|_\infty := C \sup_{z \in \Sigma_\sigma} |f(z)|, \qquad f \in H_0^\infty(\Sigma_\sigma).$$

The infimum of all σ such that $(A, D(A))$ admits a bounded $H^\infty(\Sigma_\sigma)$-calculus is called *angle* of the calculus. The following two results are mentioned in [122].

Theorem 2.69 ([71, Corollary 5.2]). *Let $G \subseteq \mathbb{R}^d$ be an arbitrary domain and let $p \in [1, \infty)$. If $(-A, D(-A))$ is the generator of a positive analytic contraction semigroup on $L_p(G)$, then, $(A, D(A))$ admits a bounded H^∞-calculus of angle less than $\pi/2$.*

The next result can be derived from [60, Corollary 3.5.7].

Theorem 2.70. *If $0 \in \rho(A)$ and $(A, D(A))$ admits a bounded H^∞-calculus of angle less than $\pi/2$, then, A has bounded imaginary powers and $\sup_{t \in [-1,1]} \|A^{it}\|_{\mathcal{L}(E)} < \infty$.*

Variatonal operators

Let $(V, \langle \cdot, \cdot \rangle_V)$ be a separable real Hilbert space. Furthermore, let

$$a(\cdot, \cdot) : V \times V \to \mathbb{R}$$

be a continuous, symmetric and elliptic bilinear form. This means that there exist two constants $\delta_{\mathrm{ell}}, K_{\mathrm{ell}} > 0$, such that for arbitrary $u, v \in V$, the bilinear form fulfils the following conditions:

$$\delta_{\mathrm{ell}} \|u\|_V^2 \le a(u, u), \qquad a(u, v) = a(v, u), \qquad |a(u, v)| \le K_{\mathrm{ell}} \|u\|_V \|v\|_V. \tag{2.52}$$

Then, by the Lax-Milgram theorem, the operator

$$\begin{aligned} A : V &\to V^* \\ v &\mapsto Av := a(v, \cdot) \end{aligned} \tag{2.53}$$

is an isomorphism between V and its dual space V^*. Let us now assume that V is densely embedded into a real Hilbert space $(E, \langle \cdot, \cdot \rangle_E)$ via a linear embedding j. Then, the adjoint map $j^* : E^* \to V^*$ of j embeds E^* densely into the dual V^* of V. If we identify the Hilbert space E with its dual E^* via Riesz's isometric isomorphism $E \ni u \mapsto \Psi u := \langle u, \cdot \rangle_E \in E^*$, we obtain a so called *Gelfand triple* (V, E, V^*),

$$V \overset{j}{\hookrightarrow} E \overset{\Psi}{\cong} E^* \overset{j^*}{\hookrightarrow} V^*. \tag{2.54}$$

We have

$$\langle j(v_1), j(v_2) \rangle_E = \langle j^* \Psi j(v_1), v_2 \rangle_{V^* \times V} \qquad \text{for all } v_1, v_2 \in V. \tag{2.55}$$

It is worth noting that, although V is a Hilbert space, at this point we do *not* identify V and its dual V^* via the Riesz isomorphism in V. This would not match with (2.54) and (2.55). Here, the vector space V is considered as a subspace of V^* by means of the embedding $j^* \Psi j$, where Ψ is the Riesz isomorphism for E and not for V. In this setting, we can consider the operator

$A : V \to V^*$ as an (in general) unbounded operator on the intermediate space E. Therefore, we set

$$D(A) := D(A; E) := \{u \in V \ : \ Au \in j^*\Psi(E)\},$$

and define the operator

$$\tilde{A} : D(\tilde{A}) := j(D(A; E)) \subseteq E \to E$$
$$u \mapsto \tilde{A}u := \Psi^{-1} j^{*-1} A \, j^{-1} u.$$

The (unbounded) linear operator $-\tilde{A}$ with domain $D(-\tilde{A}) := D(\tilde{A})$ is sometimes called *variational*. It is densely defined, since E^* is densely embedded in V^* and A is isomorphic. Furthermore, the symmetry of the bilinear form $a(\cdot, \cdot)$ implies that $\tilde{A}$, and therefore also $-\tilde{A}$, is self-adjoint. That is, $(\tilde{A}, D(\tilde{A})) = (\tilde{A}^*, D(\tilde{A}^*))$, where $\tilde{A}^* : D(\tilde{A}^*) \subseteq E \to E$ denotes the *adjoint* operator defined by

$$D(\tilde{A}^*) := \big\{u_1 \in E : \exists u_2 \in E : \langle \tilde{A}u, u_1 \rangle_E = \langle u, u_2 \rangle_E \text{ for all } u \in D(\tilde{A})\big\},$$
$$\tilde{A}^* u_1 := u_2, \qquad u_1 \in D(\tilde{A}^*),$$

where $u_2 \in E$ fulfils $\langle \tilde{A}u, u_1 \rangle_E = \langle u, u_2 \rangle_E$ for all $u \in D(\tilde{A})$ and is unique by the density of $D(\tilde{A})$ in E. At the same time, since $A : V \to V^*$ is an isomorphism, the operator $(-\tilde{A})^{-1} : (E, \|\cdot\|_E) \to (D(-\tilde{A}), \|\cdot\|_E)$, defined by $(-\tilde{A})^{-1} := j(-A)^{-1}j^*\Psi$ is the bounded inverse of $-\tilde{A}$. Thus, $0 \in \rho(-\tilde{A})$ and, therefore, $(-\tilde{A}, D(-\tilde{A}))$ is a closed operator on E. Moreover, by (2.52) and the definition of $\tilde{A}$, for arbitrary $\lambda > 0$ and $u \in D(-\tilde{A})$,

$$\|(\lambda \mathrm{Id} - (-\tilde{A}))u\|_E \geq \langle \lambda u, u \rangle_E + a(u, u) \geq \lambda \|u\|_E,$$

i.e., $-\tilde{A}$ is *dissipative*, see [102, Theorem 4.2]. Therefore, by the Lumer-Philips Theorem, see in particular [102, Corollary 4.4], $(-\tilde{A}, D(-\tilde{A}))$ is the generator of a contraction semigroup on E. By making slight abuse of notation, we sometimes write A instead of $\tilde{A}$, especially when $j = \mathrm{Id}$.

Chapter 3

Starting point: Linear SPDEs in weighted Sobolev spaces

In this chapter we present and discuss the main results from the L_p-theory of SPDEs on non-smooth domains as developed in [75] within the analytic approach. It is the starting point for our regularity analysis, providing existence and uniqueness of solutions for a wide class of linear SPDEs on general bounded Lipschitz domains $\mathcal{O} \subset \mathbb{R}^d$. The solutions are elements of certain classes $\mathfrak{H}_{p,\theta}^{\gamma}(\mathcal{O}, T)$ of predictable p-Bochner integrable $H_{p,\theta-p}^{\gamma}(\mathcal{O})$-valued stochastic processes. Since in the next chapter we will be able to prove a general embedding of weighted Sobolev spaces into Besov spaces from the scale (∗), this L_p-theory turns out to be tailor-made for our regularity analysis in the non-linear approximation scale (∗). A combination of the existence results from this section with the aforementioned embedding will lead to a statement about the spatial Besov regularity for linear SPDEs, as stated and proven in Section 5.1. In order to obtain similar spatial regularity results for semi-linear equations, in Section 5.2 we will also extend the L_p-theory from [75] to a class of semi-linear SPDEs. Furthermore, while analysing the Hölder regularity of the paths of the solution process in Chapter 6, we present an extension to an $L_q(L_p)$-theory for the heat equation on bounded Lipschitz domains. That is, we prove the existence of a solution in certain classes $\mathfrak{H}_{p,\theta}^{\gamma,q}(\mathcal{O}, T)$ of predictable q-Bochner integrable $H_{p,\theta-p}^{\gamma}(\mathcal{O})$-valued stochastic processes, explicitly allowing the summability parameter q in time (and with respect to $\omega \in \Omega$) to be greater than the summability parameter p used to measure the smoothness in space direction.

We split this chapter in two parts: In Section 3.1 we discuss the spaces $\mathfrak{H}_{p,\theta}^{\gamma,q}(\mathcal{O}, T)$, whereas Section 3.2 is concerned with those fragments from the L_p-regularity of SPDEs developed in [75] which are relevant for our analysis.

Before we start our exposition, we fix some notation and specify the class of equations considered in this chapter. Let $\mathcal{O}$ be a bounded Lipschitz domain in $\mathbb{R}^d$. $(\Omega, \mathcal{F}, \mathbb{P})$ denotes a complete probability space and $T > 0$ is a finite time horizon. Moreover, $(w_t^k)_{t \in [0,T]}$, $k \in \mathbb{N}$, is a sequence of independent real-valued standard Brownian motions with respect to a normal filtration $(\mathcal{F}_t)_{t \in [0,T]}$ on $(\Omega, \mathcal{F}, \mathbb{P})$. We write shorthand Ω_T instead of $\Omega \times [0,T]$. We consider equations of the form

$$\left. \begin{aligned} \mathrm{d}u &= \left(a^{ij}u_{x^i x^j} + b^i u_{x^i} + cu + f\right)\mathrm{d}t + \left(\sigma^{ik}u_{x^i} + \mu^k u + g^k\right)\mathrm{d}w_t^k \quad \text{on } \Omega_T \times \mathcal{O}, \\ u(0) &= u_0 \quad \text{on } \Omega \times \mathcal{O}, \end{aligned} \right\} \tag{3.1}$$

where the coefficients a^{ij}, b^i, c, $\sigma^{i,k}$ and μ^k, for $i, j \in \{1, \ldots, d\}$ and $k \in \mathbb{N}$, are assumed to fulfil certain assumptions. We want to emphasize that in this thesis, for a better readability, we use the so-called summation convention on the repeated indices i, j, k when writing down equations, see also Remark 3.12(i) as well as Remark 5.12(i). In order to state the assumptions on the coefficients, we need some additional notation. For $x, y \in \mathcal{O}$ we write $\rho(x, y) := \rho(x) \wedge \rho(y)$, $\rho(x)$ being the distance of a point $x \in \mathcal{O}$ to the boundary $\partial \mathcal{O}$, i.e., $\rho(x) = \operatorname{dist}(x, \partial \mathcal{O})$. For $\alpha \in \mathbb{R}$, $\delta \in (0, 1]$ and $m \in \mathbb{N}_0$ we set

$$[f]_m^{(\alpha)} := \sup_{x \in \mathcal{O}} \rho^{m+\alpha}(x) |D^m f(x)|,$$

$$[f]_{m+\delta}^{(\alpha)} := \sup_{\substack{x, y \in \mathcal{O} \\ |\beta| = m}} \rho^{m+\alpha}(x, y) \frac{|D^\beta f(x) - D^\beta f(y)|}{|x - y|^\delta},$$

$$|f|_m^{(\alpha)} := \sum_{l=0}^{m} [f]_l^{(\alpha)} \qquad \text{and} \qquad |f|_{m+\delta}^{(\alpha)} := |f|_m^{(\alpha)} + [f]_{m+\delta}^{(\alpha)},$$

whenever it makes sense. We use the same notations for ℓ_2-valued functions (just replace the absolute values in the above definitions by the ℓ_2-norms). Furthermore, let us fix an arbitrary function

$$\eta : [0, \infty) \to [0, \infty),$$

vanishing only on the set of non-negative integers, i.e., $\eta(j) = 0$ if, and only if, $j \in \mathbb{N}_0$. We set

$$t_+ := t + \eta(t).$$

These notations at hand, we present the assumptions on the coefficients of Eq. (3.1), which are identical with the ones in [75, Assumption 2.10], see also [73, Assumptions 2.5 and 2.6]. The precise solution concept for equations of the type (3.1) fulfilling these assumptions will be specified in Definition 3.10.

Assumption 3.1. **(i)** For any fixed $x \in \mathcal{O}$, the coefficients

$$a^{ij} (\cdot, \cdot, x), b^i (\cdot, \cdot, x), c (\cdot, \cdot, x), \sigma^{ik} (\cdot, \cdot, x), \mu^k (\cdot, \cdot, x) : \Omega_T \to \mathbb{R}$$

are predictable processes with respect to the given normal filtration $(\mathcal{F}_t)_{t \in [0, T]}$.

(ii) (Stochastic parabolicity) There are constants $\delta_0, K \in (0, \infty)$, such that for all $(\omega, t, x) \in \Omega_T \times \mathcal{O}$ and $\lambda \in \mathbb{R}^d$,

$$\delta_0 |\lambda|^2 \le \sum_{i,j=1}^{d} \tilde{a}^{ij}(\omega, t, x) \lambda_i \lambda_j \le K |\lambda|^2,$$

where $\tilde{a}^{ij} := a^{ij} - \frac{1}{2} \langle \sigma^{i \cdot}, \sigma^{j \cdot} \rangle_{\ell_2}$ for $i, j \in \{1, \ldots, d\}$.

(iii) For all $(\omega, t) \in \Omega_T$,

$$|a^{ij}(\omega, t, \cdot)|_{|\gamma|+}^{(0)} + |b^i(\omega, t, \cdot)|_{|\gamma|+}^{(1)} + |c(\omega, t, \cdot)|_{|\gamma|+}^{(2)}$$
$$+ |(\sigma^{ik}(\omega, t, \cdot))_{k \in \mathbb{N}}|_{|\gamma+1|+}^{(0)} + |(\mu^k(\omega, t, \cdot))_{k \in \mathbb{N}}|_{|\gamma+1|+}^{(1)} \le K.$$

(iv) The coefficients a^{ij} and $\sigma^{i \cdot}$ are uniformly continuous in $x \in \mathcal{O}$, i.e., for any $\varepsilon > 0$ there is a $\delta = \delta(\varepsilon) > 0$, such that

$$\left| a^{ij}(\omega, t, x) - a^{ij}(\omega, t, y) \right| + \left| \sigma^{i \cdot}(\omega, t, x) - \sigma^{i \cdot}(\omega, t, y) \right|_{\ell_2} \le \varepsilon,$$

for all $(\omega, t) \in \Omega_T$, whenever $x, y \in \mathcal{O}$ with $|x - y| \le \delta$.

(v) The behaviour of the coefficients b^i, c and μ can be controlled near the boundary of $\mathcal{O}$ in the following way:

$$\lim_{\substack{\rho(x)\to 0 \\ x\in\mathcal{O}}} \sup_{\substack{\omega\in\Omega \\ t\in[0,T]}} \left\{ \rho(x)|b^i(\omega,t,x)| + \rho^2(x)|c(\omega,t,x)| + \rho(x)|\mu(\omega,t,x)|_{\ell_2} \right\} = 0.$$

3.1   Stochastic parabolic weighted Sobolev spaces $\mathfrak{H}^{\gamma,q}_{p,\theta}(G,T)$

The analysis of SPDEs in the analytic approach takes place in the spaces $\mathfrak{H}^{\gamma,q}_{p,\theta}(G,T)$ consisting of certain predictable q-Bochner integrable $H^{\gamma}_{p,\theta-p}(G)$-valued stochastic processes. In this section we present the precise definition of these spaces. Furthermore, we collect some of their properties which are relevant for our analysis later on. We start with common notations for predictable processes taking values in weighted Sobolev spaces, which are frequently used in the analytic approach. In the sequel, we write $\mathcal{P}_T$ for the $(\mathcal{F}_t)_{t\in[0,T]}$-predictable σ-field on Ω_T and $\mathbb{P}_T := \mathbb{P}\otimes dt$. Unless explicitly stated otherwise, G denotes an arbitrary domain in $\mathbb{R}^d$ with non-empty boundary.

Definition 3.2. Let G be a domain in $\mathbb{R}^d$ with non-empty boundary. For $p,q \in (1,\infty)$ and $\gamma,\theta \in \mathbb{R}$ we define

$$\mathbb{H}^{\gamma,q}_{p,\theta}(G,T) := L_q(\Omega_T, \mathcal{P}_T, \mathbb{P}_T; H^{\gamma}_{p,\theta}(G)),$$
$$\mathbb{H}^{\gamma,q}_{p,\theta}(G,T;\ell_2) := L_q(\Omega_T, \mathcal{P}_T, \mathbb{P}_T; H^{\gamma}_{p,\theta}(G;\ell_2)),$$
$$U^{\gamma,q}_{p,\theta}(G) := L_q(\Omega, \mathcal{F}_0, \mathbb{P}; H^{\gamma-2/q}_{p,\theta-(1-2/q)p}(G)).$$

If $p=q$ we also write $\mathbb{H}^{\gamma}_{p,\theta}(G,T)$, $\mathbb{H}^{\gamma}_{p,\theta}(G,T;\ell_2)$ and $U^{\gamma}_{p,\theta}(G)$ instead of $\mathbb{H}^{\gamma,p}_{p,\theta}(G,T)$, $\mathbb{H}^{\gamma,p}_{p,\theta}(G,T;\ell_2)$ and $U^{\gamma,p}_{p,\theta}(G)$ respectively.

Unless explicitly stated otherwise, from now on we assume that

$$\boxed{p \in [2,\infty), \quad q \in [2,\infty), \quad \gamma \in \mathbb{R}, \quad \theta \in \mathbb{R}.}$$

Definition 3.3. Let G be a domain in $\mathbb{R}^d$ with non-empty boundary, $p,q \in [2,\infty)$ and $\gamma,\theta \in \mathbb{R}$. We write $u \in \mathfrak{H}^{\gamma,q}_{p,\theta}(G,T)$ if, and only if, $u \in \mathbb{H}^{\gamma,q}_{p,\theta-p}(G,T)$, $u(0) \in U^{\gamma,q}_{p,\theta}(G)$, and there exist some $f \in \mathbb{H}^{\gamma-2,q}_{p,\theta+p}(G,T)$ and $g \in \mathbb{H}^{\gamma-1,q}_{p,\theta}(G,T;\ell_2)$ such that

$$du = f\,dt + g^k\,dw^k_t$$

in the sense of distributions. That is, for any $\varphi \in \mathcal{C}^\infty_0(G)$, with probability one, the equality

$$(u(t,\cdot),\varphi) = (u(0,\cdot),\varphi) + \int_0^t (f(s,\cdot),\varphi)\,ds + \sum_{k=1}^{\infty} \int_0^t (g^k(s,\cdot),\varphi)\,dw^k_s \tag{3.2}$$

holds for all $t \in [0,T]$, where the series is assumed to converge uniformly on $[0,T]$ in probability. In this situation we write

$$\mathbb{D}u := f \quad \text{and} \quad \mathbb{S}u := g$$

for the *deterministic* and for the *stochastic part* of u, respectively. The norm in $\mathfrak{H}^{\gamma,q}_{p,\theta}(G,T)$ is defined as

$$\|u\|_{\mathfrak{H}^{\gamma,q}_{p,\theta}(G,T)} := \|u\|_{\mathbb{H}^{\gamma,q}_{p,\theta-p}(G,T)} + \|\mathbb{D}u\|_{\mathbb{H}^{\gamma-2,q}_{p,\theta+p}(G,T)} + \|\mathbb{S}u\|_{\mathbb{H}^{\gamma-1,q}_{p,\theta}(G,T;\ell_2)} + \|u(0)\|_{U^{\gamma,q}_{p,\theta}(G)}. \tag{3.3}$$

If $p=q$ we also write $\mathfrak{H}^{\gamma}_{p,\theta}(G,T)$ instead of $\mathfrak{H}^{\gamma,p}_{p,\theta}(G,T)$.

Remark 3.4.   (i) The phrase "…, with probability one, the equality (3.2) holds for all $t \in [0, T]$, …" in the definition above, means: There exists a set $\Omega_0 \in \mathcal{F}$ with $\mathbb{P}(\Omega_0) = 1$, such that for any fixed $\omega \in \Omega_0$, Equality (3.2) is fulfilled for all $t \in [0, T]$. In particular, the $\mathbb{P}$-null set where (3.2) might not hold, does *not* depend on $t \in [0, T]$.

(ii) Replacing G by $\mathbb{R}^d$ and omitting the weight parameters in the definitions above, we obtain the spaces $\mathbb{H}_p^{\gamma,q}(T) = \mathbb{H}_p^{\gamma,q}(\mathbb{R}^d, T)$, $\mathbb{H}_p^{\gamma,q}(T; \ell_2) = \mathbb{H}_p^{\gamma,q}(\mathbb{R}^d, T; \ell_2)$, $U_p^{\gamma,q} = U_p^{\gamma,q}(\mathbb{R}^d)$, and $\mathscr{H}_p^{\gamma,q}(T)$ as introduced in [83, Definition 3.5]. The latter are denoted by $\mathcal{H}_p^{\gamma,q}(T)$ in [82]; if $q = p$ they coincide with the spaces $\mathcal{H}_p^\gamma(T)$ introduced in [80, Definition 3.1].

Lemma 3.5. *Let G be a domain in $\mathbb{R}^d$ with non-empty boundary, $p, q \in [2, \infty)$ and $\gamma, \theta \in \mathbb{R}$.*

(i) *If $g \in \mathbb{H}_{p,\theta}^{\gamma,q}(G, T; \ell_2)$ then, for any $\varphi \in \mathcal{C}_0^\infty(G)$, the series*

$$\sum_{k=1}^\infty \int_0^\cdot (g^k(s, \cdot), \varphi) \, \mathrm{d}w_s^k \tag{3.4}$$

from (3.2) converges in $L_2(\Omega; \mathcal{C}([0, T]; \mathbb{R}))$.

(ii) *The pair $(\mathbb{D}u, \mathbb{S}u) \in \mathbb{H}_{p,\theta+p}^{\gamma-2,q}(G, T) \times \mathbb{H}_{p,\theta}^{\gamma-1,q}(G, T; \ell_2)$ in Definition 3.3 is uniquely determined by $u \in \mathfrak{H}_{p,\theta}^{\gamma,q}(G, T)$.*

(iii) *$\mathfrak{H}_{p,\theta}^{\gamma,q}(G, T)$ is a Banach space.*

Proof. (i) The convergence of the sum (3.4) in $L_2(\Omega; \mathcal{C}([0, T]; \mathbb{R}))$ has been proven in [91, Section 3.5]. However, we need to correct a minor mistake in the first equality in the last estimate on page 91 in [91].

Let $\{\xi_n : n \in \mathbb{Z}\} \subseteq \mathcal{C}_0^\infty(G)$ with $\sum_{n \in \mathbb{Z}} \xi_n = 1$ on G fulfil (2.31) for some $c > 1$ and $k_0 > 0$. Furthermore, fix a sequence $\{\tilde{\xi}_n : n \in \mathbb{Z}\} \subseteq \mathcal{C}_0^\infty(G)$, also fulfilling (2.31)—with a possibly different $k_0 > 0$—, such that

$$\tilde{\xi}_n \Big|_{\mathrm{supp}\,\{\xi_n\}} = 1 \quad \text{for all } n \in \mathbb{Z}.$$

By Remark 2.48(ii) and (iii), it is clear that we can construct such sequences. Now, by mimicking, the proof in [91, Section 3.5] with $g_n^\kappa := \xi_n(c^n \cdot) g^\kappa(c^n \cdot)$ and $\phi_n := \tilde{\xi}_n(c^n \cdot) \phi(c^n \cdot)$ for $n \in \mathbb{Z}$ and $\kappa \in \mathbb{N}$, the assertion follows.

(ii) This assertion follows by using the arguments from [80, Remark 3.3].

(iii) By (ii) we know that the norm (3.3) is well-defined. The completeness can be proven by following the lines of [83, Remark 3.8] with $\mathbb{R}_+^d$ replaced by G. $\quad\square$

Proposition 3.6. *Let G be a domain in $\mathbb{R}^d$ with non-empty boundary, $p, q \in [2, \infty)$ and $\gamma, \theta \in \mathbb{R}$. Fix $g \in \mathbb{H}_{p,\theta}^{\gamma,q}(G, T; \ell_2)$ and let $\Phi : H_{p,\theta}^\gamma(G; \ell_2) \to \Gamma(\ell_2, H_{p,\theta}^\gamma(G))$ be the isomorphism introduced in Theorem 2.54. Then, the $\Gamma(\ell_2, H_{p,\theta}^\gamma(G))$-valued stochastic process*

$$\Phi_g := \Phi \circ g \tag{3.5}$$

is L_q-stochastically integrable with respect to the ℓ_2-cylindrical Brownian motion

$$\ell_2 \ni h \mapsto W_{\ell_2}(t)h := \sum_{k=1}^\infty w_t^k \langle e_k, h \rangle_{\ell_2} \in L_2(\Omega), \qquad t \in [0, T].$$

Moreover,

$$\sum_{k=1}^\infty \int_0^\cdot (g^k(s, \cdot), \varphi) \, \mathrm{d}w_s^k = \left(\int_0^\cdot \Phi_g(s) \, \mathrm{d}W_{\ell_2}(s), \varphi \right) \qquad \mathbb{P}\text{-a.s.} \tag{3.6}$$

in $\mathcal{C}([0, T]; \mathbb{R})$.

Proof. Fix $g \in \mathbb{H}_{p,\theta}^{\gamma,q}(G,T;\ell_2) = L_q(\Omega_T, \mathcal{P}_T, \mathbb{P}_T; H_{p,\theta}^{\gamma}(G;\ell_2))$. Then, since Φ is bounded from $H_{p,\theta}^{\gamma}(G;\ell_2)$ to $\Gamma(\ell_2, H_{p,\theta}^{\gamma}(G))$, we have

$$\Phi_g \in L_q(\Omega_T, \mathcal{P}_T, \mathbb{P}_T; \Gamma(\ell_2, H_{p,\theta}^{\gamma}(G))). \tag{3.7}$$

In particular, Φ_g is an ℓ_2-strongly measurable $(\mathcal{F}_t)_{t\in[0,T]}$-adapted process. Also, Φ_g belongs to $L_q(\Omega; L_2([0,T],H))$ scalarly, and, since $H_{p,\theta}^{\gamma}(G)$ is a UMD Banach space with type 2, compare Lemma 2.50, Φ_g is stochastically integrable with respect to the ℓ_2-cylindrical Brownian motion $(W_{\ell_2}(t))_{t\in[0,T]}$, cf. Theorem 2.32. Consequently, by Theorem 2.29, see also Remark 2.30, Φ_g represents a random variable $R_{\Phi_g} \in L_q^{\mathbb{F}}(\Omega; \Gamma(L_2([0,T];H), E))$. In particular, by [120, Proposition 4.3], there exists a continuous $(\mathcal{F}_t)_{t\in[0,T]}$-adapted version of the $H_{p,\theta}^{\gamma}(G)$-valued stochastic process

$$\left(\int_0^t \Phi_g(s) \, dW_{\ell_2}(s) \right)_{t\in[0,T]},$$

which, by the Burkholder-Davis-Gundy inequality proved in [120, Theorem 4.4], satisfies

$$\mathbb{E}\left[\sup_{t\in[0,T]} \left\| \int_0^t \Phi_g(s) \, dW_{\ell_2}(s) \right\|_{H_{p,\theta}^{\gamma}(G)}^q \right] \leq C \, \mathbb{E}\left[\| R_{\Phi_g} \|_{\Gamma(L_2([0,T];\ell_2), H_{p,\theta}^{\gamma}(G))}^q \right].$$

Using Embedding (2.20) and the fact that Φ is an isomorphism between the spaces $H_{p,\theta}^{\gamma}(G;\ell_2)$ and $\Gamma(\ell_2, H_{p,\theta}^{\gamma}(G))$, see Theorem 2.54, this leads to

$$\mathbb{E}\left[\sup_{t\in[0,T]} \left\| \int_0^t \Phi_g(s) \, dW_{\ell_2}(s) \right\|_{H_{p,\theta}^{\gamma}(G)}^q \right] \leq C \, \mathbb{E}\left[\int_0^T \| \Phi_g \|_{\Gamma(\ell_2, H_{p,\theta}^{\gamma}(G))}^q \, dt \right] \leq C \, \| g \|_{\mathbb{H}_{p,\theta}^{\gamma,q}(G,T;\ell_2)}^q.$$

Fix two arbitrary positive integers $m \leq n$ and set $g_{m,n} := (g_{m,n}^k)_{k\in\mathbb{N}}$ with

$$g_{m,n}^k := \begin{cases} g^k, & \text{if } k \in \{m,\dots,n\} \\ 0, & \text{else} \end{cases} \in \mathbb{H}_{p,\theta}^{\gamma,q}(G,T;\ell_2).$$

Then, by the same arguments as above, the stochastic processes

$$\left(\sum_{k=m}^n \int_0^t g^k(s,\cdot) \, dw_s^k \right)_{t\in[0,T]} \quad \text{and} \quad \left(\int_0^t \Phi_{g_{m,n}}(s) \, dW_{\ell_2}(s) \right)_{t\in[0,T]}$$

have continuous versions, which, by Theorem 2.31 coincide $\mathbb{P}$-a.s. in $\mathcal{C}([0,T]; H_{p,\theta}^{\gamma}(G))$. Moreover,

$$\mathbb{E}\left[\sup_{t\in[0,T]} \left\| \int_0^t \Phi_{g_{m,n}}(s) \, dW_{\ell_2}(s) \right\|_{H_{p,\theta}^{\gamma}(G)}^q \right] \leq C \, \| g_{m,n} \|_{\mathbb{H}_{p,\theta}^{\gamma,q}(G,T;\ell_2)}^q.$$

The right hand side converges to zero as $m,n \to \infty$, since $g \in \mathbb{H}_{p,\theta}^{\gamma,q}(G,T;\ell_2)$. Consequently, the series

$$\sum_{k=1}^{\infty} \int_0^{\cdot} g^k(s,\cdot) \, dw_s^k$$

converges in the Banach space $L_q(\Omega; \mathcal{C}([0,T]; H_{p,\theta}^{\gamma}(G)))$, and, by another application of Theorem 2.31,

$$\sum_{k=1}^{\infty} \int_0^{\cdot} g^k(s,\cdot) \, dw_s^k = \int_0^{\cdot} \Phi_g(s) \, dW_{\ell_2}(s) \qquad \mathbb{P}\text{-a.s.}$$

in $\mathcal{C}([0,T]; H^\gamma_{p,\theta}(G))$. For $\varphi \in \mathcal{C}^\infty_0(G) \subseteq (H^\gamma_{p,\theta}(G))^* \simeq H^{-\gamma}_{p',\theta'}(G)$ with $1/p + 1/p' = 1$ and $\theta/p + \theta'/p' = d$, see Lemma 2.45(viii) and (ii) together with Remark 2.49, the linear operator $(\cdot, \varphi) : H^\gamma_{p,\theta}(G) \to \mathbb{R}$, $v \mapsto (v, \varphi)$, is bounded. Thus, for any $t \in [0,T]$,

$$\left(\sum_{k=1}^\infty \int_0^t g^k(s, \cdot) \, \mathrm{d}w^k_s, \varphi \right) = \sum_{k=1}^\infty \int_0^t \left(g^k(s, \cdot), \varphi \right) \mathrm{d}w^k_s \qquad \mathbb{P}\text{-a.s.},$$

and, therefore,

$$\sum_{k=1}^\infty \int_0^\cdot \left(g^k(s, \cdot), \varphi \right) \mathrm{d}w^k_s = \left(\int_0^\cdot \Phi_g(s) \, \mathrm{d}W_{\ell_2}(s), \varphi \right) \qquad \mathbb{P}\text{-a.s.} \tag{3.8}$$

in $\mathcal{C}([0,T]; \mathbb{R})$—after possibly changing to suitable versions of the processes. $\qquad \square$

Remark 3.7. It is worth noting that similar arguments as in the proof of Proposition 3.6 yield an alternative proof of the convergence of the series (3.4) in $L_2(\Omega; \mathcal{C}([0,T]; \mathbb{R}))$, and even in $L_q(\Omega; \mathcal{C}([0,T]; \mathbb{R}))$.

Besides the analysis of the spatial regularity of solutions to SPDEs, in this thesis we are also interested in the Hölder regularity of the paths of the solution processes. Since our solutions will always be contained in $\mathfrak{H}^{\gamma,q}_{p,\theta}(G,T)$, results on the Hölder regularity of the elements of these spaces are of major interest. For $p = q \in [2,\infty)$ the following result concerning the regularity of the paths of an element of $\mathfrak{H}^\gamma_{p,\theta}(G,T)$, considered as a stochastic process with values in weighted Sobolev spaces, can be found in [75, Theorem 2.9]. Its proof strongly relies on [83, Corollary 4.12 and Remark 4.14], which are corresponding results on the whole space $\mathbb{R}^d$. Note that the statement of [75, Theorem 2.9] is formulated only for a certain class of bounded non-smooth domains. However, the arguments go through for arbitrary domains with non-empty boundary.

Theorem 3.8. *Let $G \subset \mathbb{R}^d$ be an arbitrary domain with non-empty boundary, $\gamma \in \mathbb{R}$, and $\theta \in \mathbb{R}$.*

(i) *Let $2/p < \tilde{\beta} < \beta \le 1$. Then*

$$\mathbb{E}[u]^p_{\mathcal{C}^{\tilde{\beta}/2 - 1/p}([0,T]; H^{\gamma+2-\beta}_{p,\theta-(1-\beta)p}(G))} \le C T^{(\beta-\tilde{\beta})p/2} \|u\|^p_{\mathfrak{H}^{\gamma+2}_{p,\theta}(G,T)},$$

where $C \in (0,\infty)$ is a constant independent of T and u.

(ii) *Let $p \in [2,\infty)$. Then*

$$\mathbb{E}\left[\sup_{t \in [0,T]} \|u\|^p_{H^{\gamma+1}_{p,\theta}(G)} \right] \le C \|u\|^p_{\mathfrak{H}^{\gamma+2}_{p,\theta}(G,T)},$$

where the constant C depends on d, p, γ, θ, G, and T. The function $T \mapsto C(T)$ is non-decreasing. In particular, there exists a constant $C \in (0,\infty)$, such that for any $u \in \mathfrak{H}^{\gamma+2}_{p,\theta}(G,T)$ and all $t \in [0,T]$,

$$\|u\|^p_{\mathbb{H}^{\gamma+1}_{p,\theta}(G,t)} \le C \int_0^t \|u\|^p_{\mathfrak{H}^{\gamma+2}_{p,\theta}(G,s)} \, \mathrm{d}s.$$

Remark 3.9. In Chapter 6 we will need a generalization of Theorem 3.8(i) for the paths of elements of $\mathfrak{H}^{\gamma,q}_{p,\theta}(\mathcal{O},T)$ with $p \ne q$ in order to obtain Hölder regularity of the solution to the stochastic heat equation, considered as a process taking values in the Besov spaces from the scale $(*)$; see Theorem 6.1. Its proof will require more involved arguments.

3.2 An L_p-theory of linear SPDEs on bounded Lipschitz domains

In this section we present the main existence and uniqueness result of the L_p-theory of linear SPDEs on bounded non-smooth domains developed recently in [75]. Since in this thesis we are only interested in SPDEs on bounded Lipschitz domains $\mathcal{O} \subset \mathbb{R}^d$, we will restrict ourselves to this case, and consider equations of the form (3.1) with coefficients fulfilling Assumption 3.1. We first specify the solution concept which will be used in this thesis. For $i, j \in \{1, \ldots, d\}$ and a $\mathcal{D}'(\mathcal{O})$-valued function u on Ω_T we use the common notation $u_{x^i} := u_{x^i}(\omega, t) := u(\omega, t)_{x^i}$ and $u_{x^i x^j} := u_{x^i x^j}(\omega, t) := u(\omega, t)_{x^i x^j}$, $(\omega, t) \in \Omega_T$, respectively.

Definition 3.10. Let $\mathcal{O}$ be a bounded Lipschitz domain in $\mathbb{R}^d$. Given $\gamma \in \mathbb{R}$, let a^{ij}, b^i, c, σ^{ik} and μ^k, where $i, j \in \{1, \ldots, d\}$ and $k \in \mathbb{N}$, satisfy Assumptions 3.1. A stochastic process $u \in \mathbb{H}^{\gamma, q}_{p, \theta - p}(\mathcal{O}, T)$ is called a *solution of Eq. (3.1) in the class* $\mathfrak{H}^{\gamma, q}_{p, \theta}(\mathcal{O}, T)$ if, and only if, $u \in \mathfrak{H}^{\gamma, q}_{p, \theta}(\mathcal{O}, T)$ with

$$u(0, \cdot) = u_0, \quad \mathbb{D}u = \sum_{i,j=1}^{d} a^{ij} u_{x^i x^j} + \sum_{i=1}^{d} b^i u_{x^i} + cu + f, \quad \text{and} \quad \mathbb{S}u = \left(\sum_{i=1}^{d} \sigma^{ik} u_{x^i} + \mu^k u + g^k \right)_{k \in \mathbb{N}}$$

in the sense of Definition 3.3.

Remark 3.11. In this thesis, if we call an element $u \in \mathfrak{H}^{\gamma, q}_{p, \theta}(\mathcal{O}, T)$ a *solution* of Eq. (3.1), we mean that u is a solution of Eq. (3.1) in the class $\mathfrak{H}^{\gamma, q}_{p, \theta}(\mathcal{O}, T)$.

Remark 3.12.   **(i)** As already mentioned, throughout this thesis, for a better readability, we omit the notation of the sums $\sum_{i,j}$ and $\sum_k$ when writing down equations and use the so-called summation convention on the repeated indices i, j, k. Thus the expression

$$du = \left(a^{ij} u_{x^i x^j} + b^i u_{x^i} + cu + f \right) dt + \left(\sigma^{ik} u_{x^i} + \mu^k u + g^k \right) dw_t^k$$

is shorthand for

$$du = \left(\sum_{i,j=1}^{d} a^{ij} u_{x^i x^j} + \sum_{i=1}^{d} b^i u_{x^i} + cu + f \right) dt + \left(\sum_{i=1}^{d} \sigma^{ik} u_{x^i} + \mu^k u + g^k \right) dw_t^k$$

in the sense of Definition 3.3.

(ii) The solution concept presented in Definition 3.10 is a natural generalization of the definition given in [75]. Therein, only the case $p = q$ is considered. However, we will need this generalization later on in Chapter 6.

The main existence and uniqueness results for equations on bounded Lipschitz domains proven in [75], see Theorem 2.12, Remark 2.13 as well as Theorem 2.15 therein, can be summarized as follows.

Theorem 3.13. *Let $\mathcal{O}$ be a bounded Lipschitz domain in $\mathbb{R}^d$, and $\gamma \in \mathbb{R}$. For $i, j \in \{1, \ldots, d\}$ and $k \in \mathbb{N}$, let a^{ij}, b^i, c, σ^{ik}, and μ^k be given coefficients satisfying Assumption 3.1 with suitable constants δ_0 and K.*

(i) *For $p \in [2, \infty)$, there exists a constant $\kappa_0 \in (0, 1)$, depending only on d, p, δ_0, K and $\mathcal{O}$, such that for any $\theta \in (d + p - 2 - \kappa_0, d + p - 2 + \kappa_0)$, $f \in \mathbb{H}^{\gamma}_{p, \theta + p}(\mathcal{O}, T)$, $g \in \mathbb{H}^{\gamma+1}_{p, \theta}(\mathcal{O}, T; \ell_2)$ and $u_0 \in U^{\gamma+2}_{p, \theta}(\mathcal{O})$, Eq. (3.1) has a unique solution u in the class $\mathfrak{H}^{\gamma+2}_{p, \theta}(\mathcal{O}, T)$. For this solution*

$$\|u\|^p_{\mathfrak{H}^{\gamma+2}_{p,\theta}(\mathcal{O},T)} \leq C \left(\|f\|^p_{\mathbb{H}^{\gamma}_{p,\theta+p}(\mathcal{O},T)} + \|g\|^p_{\mathbb{H}^{\gamma+1}_{p,\theta}(\mathcal{O},T;\ell_2)} + \|u_0\|^p_{U^{\gamma+2}_{p,\theta}(\mathcal{O})} \right), \tag{3.9}$$

where the constant C depends only on d, p, γ, θ, δ_0, K, T and $\mathcal{O}$.

(ii) *There exists $p_0 > 2$, such that the following statement holds: If $p \in [2, p_0)$, then there exists a constant $\kappa_1 \in (0, 1)$, depending only on d, p, δ_0, K and $\mathcal{O}$, such that for any $\theta \in (d - \kappa_1, d + \kappa_1)$, $f \in \mathbb{H}^\gamma_{p,\theta+p}(\mathcal{O}, T)$, $g \in \mathbb{H}^{\gamma+1}_{p,\theta}(\mathcal{O}, T; \ell_2)$ and $u_0 \in U^{\gamma+2}_{p,\theta}(\mathcal{O})$, Eq. (3.1) has a unique solution u in the class $\mathfrak{H}^{\gamma+2}_{p,\theta}(\mathcal{O}, T)$. For this solution, estimate (3.9) holds.*

Remark 3.14.   (i) For $p = 2$ there is no difference between (i) and (ii) in Theorem 3.13. In particular, existence of solutions in $\mathfrak{H}^\gamma_{2,d}(\mathcal{O}, T) \hookrightarrow L_2(\Omega_T; \mathring{W}^1_2(\mathcal{O}))$ is guaranteed under suitable assumptions on the data of the equation. Things are different if $p > 2$. Since we do not know the precise value of $\kappa_0 = \kappa_0(d, p, \mathcal{O})$, we can not expect that $d \in (d+p-2-\kappa_0, d+p-2+\kappa_0)$ if $p > 2$. Thus, Theorem 3.13(i) does not yield the existence of a solution $u \in \mathfrak{H}^\gamma_{p,d}(\mathcal{O}, T)$, even if the data of the equation are assumed to be arbitrarily smooth. However, Theorem 3.13(ii) guarantees that at least for certain $p > 2$, i.e., for $p \in [2, p_0)$ with some $p_0 > 2$, a solution $u \in \mathfrak{H}^\gamma_{p,d}(\mathcal{O}, T) \hookrightarrow L_p(\Omega_T; \mathring{W}^1_p(\mathcal{O}))$ exists under suitable assumptions on the data. In general, p_0 is not very high due to a counterexample of N.V. Krylov, which can be found in [75, Example 2.17]. It is shown therein that for any $p > 4$, there exists a bounded Lipschitz domain $\mathcal{O} \subset \mathbb{R}^2$ and a function $f \in L_p([0, T]; L_p(\mathcal{O}))$ such that a solution of the (deterministic) heat equation

$$\left.\begin{aligned} du &= (\Delta u + f) \, dt \quad \text{on } \Omega_T \times \mathcal{O}, \\ u(0) &= 0 \quad \text{on } \Omega \times \mathcal{O}, \end{aligned}\right\}$$

fails to be in $L_p([0, T]; L_{p,d-p}(\mathcal{O}))$, see [75, Example 2.17]. Thus, if we do not specify any further properties of the domain $\mathcal{O}$ except the fact that it is bounded and Lipschitz, the assertion of Theorem 3.13(ii) holds only with $p_0 \leq 4$.

(ii) Assume that the bounded $\mathcal{O}$ is not only Lipschitz but of class $\mathcal{C}^1_u$, see Definition 2.3. Then, if $\sigma = 0$, the statement of Theorem 3.13(i) holds for any $p \in [2, \infty)$ and $\theta \in \mathbb{R}$ fulfilling

$$d - 1 < \theta < d + p - 1; \tag{3.10}$$

see [72, Theorem 2.9 together with Remark 2.7]. That is: Let $\mathcal{O}$ be a bounded $\mathcal{C}^1_u$-domain in $\mathbb{R}^d$, and assume that a^{ij}, b^i, c, and μ^k with $i, j \in \{1, \ldots, d\}$ and $k \in \mathbb{N}$, are given coefficients satisfying Assumption 3.1 for some $\gamma \in \mathbb{R}$ with $\sigma = 0$ and suitable constants δ_0 and K. Then for any $p \in [2, \infty)$ and any $\theta \in \mathbb{R}$ fulfilling (3.10), Eq. (3.1) with $u_0 \in U^{\gamma+2}_{p,\theta}(\mathcal{O})$, $f \in \mathbb{H}^\gamma_{p,\theta+p}(\mathcal{O}, T)$ and $g \in \mathbb{H}^{\gamma+1}_{p,\theta}(\mathcal{O}, T; \ell_2)$ has a unique solution u in the class $\mathfrak{H}^{\gamma+2}_{p,\theta}(\mathcal{O}, T)$. Moreover, the estimate (3.9) holds.

(iii) As mentioned in [86, Remark 3.6], if $\mathcal{O}$ is replaced by $\mathbb{R}^d_+$, the statement of Theorem 3.13 fails to hold for $\theta \geq d + p - 1$ and $\theta \leq d - 1$. Therefore, in general, we do not expect that the κ_0 and κ_1 can be chosen to be greater than one. Explicit counterexamples on general bounded Lipschitz domains are yet to be constructed.

(iv) Remember that, as mentioned in the introduction, in this thesis we are interested in equations with zero Dirichlet boundary conditions. However, it is not immediately clear in which sense solutions in the class $\mathfrak{H}^{\gamma,q}_{p,\theta}(\mathcal{O}, T)$ fulfil such boundary conditions and therefore can be understood as solutions to Eq. (1.1). This will be clarified in Chapter 4, see in particular Remark 4.3.

(v) The statement of Theorem 3.13 is proved in [75] not only for bounded Lipschitz domains but for any bounded domain $G \subset \mathbb{R}^d$ which *admits the Hardy inequality*, i.e., for which

$$\int_G \left|\rho_G(x)^{-1}\varphi(x)\right|^2 \mathrm{d}x \le C \int_G |\varphi_x(x)|^2 \mathrm{d}x, \qquad \text{for all} \quad \varphi \in \mathcal{C}_0^\infty(G), \tag{3.11}$$

with a constant C which does not depend on $\varphi \in \mathcal{C}_0^\infty(G)$; the solution concept is analogous to the one introduced in Definition 3.10 with G instead of $\mathcal{O}$. It is known that bounded Lipschitz domains admit the Hardy inequality, see, e.g., [99] for a proof.

The analysis in [75] is done in the framework of the analytic approach. As pointed out in the introduction, alternatively, equations of the type (3.1) can be consider within a semigroup framework. Since many contributions to the regularity analysis of SPDEs use this semigroup approach, it is important to know whether the solution concept used in this thesis matches with the one(s) used therein. In what follows we present a specific setting where a solution to Eq. (3.1) in the sense given above is a weak solution of the corresponding Cauchy problem in the sense of Da Prato and Zabczyk [32, Section 5.1.1], which is the common solution concept used within the semigroup approach. We restrict ourselves to the Hilbert space case (i.e., $p = 2$) and particularly to equations in $L_2(\mathcal{O})$. A generalization to Banach spaces (i.e., $p > 2$) will be discussed in Chapter 6. We start by defining what is called a weak solution in the semigroup framework. To this end, we first fix our specific setting.

Assumption 3.15.　**(i)** The operator $(-A, D(-A))$ generates a strongly continuous semigroup $\{S(t)\}_{t \ge 0}$ in $L_2(\mathcal{O})$.

(ii) $(W^Q(t))_{t \in [0,T]}$ is a Q-Wiener process in a real Hilbert space $(H, \langle \cdot, \cdot \rangle_H)$ adapted to the given normal filtration $(\mathcal{F}_t)_{t \in [0,T]}$ with covariance operator $Q \in \mathcal{L}_1(H)$.

(iii) $f : \Omega_T \to L_2(\mathcal{O})$ is a predictable stochastic process with $\mathbb{P}$-a.s. Bochner integrable trajectories.

(iv) $B \in \mathcal{L}(H, L_2(\mathcal{O}))$.

(v) $u_0 : \Omega \to L_2(\mathcal{O})$ is an $\mathcal{F}_0$-measurable random variable.

Under these conditions we can define what is called a weak solution of the $L_2(\mathcal{O})$-valued SDE

$$\left.\begin{aligned} \mathrm{d}u(t) + Au(t)\,\mathrm{d}t &= f(t)\,\mathrm{d}t + B\,\mathrm{d}W^Q(t), \qquad t \in [0,T], \\ u(0) &= u_0, \end{aligned}\right\} \tag{3.12}$$

in the semigroup approach of Da Prato and Zabczyk [32, Section 5.1.1].

Definition 3.16. Let $(A, D(A))$, $(W^Q(t))_{t \in [0,T]}$, f, B and u_0 fulfil Assumption 3.15. Then, an $L_2(\mathcal{O})$-valued stochastic process $u = (u(t))_{t \in [0,T]}$ is a *weak solution* of Eq. (3.12), if it has the following properties:

(i) u has $\mathbb{P}$-a.s. Bochner integrable trajectories.

(ii) For all $\zeta \in D(A^*)$ and $t \in [0,T]$, we have

$$\begin{aligned} \langle u(t), \zeta \rangle_{L_2(\mathcal{O})} = \langle u_0, \zeta \rangle_{L_2(\mathcal{O})} &- \int_0^t \langle u(s), A^*\zeta \rangle_{L_2(\mathcal{O})}\,\mathrm{d}s \\ &+ \int_0^t \langle f(s), \zeta \rangle_{L_2(\mathcal{O})}\,\mathrm{d}s + \langle BW^Q(t), \zeta \rangle_{L_2(\mathcal{O})} \quad \mathbb{P}\text{-a.s.} \end{aligned} \tag{3.13}$$

Remark 3.17. **(i)** In parts of the literature such as [104] a weak solution in the sense of Definition 3.16 is called *analytically weak solution.*

(ii) Typically, if $H \hookrightarrow L_2(\mathcal{O})$ and B is the identity operator from H into $L_2(\mathcal{O})$, we omit B in (3.12) and (3.13).

Now we can prove exemplarily that in a specific setting the solution to Eq. (3.1) is a weak solution of the corresponding $L_2(\mathcal{O})$-valued SDE of the form (3.12).

Proposition 3.18. *Assume that the coefficients (a^{ij}) are constant and symmetric, i.e., they do not depend on $(\omega, t, x) \in \Omega_T \times \mathcal{O}$ and $a^{ij} = a^{ji}$ for $i, j \in \{1, \dots, d\}$. Furthermore, let Assumption 3.1 be fulfilled with vanishing b^i, c, σ^{ik}, and μ^k, for $i \in \{1, \dots, d\}$ and $k \in \mathbb{N}$. Fix $f \in \mathbb{H}^0_{2,d}(\mathcal{O}, T)$, $g \in H^1_{2,d}(\mathcal{O}; \ell_2)$ and $u_0 \in U^2_{2,d}(\mathcal{O})$. Then, the solution $u \in \mathfrak{H}^2_{2,d}(\mathcal{O}, T)$ of Eq. (3.1), which exists by Theorem 3.13, is the unique weak solution of Eq. (3.12), where B is the identity operator from $H^1_{2,d}(\mathcal{O})$ into $L_2(\mathcal{O})$,*

$$(-A, D(-A)) := \left(\sum_{i,j=1}^{d} a^{ij} u_{x^i x^j}, \left\{ u \in \mathring{W}^1_2(\mathcal{O}) : \sum_{i,j=1}^{d} a^{ij} u_{x^i x^j} \in L_2(\mathcal{O}) \right\} \right), \quad (3.14)$$

and

$$W^Q(t) := \sum_{k=1}^{\infty} g^k w^k_t, \qquad t \in [0, T], \quad (3.15)$$

is an $H^1_{2,d}(\mathcal{O})$-valued Q-Wiener process with covariance operator $Q \in \mathcal{L}_1(H^1_{2,d}(\mathcal{O}))$ given by

$$Qv = \sum_{k=1}^{\infty} \langle g^k, v \rangle_{H^1_{2,d}(\mathcal{O})} g^k, \qquad v \in H^1_{2,d}(\mathcal{O}).$$

Proof. In the given setting, by Theorem 3.13, the equation

$$\left. \begin{array}{ll} \mathrm{d}u = \left(a^{ij} u_{x^i x^j} + f \right) \mathrm{d}t + g^k \, \mathrm{d}w^k_t & \text{on } \Omega_T \times \mathcal{O}, \\ u(0) = u_0 & \text{on } \Omega \times \mathcal{O} \end{array} \right\} \quad (3.16)$$

has a unique solution $u \in \mathfrak{H}^2_{2,d}(\mathcal{O}, T)$. In particular, for all $\varphi \in \mathcal{C}^\infty_0(\mathcal{O})$, with probability one, the equality

$$(u(t, \cdot), \varphi) = (u(0, \cdot), \varphi) + \int_0^t \left(\sum_{i,j=1}^{d} a^{ij} u_{x^i x^j}(s, \cdot) + f(s, \cdot), \varphi \right) \mathrm{d}s + \sum_{k=1}^{\infty} \int_0^t (g^k, \varphi) \, \mathrm{d}w^k_s \quad (3.17)$$

holds for all $t \in [0, T]$. Fix $\zeta \in D(A^*) \subseteq \mathring{W}^1_2(\mathcal{O})$. Then, there exists a sequence $(\varphi_n)_{n \in \mathbb{N}} \subseteq \mathcal{C}^\infty_0(\mathcal{O})$ approximating ζ in $\mathring{W}^1_2(\mathcal{O})$. We fix such a sequence and show that for any $t \in [0, T]$, each side of (3.17) with φ_n instead of φ converges $\mathbb{P}$-a.s. to the corresponding side of (3.13) with A, B and W^Q as defined in (3.14) and (3.15). This obviously would prove the assertion of the theorem. We start with the left hand sides. Since $u \in \mathbb{H}^2_{2,d-2}(\mathcal{O}, T)$, by Theorem 3.8, it has a version with continuous paths, if considered as a process with state space $H^1_{2,d}(\mathcal{O})$. Consequently, with probability one, $u(t, \cdot) \in L_2(\mathcal{O})$ for all $t \in [0, T]$. Thus, with probability one,

$$\lim_{n \to \infty} (u(t, \cdot), \varphi_n) = \lim_{n \to \infty} \langle u(t, \cdot), \varphi_n \rangle_{L_2(\mathcal{O})} = \langle u(t, \cdot), \zeta \rangle_{L_2(\mathcal{O})} \quad \text{for all} \quad t \in [0, T].$$

We continue with the right hand sides. Since $u_0 \in U^2_{2,d}(\mathcal{O}) = L_2(\Omega, \mathcal{F}_0, \mathbb{P}; H^1_{2,d}(\mathcal{O}))$, $u_0 \in L_2(\mathcal{O})$ $\mathbb{P}$-a.s., and

$$\lim_{n \to \infty} (u_0, \varphi_n) = \lim_{n \to \infty} \langle u_0, \varphi_n \rangle_{L_2(\mathcal{O})} = \langle u_0, \zeta \rangle_{L_2(\mathcal{O})} \qquad \mathbb{P}\text{-a.s.}$$

Furthermore, since $f \in \mathbb{H}^0_{2,d}(\mathcal{O}, T) = L_2(\Omega_T, \mathcal{P}_T, \mathbb{P}_T; L_2(\mathcal{O}))$, using Tonelli's theorem and the dominated convergence theorem, we obtain that with probability one,

$$\lim_{n \to \infty} \int_0^t (f(s, \cdot), \varphi_n) \, ds = \int_0^t \langle f(s, \cdot), \zeta \rangle_{L_2(\mathcal{O})} \, ds$$

holds for all $t \in [0, T]$ (after possibly passing to a subsequence). Moreover, since $g \in H^1_{2,d}(\mathcal{O})$ and the Brownian motions $(w_t^k)_{t \in [0,T]}$, $k \in \mathbb{N}$, are independent, an application of Doob's inequality yields

$$\mathbb{E} \left[\sup_{t \in [0,T]} \left\| \sum_{k=1}^\infty g^k w_t^k \right\|^2_{H^1_{2,d}(\mathcal{O})} \right] \leq C \, T \, \|g\|^2_{H^1_{2,d}(\mathcal{O}; \ell_2)}.$$

In particular, the series $\sum_{k=1}^\infty g^k w_{\cdot}^k$ converges in $L_2(\Omega; \mathcal{C}([0, T]; H^1_{2,d}(\mathcal{O})))$. Thus, using the properties of Itô's one-dimensional stochastic integral, yields that with probability one,

$$\lim_{n \to \infty} \sum_{k=1}^\infty \int_0^{\cdot} (g^k, \varphi_n) \, dw_s^k = \langle \sum_{k=1}^\infty g^k w_{\cdot}^k, \zeta \rangle_{L_2(\mathcal{O})}.$$

It remains to prove that for all $t \in [0, T]$,

$$\lim_{n \to \infty} \int_0^t \Big(\sum_{i,j=1}^d a^{ij} u_{x^i x^j}(s, \cdot), \varphi_n \Big) \, ds = - \int_0^t \langle u, A^* \zeta \rangle_{L_2(\mathcal{O})} \, ds \qquad \mathbb{P}\text{-a.s.,} \qquad (3.18)$$

which can be proven by reasoning as follows: The operator $(-A, D(-A))$ in (3.14) can be introduced as the variational operator $(-\tilde{A}, D(-\tilde{A}))$ in Section 2.4 starting with the bilinear form

$$a : \mathring{W}^1_2(\mathcal{O}) \times \mathring{W}^1_2(\mathcal{O}) \to \mathbb{R}$$

$$(u, v) \mapsto a(u, v) := \int_{\mathcal{O}} \sum_{i,j=1}^d a^{ij} u_{x^i} v_{x^j} \, dx. \qquad (3.19)$$

Thus, it is a densely defined, closed, self-adjoint and dissipative operator generating a contraction semigroup $\{S(t)\}_{t \in [0,T]}$ on $L_2(\mathcal{O})$. Since $u_x \in \mathbb{H}^1_{2,d}(\mathcal{O}, T) \subseteq \mathbb{H}^0_{2,d}(\mathcal{O}, T)$, which easily follows by (2.28) and the fact that $u \in \mathbb{H}^2_{2,d-2}(\mathcal{O}, T)$, the equality

$$(a^{ij} u_{x^i x^j}, \varphi_n) = -(a^{ij} u_{x^i}, (\varphi_n)_{x^j}) = - \int_{\mathcal{O}} a^{ij} u_{x^i} (\varphi_n)_{x^j} \, dx = -\langle a^{ij} u_{x^i}, (\varphi_n)_{x^j} \rangle_{L_2(\mathcal{O})}$$

holds $\mathbb{P}_T$-a.e. for all $i, j \in \{1, \ldots, d\}$ and $n \in \mathbb{N}$. Therefore,

$$\lim_{n \to \infty} \sum_{i,j=1}^d (a^{ij} u_{x^i x^j}, \varphi_n) = - \sum_{i,j=1}^d \langle a^{ij} u_{x^i}, \zeta_{x^j} \rangle_{L_2(\mathcal{O})} \qquad \mathbb{P}_T\text{-a.e.,}$$

and, consequently,

$$\lim_{n \to \infty} \sum_{i,j=1}^d (a^{ij} u_{x^i x^j}, \varphi_n) = -a(u, \zeta) = \langle u, -A\zeta \rangle_{L_2(\mathcal{O})} = -\langle u, A^* \zeta \rangle_{L_2(\mathcal{O})} \qquad \mathbb{P}_T\text{-a.e.,}$$

where $a(\cdot, \cdot)$ is given by (3.19). Now one can use the dominated convergence theorem to prove that (3.18) holds for all $t \in [0, T]$. In summary, u is a weak solution of the corresponding infinite-dimensional SDE of the type (3.12). The uniqueness follows from [32, Theorem 5.4]. $\square$

Chapter 4

Embeddings of weighted Sobolev spaces into Besov spaces

In this chapter we analyse the regularity within the non-linear approximation scale

$$B^\alpha_{\tau,\tau}(\mathcal{O}), \quad \frac{1}{\tau} = \frac{\alpha}{d} + \frac{1}{p}, \quad \alpha > 0, \tag{$*$}$$

of elements from the weighted Sobolev spaces $H^\gamma_{p,\theta}(\mathcal{O})$ introduced in Subsection 2.3.3. Our main goal is to prove that for $\gamma, \nu > 0$ and $p \geq 2$, the space $H^\gamma_{p,d-\nu p}(\mathcal{O})$ is embedded into the Besov spaces $B^\alpha_{\tau,\tau}(\mathcal{O})$ from ($*$) for certain $\alpha < \alpha_{\max} = \alpha_{\max}(\gamma, \nu, d)$. As before, also in this chapter, $\mathcal{O}$ denotes a bounded Lipschitz domain in $\mathbb{R}^d$.

Remember that, if we want to clarify whether adaptive wavelet methods for solving SPDEs bear the potential to be more efficient than their uniform alternatives, we need to analyse the regularity of the corresponding solution in the scale ($*$), cf. Section 1.1. In Chapter 3 we have seen that there exists a quite satisfactory solvability theory for a wide class of linear SPDEs within the spaces $\mathfrak{H}^\gamma_{p,\theta}(\mathcal{O}, T) = \mathfrak{H}^{\gamma,p}_{p,\theta}(\mathcal{O}, T)$ with suitable parameters $\gamma \in \mathbb{R}$, $p \in [2, \infty)$ and $\theta \in \mathbb{R}$ (cf. Theorem 3.13). For $q, p \in [2, \infty)$ and $\gamma, \theta \in \mathbb{R}$, the elements of $\mathfrak{H}^{\gamma,q}_{p,\theta}(\mathcal{O}, T)$ are L_q-integrable stochastic processes taking values in

$$H^\gamma_{p,\theta-p}(\mathcal{O}) = H^\gamma_{p,d-\nu p}(\mathcal{O}) \quad \text{with} \quad \nu = 1 + \frac{d-\theta}{p}.$$

Thus, a combination of the embedding mentioned above with Theorem 3.13 yields a statement about the spatial regularity of linear SPDEs within the scale ($*$) of Besov spaces (Theorem 5.2). Even more, this embedding shows that—to a certain extent—the regualrity analysis for SPDEs in terms of the scale ($*$) can be traced back to the analysis of such equations in terms of the spaces $\mathfrak{H}^{\gamma,q}_{p,\theta}(\mathcal{O}, T)$ (see Theorem 5.1).

Our results also have an impact on the regularity analysis of deterministic partial differential equations. E.g., the results from [76] on the weighted Sobolev regularity of deterministic parabolic and elliptic equations on bounded $\mathcal{C}^1_u$-domains will automatically lead to regularity results in the scale ($*$) for these equations. Using the mentioned embedding, one can also derive Besov regularity estimates for degenerate elliptic equations on bounded Lipschitz domains as considered, e.g., in [93]. Our results can also be seen as an extension of and a supplement to the Besov regularity results for elliptic equations in [38] and [34–36, 40, 63]. It is worth noting that first results on the regularity in the scale ($*$) of solutions to (deterministic) parabolic equations have been obtained in [3], see also the preparative results in [4, 5].

We choose the following outline. First, we will discuss the relationship between weighted Sobolev spaces with and without weights (Section 4.1). As we have mentioned in Subsection 2.3.3,

$H^m_{p,d-\nu p}(\mathcal{O}) = \mathring{W}^m_p(\mathcal{O})$ for $m \in \mathbb{N}_0$, see Lemma 2.51. We generalize this result and prove a general embedding of weighted Sobolev spaces into Sobolev spaces without weights (Proposition 4.1). Moreover, we enlighten the fact that, for the range of parameters γ and ν relevant for SPDEs, the elements of $H^\gamma_{p,d-\nu p}(\mathcal{O})$ have zero boundary trace. In particular, this justifies saying that the solutions considered in this thesis 'fulfil a zero Dirichlet boundary condition'. In the intermediate Section 4.2, we recall some fundamental results on the wavelet decomposition of Besov spaces. They will be used in Section 4.3, when proving the embedding mentioned above of weighted Sobolev spaces into Besov spaces from the scale $(*)$ (Theorem 4.13). The proof of this theorem is split into two parts: In Part One, we restrict ourselves to integer $\gamma \in \mathbb{N}$. In Part Two the complex interpolation method of A.P. Calderón and its extension to suitable quasi-Banach spaces by O. Mendez and M. Mitrea [96] is applied in order to prove the embedding for fractional $\gamma \in \mathbb{R}_+ \setminus \mathbb{N}$. In Section 4.4, we present an alternative proof of Theorem 4.7, which does not require any knowledge about complex interpolation in quasi-Banach spaces.

4.1 Weighted Sobolev spaces and Sobolev spaces without weights

We start with a general embedding of weighted Sobolev spaces into the closure of $\mathcal{C}^\infty_0(\mathcal{O})$ in the Sobolev spaces without weights.

Proposition 4.1. *Let* $\gamma, \nu \in (0, \infty)$ *and* $p \in [2, \infty)$. *Then the following embedding holds:*

$$H^\gamma_{p,d-\nu p}(\mathcal{O}) \hookrightarrow \mathring{W}^{\gamma \wedge \nu}_p(\mathcal{O}). \tag{4.1}$$

Proof. Since $\mathcal{C}^\infty_0(\mathcal{O})$ is densely embedded in the weighted Sobolev spaces, see Lemma 2.45(ii), it is enough to prove that $H^\gamma_{p,d-\nu p}(\mathcal{O}) \hookrightarrow W^{\gamma \wedge \nu}_p(\mathcal{O})$ for the particular parameters. We start the proof by considering the case where $\gamma = \nu$, i.e., we prove that for $\gamma > 0$ and $p \in [2, \infty)$ we have

$$H^\gamma_{p,d-\gamma p}(\mathcal{O}) \hookrightarrow W^\gamma_p(\mathcal{O}). \tag{4.2}$$

For $\gamma = m \in \mathbb{N}_0$ this follows from Lemma 2.51. In the case of fractional $\gamma \in \mathbb{R}_+ \setminus \mathbb{N}$ we argue as follows. Let $\gamma = m + \eta$ with $m \in \mathbb{N}_0$ and $\eta \in (0, 1)$. By Lemma 2.45(v),

$$H^{m+\eta}_{p,d-(m+\eta)p}(\mathcal{O}) = \left[H^m_{p,d-mp}(\mathcal{O}), H^{m+1}_{p,d-(m+1)p}(\mathcal{O}) \right]_\eta.$$

Thus, since (4.2) holds for the integer case,

$$H^{m+\eta}_{p,d-(m+\eta)p}(\mathcal{O}) \hookrightarrow \left[W^m_p(\mathcal{O}), W^{m+1}_p(\mathcal{O}) \right]_\eta.$$

By Theorem 2.66(ii) this yields

$$H^{m+\eta}_{p,d-(m+\eta)p}(\mathcal{O}) \hookrightarrow \left[F^m_{p,2}(\mathcal{O}), F^{m+1}_{p,2}(\mathcal{O}) \right]_\eta.$$

Since the Triebel-Lizorkin spaces constitute a scale of complex interpolation spaces, see, e.g., [117, Corollary 1.111], this leads to

$$H^{m+\eta}_{p,d-(m+\eta)p}(\mathcal{O}) \hookrightarrow F^{m+\eta}_{p,2}(\mathcal{O}).$$

Therefore, since $F^{m+\eta}_{p,2}(\mathcal{O}) \hookrightarrow B^{m+\eta}_{p,p}(\mathcal{O})$ by Theorem 2.66(i),

$$H^{m+\eta}_{p,d-(m+\eta)p}(\mathcal{O}) \hookrightarrow B^{m+\eta}_{p,p}(\mathcal{O}) = W^{m+\eta}_p(\mathcal{O}),$$

where the last equality follows from Theorem 2.60(i). Thus, Embedding (4.2) is proven. The embedding (4.1) for $\gamma \neq \nu$ follows now by using standard arguments. Indeed, since $\gamma \geq \gamma \wedge \nu$ we have

$$H_{p,d-\nu p}^{\gamma}(\mathcal{O}) \hookrightarrow H_{p,d-\nu p}^{\gamma \wedge \nu}(\mathcal{O}),$$

see [93, page 3]. Furthermore, the boundedness of the domain $\mathcal{O}$ and the fact that $d - \nu p \leq d - (\gamma \wedge \nu)p$ imply

$$H_{p,d-\nu p}^{\gamma \wedge \nu}(\mathcal{O}) \hookrightarrow H_{p,d-(\gamma \wedge \nu)p}^{\gamma \wedge \nu}(\mathcal{O}),$$

see Lemma 2.45(vii). A combination of these two embeddings with (4.2) finally gives the asserted Embedding (4.1). □

The following embedding is a consequence of Corollary 2.67, Theorem 2.60(i) and Lemma 2.45(ii). We use the common notation

$$\mathring{B}_{p,q}^{s}(\mathcal{O}) := \overline{\mathcal{C}_0^{\infty}(\mathcal{O})}^{\|\cdot\|_{B_{p,q}^s(\mathcal{O})}}$$

for the closure of the test functions $\mathcal{C}_0^{\infty}(\mathcal{O})$ in the Besov space $B_{p,q}^{s}(\mathcal{O})$ for $s \in \mathbb{R}$ and $p, q \in (0, \infty)$.

Corollary 4.2. *Let* $\gamma, \nu \in (0, \infty)$ *and* $p \in [2, \infty)$. *Then the following embedding holds:*

$$H_{p,d-\nu p}^{\gamma}(\mathcal{O}) \hookrightarrow \mathring{B}_{p,p}^{\gamma \wedge \nu}(\mathcal{O}).$$

Remark 4.3. Since $\mathcal{O} \subset \mathbb{R}^d$ is assumed to be a bounded Lipschitz domain, we know by [69, Chapter VIII, Theorem 2] that for $1/p < s$ the operator Tr, initially defined on $\mathcal{C}^{\infty}(\overline{\mathcal{O}})$ as the restriction to $\partial\mathcal{O}$, extends to a bounded linear operator from $B_{p,p}^{s}(\mathcal{O})$ to $B_{p,p}^{s-1/p}(\partial\mathcal{O})$, see [69] for a definition of Besov spaces on $\partial\mathcal{O}$. In this case we denote by $\mathring{B}_{p,p,0}^{s}(\mathcal{O})$ the subspace of $B_{p,p}^{s}(\mathcal{O})$ with zero boundary trace, i.e.,

$$B_{p,p,0}^{s}(\mathcal{O}) := \left\{ u \in B_{p,p}^{s}(\mathcal{O}) : \operatorname{Tr} u = 0 \right\}, \quad \frac{1}{p} < s.$$

If additionally $s < 1 + 1/p$, then, by [67, Theorem 3.12], these spaces coincide with the closure of $\mathcal{C}_0^{\infty}(\mathcal{O})$ in $B_{p,p}^{s}(\mathcal{O})$, i.e.,

$$\mathring{B}_{p,p}^{s}(\mathcal{O}) = B_{p,p,0}^{s}(\mathcal{O}) \quad \text{for} \quad \frac{1}{p} < s < 1 + \frac{1}{p}.$$

Thus, if $1/p < \gamma \wedge \nu < 1 + 1/p$, by Corollary 4.2,

$$H_{p,d-\nu p}^{\gamma}(\mathcal{O}) \hookrightarrow \mathring{B}_{p,p}^{\gamma \wedge \nu}(\mathcal{O}) = B_{p,p,0}^{\gamma \wedge \nu}(\mathcal{O}) = \left\{ u \in B_{p,p}^{\gamma \wedge \nu}(\mathcal{O}) : \operatorname{Tr} u = 0 \right\}.$$

In Section 3.2 we considered SPDEs in the setting of [75]. The solutions to these equations are stochastic processes taking values in $H_{p,d-\nu p}^{\gamma}(\mathcal{O})$ with $\nu := 1 + \frac{d-\theta}{p}$, where the value of θ never leaves the range

$$d - 1 < \theta < d + p - 1; \tag{4.3}$$

see also Remark 3.14(ii) and (iii). This condition is equivalent to $1/p < \nu < 1 + 1/p$ with ν as introduced before. Hence, if $\gamma > 1/p$ we deal with solutions fulfilling a zero Dirichlet boundary condition in the sense that they can be considered as stochastic processes taking values in $B_{p,p,0}^{\gamma \wedge \nu}(\mathcal{O})$.

4.2 Wavelet decomposition of Besov spaces on $\mathbb{R}^d$

In this section we present some fundamental results on the wavelet decomposition of Besov spaces. They will serve as a key ingredient in the proof of an embedding of weighted Sobolev spaces into Besov spaces from the non-linear approximation scale $(*)$ in the subsequent section. Our standard reference concerning wavelete decompositions of Besov spaces is the monograph [27], we also refer to the seminal works [47, 56, 64, 97, 109, 117] for further details.

Throughout this chapter, let ϕ be a scaling function of tensor product type on $\mathbb{R}^d$ and let ψ_i, $i = 1, \ldots, 2^d - 1$, be corresponding multivariate mother wavelets such that, for a given $r \in \mathbb{N}$ and some $M > 0$, the following locality, smoothness and vanishing moment conditions hold. For all $i = 1, \ldots, 2^d - 1$,

$$\operatorname{supp} \phi, \operatorname{supp} \psi_i \subseteq [-M, M]^d, \tag{4.4}$$

$$\phi, \psi_i \in \mathcal{C}^r(\mathbb{R}^d), \tag{4.5}$$

$$\int_{\mathbb{R}^d} x^\alpha \psi_i(x)\, dx = 0 \quad \text{for all } \alpha \in \mathbb{N}_0^d \text{ with } |\alpha| \leq r. \tag{4.6}$$

For the dyadic shifts and dilations of the scaling function and the corresponding wavelets we use the abbreviations

$$\phi_k(x) := \phi(x - k), \ x \in \mathbb{R}^d, \qquad \text{for } k \in \mathbb{Z}^d, \text{ and} \tag{4.7}$$

$$\psi_{i,j,k}(x) := 2^{jd/2}\psi_i(2^j x - k), \ x \in \mathbb{R}^d, \qquad \text{for } (i, j, k) \in \{1, \ldots, 2^d - 1\} \times \mathbb{N}_0 \times \mathbb{Z}^d, \tag{4.8}$$

and assume that

$$\left\{ \phi_k, \psi_{i,j,k} : (i, j, k) \in \{1, \ldots, 2^d - 1\} \times \mathbb{N}_0 \times \mathbb{Z}^d \right\}$$

is a Riesz basis of $L_2(\mathbb{R}^d)$. Further, we assume that there exists a dual Riesz basis satisfying the same requirements. That is, there exist functions $\widetilde{\phi}$ and $\widetilde{\psi}_i$, $i = 1, \ldots, 2^d - 1$, such that conditions (4.4), (4.5) and (4.6) hold if ϕ and ψ_i are replaced by $\widetilde{\phi}$ and $\widetilde{\psi}_i$, and such that the biorthogonality relations

$$\langle \widetilde{\phi}_k, \psi_{i,j,k} \rangle = \langle \widetilde{\psi}_{i,j,k}, \phi_k \rangle = 0, \quad \langle \widetilde{\phi}_k, \phi_l \rangle = \delta_{k,l}, \quad \langle \widetilde{\psi}_{i,j,k}, \psi_{u,v,l} \rangle = \delta_{i,u}\, \delta_{j,v}\, \delta_{k,l},$$

are fulfilled. Here we use analogous abbreviations to (4.7) and (4.8) for the dyadic shifts and dilations of $\widetilde{\phi}$ and $\widetilde{\psi}_i$, and $\delta_{k,l}$ denotes the Kronecker symbol. We refer to [27, Chapter 2] for the construction of biorthogonal wavelet bases, see also [45] and [30]. To keep notation simple, we will write

$$\psi_{i,j,k,p} := 2^{jd\left(\frac{1}{p} - \frac{1}{2}\right)} \psi_{i,j,k} \qquad \text{and} \qquad \widetilde{\psi}_{i,j,k,p'} := 2^{jd\left(\frac{1}{p'} - \frac{1}{2}\right)} \widetilde{\psi}_{i,j,k},$$

for the L_p-normalized wavelets and the correspondingly modified duals, with $p' := p/(p-1)$ if $p \in (0, \infty)$, $p \neq 1$, and $p' := \infty$, $1/p' := 0$ if $p = 1$.

The following theorem shows how Besov spaces on $\mathbb{R}^d$ can be described by decay properties of the wavelet coefficients, if the parameters fulfil certain conditions.

Theorem 4.4. *Let $p, q \in (0, \infty)$ and $s > \max\{0, d\,(1/p - 1)\}$. Choose $r \in \mathbb{N}$ such that $r > s$ and construct a biorthogonal wavelet Riesz basis as described above. Then a locally integrable function $f : \mathbb{R}^d \to \mathbb{R}$ is in the Besov space $B_{p,q}^s(\mathbb{R}^d)$ if, and only if,*

$$f = \sum_{k \in \mathbb{Z}^d} \langle f, \widetilde{\phi}_k \rangle\, \phi_k + \sum_{i=1}^{2^d - 1} \sum_{j \in \mathbb{N}_0} \sum_{k \in \mathbb{Z}^d} \langle f, \widetilde{\psi}_{i,j,k,p'} \rangle\, \psi_{i,j,k,p} \tag{4.9}$$

(convergence in $\mathcal{D}'(\mathbb{R}^d)$) with

$$\left(\sum_{k\in\mathbb{Z}^d}|\langle f,\tilde{\phi}_k\rangle|^p\right)^{\frac{1}{p}}+\left(\sum_{i=1}^{2^d-1}\sum_{j\in\mathbb{N}_0}2^{jsq}\left(\sum_{k\in\mathbb{Z}^d}|\langle f,\tilde{\psi}_{i,j,k,p'}\rangle|^p\right)^{\frac{q}{p}}\right)^{\frac{1}{q}}<\infty, \qquad (4.10)$$

and (4.10) is an equivalent (quasi-)norm for $B^s_{p,q}(\mathbb{R}^d)$.

Remark 4.5. A proof of this theorem for the case $p\geq 1$ can be found in [97, §10 of Chapter 6]. For the general case see for example [89] or [27, Theorem 3.7.7]. Of course, if (4.10) holds then the infinite sum in (4.9) converges also in $B^s_{p,q}(\mathbb{R}^d)$. If $s>\max\{0,d(1/p-1)\}$ we have the embedding $B^s_{p,q}(\mathbb{R}^d)\hookrightarrow L_{s_0}(\mathbb{R}^d)$ for some $s_0>1$, see, e.g., [117, Theorem 1.73(i)].

A simple computation gives us the following characterization of Besov spaces $B^\alpha_{\tau,\tau}(\mathbb{R}^d)$, in the case that the parameters α and τ are linked as in the scale $(*)$.

Corollary 4.6. *Let $p\in(1,\infty)$, $\alpha>0$ and $\tau\in\mathbb{R}$ such that $1/\tau=\alpha/d+1/p$. Choose $r\in\mathbb{N}$ such that $r>\alpha$ and construct a biorthogonal wavelet Riesz basis as described above. Then a locally integrable function $f:\mathbb{R}^d\to\mathbb{R}$ is in the Besov space $B^\alpha_{\tau,\tau}(\mathbb{R}^d)$ if, and only if,*

$$f=\sum_{k\in\mathbb{Z}^d}\langle f,\tilde{\phi}_k\rangle\phi_k+\sum_{i=1}^{2^d-1}\sum_{j\in\mathbb{N}_0}\sum_{k\in\mathbb{Z}^d}\langle f,\tilde{\psi}_{i,j,k,p'}\rangle\psi_{i,j,k,p} \qquad (4.11)$$

(convergence in $\mathcal{D}'(\mathbb{R}^d)$) with

$$\left(\sum_{k\in\mathbb{Z}^d}|\langle f,\tilde{\phi}_k\rangle|^\tau\right)^{\frac{1}{\tau}}+\left(\sum_{i=1}^{2^d-1}\sum_{j\in\mathbb{N}_0}\sum_{k\in\mathbb{Z}^d}|\langle f,\tilde{\psi}_{i,j,k,p'}\rangle|^\tau\right)^{\frac{1}{\tau}}<\infty, \qquad (4.12)$$

and (4.12) is an equivalent (quasi-)norm for $B^\alpha_{\tau,\tau}(\mathbb{R}^d)$.

4.3 Weighted Sobolev spaces and the non-linear approximation scale

In this section we prove our main result concerning the relationship between weighted Sobolev spaces and the Besov spaces from the non-linear approximation scale $(*)$. That is, we prove the following embedding.

Theorem 4.7. *Let $\mathcal{O}$ be a bounded Lipschitz domain in $\mathbb{R}^d$. Let $p\in[2,\infty)$, and $\gamma,\nu\in(0,\infty)$. Then*

$$H^\gamma_{p,d-\nu p}(\mathcal{O})\hookrightarrow B^\alpha_{\tau,\tau}(\mathcal{O}), \qquad \frac{1}{\tau}=\frac{\alpha}{d}+\frac{1}{p}, \qquad \text{for all} \qquad 0<\alpha<\min\left\{\gamma,\nu\frac{d}{d-1}\right\}. \qquad (4.13)$$

Before we start proving this result, we make some notes on our strategy. We split our proof into two parts. In the first part we assume that γ is an integer, i.e., $\gamma\in\mathbb{N}$. In this particular case we follow the lines of the proof of [38, Theorem 3.2]. This theorem can be restated as follows: If a harmonic function u lies in the Besov space $B^\nu_{p,p}(\mathcal{O})$ for some $p\in(1,\infty)$ and $\nu>0$, then it is contained in the Besov spaces

$$B^\alpha_{\tau,\tau}(\mathcal{O}), \qquad \frac{1}{\tau}=\frac{\alpha}{d}+\frac{1}{p}, \qquad \text{for all} \qquad 0<\alpha<\nu\frac{d}{d-1}. \qquad (4.14)$$

In order to prove this statement, the authors of [38] use fundamental results on extension operators [110] and on wavelet characterizations of Besov spaces—as presented in the previous section—and estimate the (weighted) ℓ_τ-norm of suitable wavelet coefficients. In this way they prove that any harmonic function $u \in B_{p,p}^\nu(\mathcal{O})$ fulfils the estimate

$$\|u\|_{B_{\tau,\tau}^\alpha(\mathcal{O})} \leq C \|u\|_{B_{p,p}^\nu(\mathcal{O})}, \qquad \frac{1}{\tau} = \frac{\alpha}{d} + \frac{1}{p}, \qquad \text{for all} \qquad 0 < \alpha < \nu \frac{d}{d-1}, \qquad (4.15)$$

where the constant C does not depend on u. A close look at the proof reveals that two facts are proven and combined in order to show that (4.15) holds in the prescribed setting. First, without making use of the harmonicity of the considered function $u \in B_{p,p}^\nu(\mathcal{O})$, it is proven that

$$\|u\|_{B_{\tau,\tau}^\alpha(\mathcal{O})} \leq C \big(\|u\|_{B_{p,p}^\nu(\mathcal{O})} + |u|_{H_{p,d-\nu p}^\gamma(\mathcal{O})} \big), \qquad \frac{1}{\tau} = \frac{\alpha}{d} + \frac{1}{p}, \qquad \text{for all} \quad 0 < \alpha < \gamma \wedge \nu \frac{d}{d-1}, \quad (4.16)$$

provided the semi-norm

$$|u|_{H_{p,d-\nu p}^\gamma(\mathcal{O})}^p = \sum_{|\alpha|=\gamma} \int_{\mathcal{O}} |\rho(x)^{|\alpha|} D^\alpha u(x)|^p \rho(x)^{-\nu p} \, \mathrm{d}x, \qquad (4.17)$$

is finite for some $\gamma \in \mathbb{N}$ with $\gamma > \nu$. Then, it is proven in the same publication, see [38, Theorem 3.1], that for any harmonic function we have

$$|u|_{H_{p,d-\nu p}^\gamma(\mathcal{O})} \leq C \|u\|_{B_{p,p}^\nu(\mathcal{O})}, \qquad \text{for all} \qquad 0 < \nu < \gamma \in \mathbb{N}. \qquad (4.18)$$

Finally, a combination of (4.16) and (4.18) yields (4.15) for harmonic functions. However, as already mentioned, (4.16) can be proven without assuming that u is harmonic. Thus, if we assume that $u \in H_{p,d-\nu p}^\gamma(\mathcal{O}) \cap B_{p,p}^\nu(\mathcal{O})$ with $\gamma \in \mathbb{N}$, the same strategy yields the estimate

$$\|u\|_{B_{\tau,\tau}^\alpha(\mathcal{O})} \leq C \big(\|u\|_{B_{p,p}^\nu(\mathcal{O})} + \|u\|_{H_{p,d-\nu p}^\gamma(\mathcal{O})} \big), \qquad \frac{1}{\tau} = \frac{\alpha}{d} + \frac{1}{p}, \qquad \text{for all} \quad 0 < \alpha < \min\left\{\gamma, \nu \frac{d}{d-1}\right\}.$$

Since, as proven in Corollary 4.2, $H_{p,d-\nu p}^\gamma(\mathcal{O}) \hookrightarrow B_{p,p}^\nu(\mathcal{O})$ for $p \in [2,\infty)$ and $\nu < \gamma$, this leads to

$$\|u\|_{B_{\tau,\tau}^\alpha(\mathcal{O})} \leq C \|u\|_{H_{p,d-\nu p}^\gamma(\mathcal{O})}, \qquad \frac{1}{\tau} = \frac{\alpha}{d} + \frac{1}{p}, \qquad \text{for all} \quad 0 < \alpha < \gamma \wedge \nu \frac{d}{d-1},$$

and all $u \in H_{p,d-\nu p}^\gamma(\mathcal{O})$, if $p \in [2,\infty)$ and $0 < \nu \leq \gamma \in \mathbb{N}$. But this is exactly our assertion for $\gamma \in \mathbb{N}$ (and $\nu \leq \gamma$).

Now we present this proof strategy in detail. The case $\gamma \in \mathbb{R}_+ \setminus \mathbb{N}$ will be considered thereafter in Part Two.

Proof of Theorem 4.7 (Part One). In this first part, we prove that the assertion holds for $\gamma \in \mathbb{N}$. We fix $p \in [2,\infty)$ and start with the case $\nu > \gamma$. Then, by Corollary 4.2 and Theorem 2.61(iii), for any $0 < \alpha < \gamma$, we have

$$H_{p,d-\nu p}^\gamma(\mathcal{O}) \hookrightarrow B_{p,p}^\gamma(\mathcal{O}) \hookrightarrow B_{\tau,\tau}^\alpha(\mathcal{O}), \qquad \frac{1}{\tau} = \frac{\alpha}{d} + \frac{1}{p}.$$

Therefore, in this case the assertion follows immediately. From now on, let us assume that $0 < \nu \leq \gamma \in \mathbb{N}$. We fix α and τ as stated in the theorem and choose a wavelet Riesz basis

$$\left\{ \phi_k, \psi_{i,j,k} : (i,j,k) \in \{1,\ldots,2^d-1\} \times \mathbb{N}_0 \times \mathbb{Z}^d \right\}$$

of $L_2(\mathbb{R}^d)$ which satisfies the assumptions from Section 4.2 with $r > \gamma$ and some arbitrary $M > 0$. (Later on, without loss of generality, $2M \in \mathbb{N}$ will be assumed.) Given $(j,k) \in \mathbb{N}_0 \times \mathbb{Z}^d$ let

$$Q_{j,k} := 2^{-j}k + 2^{-j}[-M, M]^d,$$

so that supp $\psi_{i,j,k} \subset Q_{j,k}$ for all $i \in \{1, \dots, 2^d-1\}$ and supp $\phi_k \subset Q_{0,k}$ for all $k \in \mathbb{Z}^d$. Remember that the supports of the corresponding dual basis meet the same requirements. For our purpose the set of all indices associated with those wavelets and scaling functions that may have common support with the domain $\mathcal{O}$ will play an important role and we denote them by

$$\Lambda := \left\{ (i,j,k) \in \{1, \dots, 2^d - 1\} \times \mathbb{N}_0 \times \mathbb{Z}^d \, : \, Q_{j,k} \cap \mathcal{O} \neq \emptyset \right\},$$

and

$$\Gamma := \left\{ k \in \mathbb{Z}^d \, : \, Q_{0,k} \cap \mathcal{O} \neq \emptyset \right\}.$$

After these preparations, we fix $u \in H^\gamma_{p,d-\nu p}(\mathcal{O})$. Due to Corollary 4.2 we have $u \in B^\nu_{p,p}(\mathcal{O})$. As $\mathcal{O}$ is a Lipschitz domain there exists a linear and bounded extension operator $\mathcal{E} : B^\nu_{p,p}(\mathcal{O}) \to B^\nu_{p,p}(\mathbb{R}^d)$, i.e., there exists a constant $C > 0$, such that

$$\mathcal{E}u\big|_{\mathcal{O}} = u \qquad \text{and} \qquad \|\mathcal{E}u\|_{B^\nu_{p,p}(\mathbb{R}^d)} \leq C\|u\|_{B^\nu_{p,p}(\mathcal{O})}, \tag{4.19}$$

see, e.g., [110]. The constant in (4.19) as well as all the constants C appearing in the rest of this proof do not depend on u. In the sequel we will omit the $\mathcal{E}$ in our notation and write u instead of $\mathcal{E}u$. Theorem 4.4 tells us that the following equality holds on the domain $\mathcal{O}$:

$$u = \sum_{k \in \Gamma} \langle u, \widetilde{\phi}_k \rangle \phi_k + \sum_{(i,j,k) \in \Lambda} \langle u, \widetilde{\psi}_{i,j,k,p'} \rangle \psi_{i,j,k,p},$$

where the sums converge unconditionally in $B^\nu_{p,p}(\mathbb{R}^d)$. Furthermore, cf. Corollary 4.6, we have

$$\|u\|^\tau_{B^\alpha_{\tau,\tau}(\mathcal{O})} \leq C \left(\sum_{k \in \Gamma} |\langle u, \widetilde{\phi}_k \rangle|^\tau + \sum_{(i,j,k) \in \Lambda} |\langle u, \widetilde{\psi}_{i,j,k,p'} \rangle|^\tau \right).$$

Hence, in order to prove Embedding (4.13), it is enough to prove that

$$\sum_{k \in \Gamma} |\langle u, \widetilde{\phi}_k \rangle|^\tau \leq C \, \|u\|^\tau_{B^\nu_{p,p}(\mathcal{O})} \tag{4.20}$$

and

$$\sum_{(i,j,k) \in \Lambda} |\langle u, \widetilde{\psi}_{i,j,k,p'} \rangle|^\tau \leq C \left(\|u\|_{H^\gamma_{p,d-\nu p}(\mathcal{O})} + \|u\|_{B^\nu_{p,p}(\mathcal{O})} \right)^\tau, \tag{4.21}$$

cf. Corollary 4.2.

We start with (4.20). The index set Γ introduced above is finite because of the boundedness of $\mathcal{O}$, so that we can use Jensen's inequality followed by Theorem 4.4 together with the boundedness of the extension operator to obtain

$$\sum_{k \in \Gamma} |\langle u, \widetilde{\phi}_k \rangle|^\tau \leq C \left(\left(\sum_{k \in \Gamma} |\langle u, \widetilde{\phi}_k \rangle|^p \right)^{\frac{1}{p}} \right)^\tau \leq C \, \|u\|^\tau_{B^\nu_{p,p}(\mathcal{O})}.$$

To prove (4.21), we introduce the following notation:

$$\rho_{j,k} := \operatorname{dist}(Q_{j,k}, \partial\mathcal{O}) = \inf_{x \in Q_{j,k}} \rho(x),$$

$$\Lambda_j := \left\{ (i,l,k) \in \Lambda : l = j \right\},$$

$$\Lambda_{j,m} := \left\{ (i,j,k) \in \Lambda_j : m2^{-j} \le \rho_{j,k} < (m+1)2^{-j} \right\},$$

$$\Lambda_j^0 := \Lambda_j \setminus \Lambda_{j,0},$$

$$\Lambda^0 := \bigcup_{j \in \mathbb{N}_0} \Lambda_j^0,$$

where $j, m \in \mathbb{N}_0$ and $k \in \mathbb{Z}^d$. We split the expression on the left hand side of (4.21) into

$$\sum_{(i,j,k) \in \Lambda^0} |\langle u, \widetilde{\psi}_{i,j,k,p'}\rangle|^\tau + \sum_{(i,j,k) \in \Lambda \setminus \Lambda^0} |\langle u, \widetilde{\psi}_{i,j,k,p'}\rangle|^\tau =: I + I\!I \tag{4.22}$$

and estimate each term separately.

Let us begin with I, i.e., with the coefficients corresponding to wavelets with support guaranteed to be completely contained in the domain $\mathcal{O}$. Recall that in this thesis we write A° for the interior of a set $A \subseteq \mathbb{R}^d$. Fix $(i,j,k) \in \Lambda^0$. In this case, the semi-norm

$$|u|_{W_p^\gamma(Q_{j,k}^\circ)} := \sup_{|\alpha| = \gamma} \|D^\alpha u\|_{L_p(Q_{j,k})}$$

is finite, since $\rho_{j,k} = \operatorname{dist}(Q_{j,k}, \partial\mathcal{O}) > 0$ and since $u \in H_{p,d-\nu p}^\gamma(\mathcal{O})$, which implies

$$|u|_{H_{p,d-\nu p}^\gamma(\mathcal{O})}^p = \sum_{|\alpha| = \gamma} \int_\mathcal{O} |\rho(x)^{|\alpha|} D^\alpha u(x)|^p \rho(x)^{-\nu p} \, dx < \infty,$$

cf. Remark 2.44. By a Whitney-type inequality, also known as the Deny-Lions lemma, see, e.g., [48, Theorem 3.4], there exists a polynomial $P_{j,k}$ of total degree less than γ, and a constant C, which does not depend on j or k, such that

$$\|u - P_{j,k}\|_{L_p(Q_{j,k})} \le C \, 2^{-j\gamma} |u|_{W_p^\gamma(Q_{j,k}^\circ)}.$$

Since $\widetilde{\psi}_{i,j,k,p'}$ is orthogonal to every polynomial of total degree less than γ, we have

$$\begin{aligned}
|\langle u, \widetilde{\psi}_{i,j,k,p'}\rangle| &= |\langle u - P_{j,k}, \widetilde{\psi}_{i,j,k,p'}\rangle| \\
&\le \|u - P_{j,k}\|_{L_p(Q_{j,k})} \|\widetilde{\psi}_{i,j,k,p'}\|_{L_{p'}(Q_{j,k})} \\
&\le C \, 2^{-j\gamma} |u|_{W_p^\gamma(Q_{j,k}^\circ)}.
\end{aligned}$$

The constant C does not depend on j or k, since $\|\widetilde{\psi}_{i,j,k,p'}\|_{L_{p'}(Q_{j,k})} = \|\widetilde{\psi}_i\|_{L_{p'}(\mathbb{R}^d)}$. Inserting the definition of the semi-norm on the right hand side and putting $1 = \rho(x)^{\gamma-\nu}\rho(x)^{\nu-\gamma}$ into the integrals, yields

$$\begin{aligned}
|\langle u, \widetilde{\psi}_{i,j,k,p'}\rangle| &\le C \, 2^{-j\gamma} \sup_{|\alpha| = \gamma} \left(\int_{Q_{j,k}} |D^\alpha u(x)|^p \, dx \right)^{\frac{1}{p}} \\
&\le C \, 2^{-j\gamma} \rho_{j,k}^{\nu-\gamma} \sup_{|\alpha| = \gamma} \left(\int_{Q_{j,k}} |\rho(x)^{\gamma-\nu} D^\alpha u(x)|^p \, dx \right)^{\frac{1}{p}} \\
&=: C \, 2^{-j\gamma} \rho_{j,k}^{\nu-\gamma} \mu_{j,k}.
\end{aligned}$$

Fix $j \in \mathbb{N}_0$. Summing over all indices $(i,j,k) \in \Lambda_j^0$ and applying Hölder's inequality with exponents $p/\tau > 1$ and $p/(p-\tau)$ one finds

$$
\sum_{(i,j,k)\in\Lambda_j^0} \left| \langle u, \widetilde{\psi}_{i,j,k,p'} \rangle \right|^\tau \leq C \sum_{(i,j,k)\in\Lambda_j^0} 2^{-j\gamma\tau} \rho_{j,k}^{(\nu-\gamma)\tau} \mu_{j,k}^\tau
$$

$$
\leq C \left(\sum_{(i,j,k)\in\Lambda_j^0} \mu_{j,k}^p \right)^{\frac{\tau}{p}} \left(\sum_{(i,j,k)\in\Lambda_j^0} 2^{-j\frac{\gamma\tau p}{p-\tau}} \rho_{j,k}^{\frac{(\nu-\gamma)\tau p}{p-\tau}} \right)^{\frac{p-\tau}{p}}, \tag{4.23}
$$

where C does not depend on j. In order to estimate the first sum in the product on the right hand side of (4.23) we regroup the cubes $Q_{j,k}$, $k \in \mathbb{Z}^d$, in the following way. Without loss of generality, we assume that $2M \in \mathbb{N}$. Let a_n, $n = 1, \ldots, (2M)^d$, be an arbitrary arrangement of the d-tuples from $\{0, 1, \ldots, 2M-1\}^d$, and define

$$
R_{j,n} := \left\{ Q_{j,k} : k \in a_n + 2M\mathbb{Z}^d \right\}, \qquad n \in \{1, \ldots, (2M)^d\}. \tag{4.24}
$$

Then, one can check that $\{R_{j,n} : n \in \{1, \ldots, (2M)^d\}\}$ is a finite partition of $\{Q_{j,k} : k \in \mathbb{Z}^d\}$, i.e.,

$$
\bigcup_{n=1}^{(2M)^d} R_{j,n} = \left\{ Q_{j,k} : k \in \mathbb{Z}^d \right\} \quad \text{and} \quad R_{j,n} \cap R_{j,n'} = \emptyset \quad \text{for} \quad n \neq n'. \tag{4.25}
$$

Furthermore, for any fixed $n \in \{1, \ldots, (2M)^d\}$,

$$
\text{if} \quad Q_{j,k}, Q_{j,k'} \in R_{j,n} \quad \text{for some} \quad k \neq k', \quad \text{then} \quad Q_{j,k}^\circ \cap Q_{j,k'}^\circ = \emptyset. \tag{4.26}
$$

Thus, setting

$$
R_{j,n}^0 := \left\{ (i,j,k) \in \Lambda_j^0 : Q_{j,k} \in R_{j,n} \right\}, \qquad n \in \{1, \ldots, (2M)^d\},
$$

and using (4.25), we obtain

$$
\left(\sum_{(i,j,k)\in\Lambda_j^0} \mu_{j,k}^p \right)^{\frac{\tau}{p}} = \left(\sum_{(i,j,k)\in\Lambda_j^0} \sup_{|\alpha|=\gamma} \int_{Q_{j,k}} \left| \rho(x)^{\gamma-\nu} D^\alpha u(x) \right|^p dx \right)^{\frac{\tau}{p}}
$$

$$
= \left(\sum_{n=1}^{(2M)^d} \sum_{(i,j,k)\in R_{j,n}^0} \sup_{|\alpha|=\gamma} \int_{Q_{j,k}} \left| \rho(x)^{\gamma-\nu} D^\alpha u(x) \right|^p dx \right)^{\frac{\tau}{p}}.
$$

Together with (4.26) and using the norm equivalence (2.28) together with some standard computations, this yields

$$
\left(\sum_{(i,j,k)\in\Lambda_j^0} \mu_{j,k}^p \right)^{\frac{\tau}{p}} \leq C \left(\sum_{|\alpha|=\gamma} \int_{\mathcal{O}} \left| \rho(x)^{\gamma-\nu} D^\alpha u(x) \right|^p dx \right)^{\frac{\tau}{p}} \leq C \|u\|_{H^\gamma_{p,d-\nu p}(\mathcal{O})}^\tau, \tag{4.27}
$$

with a constant C, which does not depend on j. In order to estimate the second sum on the right hand side of (4.23) we use the Lipschitz character of the domain $\mathcal{O}$ which implies that

$$
|\Lambda_{j,m}| \leq C \, 2^{j(d-1)} \qquad \text{for all} \quad j, m \in \mathbb{N}_0. \tag{4.28}
$$

Moreover, the boundedness of $\mathcal{O}$ yields $\Lambda_{j,m} = \emptyset$ for all $j, m \in \mathbb{N}_0$ with $m \geq C2^j$, where the constant C does not depend on j or m. Consequently,

$$
\left(\sum_{(i,j,k) \in \Lambda_j^0} 2^{-j \frac{\gamma p \tau}{p-\tau}} \rho_{j,k}^{\frac{(\nu-\gamma)p\tau}{p-\tau}} \right)^{\frac{p-\tau}{p}} \leq \left(\sum_{m=1}^{C2^j} \sum_{(i,j,k) \in \Lambda_{j,m}} 2^{-j \frac{\gamma p \tau}{p-\tau}} \rho_{j,k}^{\frac{(\nu-\gamma)p\tau}{p-\tau}} \right)^{\frac{p-\tau}{p}}
$$

$$
\leq C \left(\sum_{m=1}^{C2^j} 2^{j(d-1)} 2^{-j \frac{\gamma p \tau}{p-\tau}} (m \, 2^{-j})^{\frac{(\nu-\gamma)p\tau}{p-\tau}} \right)^{\frac{p-\tau}{p}} \tag{4.29}
$$

$$
\leq C \left(2^{j \left(d-1-\frac{\nu p \tau}{p-\tau} \right)} + 2^{j \left(d - \frac{\gamma p \tau}{p-\tau} \right)} \right)^{\frac{p-\tau}{p}}.
$$

Now, let us sum over all $j \in \mathbb{N}_0$. Inequalities (4.29) together with (4.27) and (4.23) imply

$$
\sum_{(i,j,k) \in \Lambda^0} \left| \langle u, \widetilde{\psi}_{i,j,k,p'} \rangle \right|^\tau \leq C \sum_{j \in \mathbb{N}_0} \left(2^{j \left(d-1-\frac{\nu p \tau}{p-\tau} \right)} + 2^{j \left(d - \frac{\gamma p \tau}{p-\tau} \right)} \right)^{\frac{p-\tau}{p}} \| u \|_{H_{p,d-\nu p}^\gamma(\mathcal{O})}^\tau.
$$

Obviously, the sums on the right hand side converge if, and only if, $\alpha \in \left(0, \gamma \wedge \nu \frac{d}{d-1} \right)$. Finally,

$$
\sum_{(i,j,k) \in \Lambda^0} \left| \langle u, \widetilde{\psi}_{i,j,k,p'} \rangle \right|^\tau \leq C \| u \|_{H_{p,d-\nu p}^\gamma(\mathcal{O})}^\tau.
$$

Now we estimate the second term $I\!I$ in (4.22). First we fix $j \in \mathbb{N}_0$ and use Hölder's inequality and (4.28) to obtain

$$
\sum_{(i,j,k) \in \Lambda_{j,0}} \left| \langle u, \widetilde{\psi}_{i,j,k,p'} \rangle \right|^\tau \leq C \, 2^{j(d-1)\frac{p-\tau}{p}} \left(\sum_{(i,j,k) \in \Lambda_{j,0}} \left| \langle u, \widetilde{\psi}_{i,j,k,p'} \rangle \right|^p \right)^{\frac{\tau}{p}},
$$

with a constant C which does not depend on j. Summing over all $j \in \mathbb{N}_0$ and using Hölder's inequality again yields

$$
\sum_{(i,j,k) \in \Lambda \backslash \Lambda^0} \left| \langle u, \widetilde{\psi}_{i,j,k,p'} \rangle \right|^\tau = \sum_{j \in \mathbb{N}_0} \sum_{(i,j,k) \in \Lambda_{j,0}} \left| \langle u, \widetilde{\psi}_{i,j,k,p'} \rangle \right|^\tau
$$

$$
\leq C \sum_{j \in \mathbb{N}_0} \left(2^{j(d-1)\frac{p-\tau}{p}} \left(\sum_{(i,j,k) \in \Lambda_{j,0}} \left| \langle u, \widetilde{\psi}_{i,j,k,p'} \rangle \right|^p \right)^{\frac{\tau}{p}} \right)
$$

$$
\leq C \left(\sum_{j \in \mathbb{N}_0} 2^{j \left(\frac{(d-1)(p-\tau)}{p} - \nu \tau \right) \frac{p}{p-\tau}} \right)^{\frac{p-\tau}{p}} \left(\sum_{j \in \mathbb{N}_0} \sum_{(i,j,k) \in \Lambda_{j,0}} 2^{j \nu p} \left| \langle u, \widetilde{\psi}_{i,j,k,p'} \rangle \right|^p \right)^{\frac{\tau}{p}}.
$$

Using Theorem 4.4 and the boundedness of the extension operator, we obtain

$$
\sum_{(i,j,k) \in \Lambda \backslash \Lambda^0} \left| \langle u, \widetilde{\psi}_{i,j,k,p'} \rangle \right|^\tau \leq C \| u \|_{B_{p,p}^\nu(\mathcal{O})}^\tau \left(\sum_{j \in \mathbb{N}_0} 2^{j \left(\frac{(d-1)(p-\tau)}{p} - \nu \tau \right) \frac{p}{p-\tau}} \right)^{\frac{p-\tau}{p}}.
$$

The series on the right hand side converges if, and only if, $\alpha \in \left(0, \nu \frac{d}{d-1} \right)$. We thus have

$$
\sum_{(i,j,k) \in \Lambda \backslash \Lambda^0} \left| \langle u, \widetilde{\psi}_{i,j,k,p'} \rangle \right|^\tau \leq C \| u \|_{B_{p,p}^\nu(\mathcal{O})}^\tau. \qquad \square
$$

So far we have proven the assertion of Theorem 4.7 provided the smoothness parameter γ is an integer. In what follows, we consider the complementary case of fractional $\gamma \in \mathbb{R}_+ \setminus \mathbb{N}$. Our strategy for proving Embedding (4.13) in this case relies on a combination of the already proven assertion for $\gamma \in \mathbb{N}$ and suitable applications of the complex interpolation method of A.P. Calderón. This method is initially defined only for Banach spaces and its extension to quasi-Banach spaces is not a trivial task. However, in our approach we need such an extension, which preserves the interpolation property and is applicable to compatible couples of Besov spaces $(B_{p_1,q_1}^{s_1}(\mathcal{O}), B_{p_2,q_2}^{s_2}(\mathcal{O}))$ on bounded Lipschitz domains $\mathcal{O} \subset \mathbb{R}^d$ with $s_1, s_2 \in \mathbb{R}$ and $0 < p_1, p_2, q_1, q_2 < \infty$. Fortunately, such a method has been developed in [96], see also [70] for more details. We use the notation

$$\left[B_{p_1,q_1}^{s_1}(\mathcal{O}), B_{p_2,q_2}^{s_2}(\mathcal{O}) \right]_\eta$$

for the (extended) complex interpolation method from [70, 96] applied to a compatible couple $(B_{p_1,q_1}^{s_1}(\mathcal{O}), B_{p_2,q_2}^{s_2}(\mathcal{O}))$ of Besov spaces. Then, since the interpolation property is preserved,

$$B_{p_1,p_1}^{s_1}(\mathcal{O}) \hookrightarrow E_1 \quad \text{and} \quad B_{p_2,p_2}^{s_2}(\mathcal{O}) \hookrightarrow E_2 \quad \text{imply} \quad \left[B_{p_1,p_1}^{s_1}(\mathcal{O}), B_{p_2,p_2}^{s_2}(\mathcal{O}) \right]_\eta \hookrightarrow [E_1, E_2]_\eta$$

for a compatible couple (E_1, E_2) of Banach spaces ($\eta \in (0, 1)$). The following result concerning the complex interpolation of Besov spaces is an immediate consequence of [117, Proposition 1.114], see also [70, Theorem 9.4]. It is a major ingredient in Part Two of the proof of Theorem 4.7.

Theorem 4.8. *Let $\mathcal{O}$ be a bounded Lipschitz domain in $\mathbb{R}^d$, and $p \in [2, \infty)$. Furthermore, let $0 \leq \alpha_0 < \alpha_1 < \infty$ and $\tau_0, \tau_1 \in (0, \infty)$ be such that*

$$\frac{1}{\tau_0} = \frac{\alpha_0}{d} + \frac{1}{p} \quad \text{and} \quad \frac{1}{\tau_1} = \frac{\alpha_1}{d} + \frac{1}{p}. \tag{4.30}$$

Then, for any $\eta \in (0, 1)$,

$$\left[B_{\tau_0,\tau_0}^{\alpha_0}(\mathcal{O}), B_{\tau_1,\tau_1}^{\alpha_1}(\mathcal{O}) \right]_\eta = B_{\tau,\tau}^\alpha(\mathcal{O}), \tag{4.31}$$

where

$$\alpha = (1 - \eta)\alpha_0 + \eta\alpha_1 \quad \text{and} \quad \frac{1}{\tau} = \frac{\alpha}{d} + \frac{1}{p}.$$

Proof. By [117, Proposition 1.114], equality (4.31) holds with

$$\frac{1}{\tau} = \frac{1 - \eta}{\tau_0} + \frac{\eta}{\tau_1}.$$

Inserting (4.30), the assertion follows. $\qquad\square$

This result at hand, we are ready to prove Embedding 4.13 for $\gamma \in \mathbb{R}_+ \setminus \mathbb{N}$.

Proof of Theorem 4.7 (Part Two). Let $\gamma \in \mathbb{R}_+ \setminus \mathbb{N}$, $\nu \in (0, \infty)$ and $p \in [2, \infty)$. If $\nu > \gamma$, (4.13) follows with the same arguments as in Part One. Thus, in what follows, we assume that $0 < \nu \leq \gamma$. We distinguish five cases.

Case 1. Let $\gamma := m + \eta$, with $m \in \mathbb{N}$, $\eta \in (0, 1)$, and $\nu \frac{d}{d-1} \geq m + 1$. Then, we can argue as follows. Fix an arbitrary $\varepsilon > 0$ with $\varepsilon \leq \eta$. Set $\alpha_0 := m - \varepsilon$, $\alpha_1 := m + 1 - \varepsilon$ and let $\tau_0, \tau_1 \in (0, \infty)$ be given by (4.30). Then, by Part One, $H_{p,d-\nu p}^{m+1}(\mathcal{O}) \hookrightarrow B_{\tau_1,\tau_1}^{\alpha_1}(\mathcal{O})$ and $H_{p,d-\nu p}^m(\mathcal{O}) \hookrightarrow B_{\tau_0,\tau_0}^{\alpha_0}(\mathcal{O})$. Thus,

$$\left[H_{p,d-\nu p}^m(\mathcal{O}), H_{p,d-\nu p}^{m+1}(\mathcal{O}) \right]_\eta \hookrightarrow \left[B_{\tau_0,\tau_0}^{\alpha_0}(\mathcal{O}), B_{\tau_1,\tau_1}^{\alpha_1}(\mathcal{O}) \right]_\eta.$$

Hence, by Theorem 4.8 and Lemma 2.45(v),

$$H^{m+\eta}_{p,d-\nu p}(\mathcal{O}) \hookrightarrow B^{\alpha^*}_{\tau^*,\tau^*}(\mathcal{O}),$$

with $\alpha^* = m + \eta - \varepsilon$ and $1/\tau^* = \alpha^*/d + 1/p$. This is true for arbitrary $\varepsilon \in (0,\eta]$. We therefore obtain

$$H^{\gamma}_{p,d-\nu p}(\mathcal{O}) = H^{m+\eta}_{p,d-\nu p}(\mathcal{O}) \hookrightarrow B^{\alpha}_{\tau,\tau}(\mathcal{O}), \quad \frac{1}{\tau} = \frac{\alpha}{d} + \frac{1}{p}, \quad \text{for all } 0 < \alpha < m + \eta = \min\left\{\gamma, \nu\frac{d}{d-1}\right\},$$

by simply applying the first embedding from Theorem 2.61(ii).

Case 2. Let $\gamma := \eta \in (0,1)$ and $\nu\frac{d}{d-1} \geq 1$. Then, since $L_p(\mathcal{O}) \hookrightarrow B^0_{p,p}(\mathcal{O})$ for $p \geq 2$, see Corollary 2.67, we also have $H^0_{p,d-\nu p}(\mathcal{O}) = L_{p,d-\nu p}(\mathcal{O}) \hookrightarrow B^0_{p,p}(\mathcal{O})$ for any $\nu > 0$. Simultaneously, by Part One, for any $\varepsilon \in (0,1)$, $H^1_{p,d-\nu p}(\mathcal{O}) \hookrightarrow B^{1-\varepsilon}_{\tau_1,\tau_1}(\mathcal{O})$ with $1/\tau_1 = (1-\varepsilon)/d + 1/p$. Thus, using again Theorem 4.8 and Lemma 2.45(v) we obtain

$$H^{\eta}_{p,d-\nu p}(\mathcal{O}) \hookrightarrow B^{(1-\varepsilon)\eta}_{\tau^*,\tau^*}(\mathcal{O}), \quad \frac{1}{\tau^*} = \frac{(1-\varepsilon)\eta}{d} + \frac{1}{p},$$

for any $\varepsilon \in (0,1)$, and therefore

$$H^{\gamma}_{p,d-\nu p}(\mathcal{O}) = H^{\eta}_{p,d-\nu p}(\mathcal{O}) \hookrightarrow B^{\alpha}_{\tau,\tau}(\mathcal{O}), \quad \frac{1}{\tau} = \frac{\alpha}{d} + \frac{1}{p}, \quad \text{for all } 0 < \alpha < \eta = \min\left\{\gamma, \nu\frac{d}{d-1}\right\}.$$

Case 3. Let $\gamma := m + \eta$, with $m \in \mathbb{N}$, $\eta \in (0,1)$, and $\nu\frac{d}{d-1} \leq m$. Since in this case, $H^{\gamma}_{p,d-\nu p}(\mathcal{O}) = H^{m+\eta}_{p,d-\nu p}(\mathcal{O}) \hookrightarrow H^m_{p,d-\nu p}(\mathcal{O})$, the embedding (4.13) holds due to the the fact that, by Part One,

$$H^m_{p,d-\nu p}(\mathcal{O}) \hookrightarrow B^{\alpha}_{\tau,\tau}(\mathcal{O}), \quad \frac{1}{\tau} = \frac{\alpha}{d} + \frac{1}{p}, \quad \text{for all } 0 < \alpha < \nu\frac{d}{d-1} = \min\left\{\gamma, \nu\frac{d}{d-1}\right\}.$$

Case 4. Let $\gamma := m + \eta$, with $m \in \mathbb{N}_0$, $\eta \in (0,1)$, and $m < \nu\frac{d}{d-1} \leq m + \eta$. Fix $\eta_0 \in (0,1)$ with $\eta_0 \leq \eta$, such that $\nu\frac{d}{d-1} = m + \eta_0$. Also, let $\varepsilon \in (0,m)$, and let $\alpha_0 := m - \varepsilon$, $\alpha_1 := m + 1 - \varepsilon$, and $\tau_0, \tau_1 \in (0,\infty)$ be given by (4.30). (If $m = 0$, set $\alpha_0 := 0$.) By Part One,

$$H^m_{p,d-m\frac{d-1}{d}p}(\mathcal{O}) \hookrightarrow B^{\alpha_0}_{\tau_0,\tau_0}(\mathcal{O}) \quad \text{and} \quad H^{m+1}_{p,d-(m+1)\frac{d-1}{d}p}(\mathcal{O}) \hookrightarrow B^{\alpha_1}_{\tau_1,\tau_1}(\mathcal{O}).$$

Therefore, by Theorem 4.8 and Lemma 2.45(v),

$$H^{m+\eta_0}_{p,d-\nu p}(\mathcal{O}) = H^{m+\eta_0}_{p,d-(m+\eta_0)\frac{d-1}{d}p}(\mathcal{O}) \hookrightarrow B^{\alpha^*}_{\tau^*,\tau^*}(\mathcal{O}), \quad \frac{1}{\tau^*} = \frac{\alpha^*}{d} + \frac{1}{p},$$

with $\alpha^* = m + \eta_0 - \varepsilon$ ($\alpha^* = \eta_0 - \varepsilon\eta_0$, if $m = 0$). Since $\varepsilon \in (0,m)$ is arbitrary, and $H^{\gamma}_{p,d-\nu p}(\mathcal{O}) = H^{m+\eta}_{p,d-\nu p}(\mathcal{O}) \hookrightarrow H^{m+\eta_0}_{p,d-\nu p}(\mathcal{O})$, we obtain

$$H^{\gamma}_{p,d-\nu p}(\mathcal{O}) \hookrightarrow B^{\alpha}_{\tau,\tau}(\mathcal{O}), \quad \frac{1}{\tau} = \frac{\alpha}{d} + \frac{1}{p}, \quad \text{for all } 0 < \alpha < m + \eta_0 = \nu\frac{d}{d-1} = \min\left\{\gamma, \nu\frac{d}{d-1}\right\}.$$

Case 5. Finally, let $\gamma := m + \eta$, with $m \in \mathbb{N}_0$, $\eta \in (0,1)$, and $m + \eta \leq \nu\frac{d}{d-1} \leq m + 1$. Following the lines of Case 5 with η instead of η_0, we obtain

$$H^{m+\eta}_{p,d-(m+\eta)\frac{d-1}{d}p}(\mathcal{O}) \hookrightarrow B^{\alpha}_{\tau,\tau}(\mathcal{O}), \quad \frac{1}{\tau} = \frac{\alpha}{d} + \frac{1}{p}, \quad \text{for all } 0 < \alpha < m + \eta = \min\left\{\gamma, \nu\frac{d}{d-1}\right\}.$$

Since $(m+\eta)\frac{d-1}{d} \leq \nu$, and therefore $d - \nu p \leq d - (m+\eta)\frac{d-1}{d}p$, we have

$$H^{m+\eta}_{p,d-\nu p}(\mathcal{O}) \hookrightarrow H^{m+\eta}_{p,d-(m+\eta)\frac{d-1}{d}p}(\mathcal{O}).$$

These two embeddings prove (4.13) also for this particular case.   □

4.4 An alternative proof of Theorem 4.7

In this section we present an alternative proof of Theorem 4.7 for arbitrary $\gamma \in \mathbb{R}$, which does not require any knowledge about complex interpolation of quasi-Banach spaces. However, the arguments are quite involved and not as elegant as in the section before. We use the same notation as in the previous sections of this chapter.

A close look at Part One of the proof of Theorem 4.7 presented in the previous section reveals that the restriction $\gamma \in \mathbb{N}$ is required only when estimating the series

$$I = \sum_{(i,j,k) \in \Lambda^0} \left| \langle u, \widetilde{\psi}_{i,j,k,p'} \rangle \right|^\tau$$

from (4.22). Let us be more detailed: The restriction $\gamma \in \mathbb{N}$ is needed for the first time when applying the Deny-Lions lemma [48, Theorem 3.4], which yields the existence of a polynomial $P_{j,k}$ of total degree less than γ, and a constant C, which does not depend on j or k, such that

$$\|u - P_{j,k}\|_{L_p(Q_{j,k})} \leq C \, 2^{-j\gamma} |u|_{W_p^\gamma(Q_{j,k}^\circ)}. \tag{4.32}$$

Using this and the orthogonality of $\widetilde{\psi}_{i,j,k,p'}$ to every polynomial of total degree less than γ, we obtain

$$\left| \langle u, \widetilde{\psi}_{i,j,k,p'} \rangle \right| \leq C \, 2^{-j\gamma} |u|_{W_p^\gamma(Q_{j,k}^\circ)},$$

which is transformed into

$$\left| \langle u, \widetilde{\psi}_{i,j,k,p'} \rangle \right| \leq C \, 2^{-j\gamma} \rho_{j,k}^{\nu-\gamma} \sup_{|\alpha|=\gamma} \left(\int_{Q_{j,k}} |\rho(x)^{\gamma-\nu} D^\alpha u(x)|^p \, \mathrm{d}x \right)^{\frac{1}{p}},$$

by putting $1 = \rho(x)^{\nu-\gamma} \rho(x)^{\gamma-\nu}$ into the integrals and using $\rho_{j,k} = \mathrm{dist}(Q_{j,k}, \partial\mathcal{O})$. Then, applying Hölder inequality, we show that

$$I \leq C \left(\sum_{(i,j,k) \in \Lambda_j^0} \sup_{|\alpha|=\gamma} \left(\int_{Q_{j,k}} |\rho(x)^{\gamma-\nu} D^\alpha u(x)|^p \, \mathrm{d}x \right) \right)^{\frac{\tau}{p}} \left(\sum_{(i,j,k) \in \Lambda_j^0} 2^{-j\frac{\gamma\tau p}{p-\tau}} \rho_{j,k}^{\frac{(\nu-\gamma)\tau p}{p-\tau}} \right)^{\frac{p-\tau}{p}}.$$

At this point we use the norm equivalence (2.28), which holds only for $\gamma \in \mathbb{N}$, in order to obtain

$$\sum_{(i,j,k) \in \Lambda_j^0} \sup_{|\alpha|=\gamma} \left(\int_{Q_{j,k}} |\rho(x)^{\gamma-\nu} D^\alpha u(x)|^p \, \mathrm{d}x \right) \leq C \, \|u\|_{H_{p,d-\nu p}^\gamma(\mathcal{O})}^p,$$

with a constant C which does not depend on j. This is the last time we use the restriction $\gamma \in \mathbb{N}$ in Part One of the proof of Theorem 4.7.

Let us now assume that we are given the setting of Theorem 4.7 with $0 < \nu \leq \gamma$ without any additional restriction on $\gamma \in (0, \infty)$. Fix $u \in H_{p,d-\nu p}^\gamma(\mathcal{O})$. The explanations above show that, in this generalized setting, if want to apply the same strategy as in the case of integer γ in order to estimate the sum I from (4.22) by the weighted Sobolev norm of u, we first need an estimate similar to (4.32). To this end we can use Corollary 4.2, which provides $H_{p,d-\nu p}^\gamma(\mathcal{O}) \hookrightarrow B_{p,p}^\nu(\mathcal{O})$ in the given setting, together with [48, Theorem 3.5], which is a generalization of the Deny-Lions lemma to arbitrary $\gamma \in (0, \infty)$. From these two facts we obtain the existence of a polynomial $P_{j,k}$ of total degree less than γ, such that

$$\|u - P_{j,k}\|_{L_p(Q_{j,k})} \leq C \, 2^{-j\gamma} |u|_{B_{p,p}^\gamma(Q_{j,k}^\circ)}, \tag{4.33}$$

where the constant C does not depend on j or k. As in the integer case, we can use the orthogonality of $\widetilde{\psi}_{i,j,k,p'}$ to every polynomial of total degree less than γ, which yields

$$\left|\langle u, \widetilde{\psi}_{i,j,k,p'}\rangle\right| \leq C\, 2^{-j\gamma}\, |u|_{B_{p,p}^{\gamma}(Q_{j,k}^{\circ})}.$$

If we multiply the right hand side with $1 = \rho_{jk}^{\nu-\gamma}\rho_{j,k}^{\gamma-\nu}$, we obtain

$$\left|\langle u, \widetilde{\psi}_{i,j,k,p'}\rangle\right| \leq C\, 2^{-j\gamma}\, \rho_{jk}^{\nu-\gamma}\rho_{j,k}^{\gamma-\nu}\, |u|_{W_p^{\gamma}(Q_{j,k}^{\circ})}.$$

Thus, an application of Hölder's inequality leads to

$$\sum_{(i,j,k)\in\Lambda^0} \left|\langle u, \widetilde{\psi}_{i,j,k,p'}\rangle\right|^{\tau} \leq C\left(\sum_{(i,j,k)\in\Lambda_j^0}\left(\rho_{j,k}^{\gamma-\nu}|u|_{B_{p,p}^{\gamma}(Q_{j,k}^{\circ})}\right)^p\right)^{\frac{\tau}{p}}\left(\sum_{(i,j,k)\in\Lambda_j^0} 2^{-j\frac{\gamma\tau p}{p-\tau}}\rho_{j,k}^{\frac{(\nu-\gamma)\tau p}{p-\tau}}\right)^{\frac{p-\tau}{p}}.$$

The following lemma shows that the first sum on the right hand side can be estimated by the weighted Sobolev norm of u times a constant C which does not depend on j or u. Using this estimate and replacing the right places in Part One of the proof presented in the previous section by the calculations above, Theorem 4.7 can be proven directly for arbitrary $\gamma \in (0,\infty)$. The details are left to the reader.

Lemma 4.9. *Let $\mathcal{O}$ be a bounded Lipschitz domain in $\mathbb{R}^d$. Let $p \in [2,\infty)$, $\gamma \in (0,\infty)$ and $\nu \in \mathbb{R}$ with $\gamma \geq \nu$. Furthermore, assume $u \in H_{p,d-\nu p}^{\gamma}(\mathcal{O})$. Then, for all $j \in \mathbb{N}_0$, the inequality*

$$\sum_{(i,j,k)\in\Lambda_j^0}\left(\rho_{j,k}^{\gamma-\nu}|u|_{B_{p,p}^{\gamma}(Q_{j,k}^{\circ})}\right)^p \leq C\,\|u\|_{H_{p,d-\nu p}^{\gamma}(\mathcal{O})}^p \tag{4.34}$$

holds, with a constant $C \in (0,\infty)$ which does not depend on j or u.

Proof. Fix $j \in \mathbb{N}_0$. Let $k_1 \geq 1$ be such that

$$2 + 2M\sqrt{d} < 2^{k_1}, \tag{4.35}$$

and construct a sequence $\{\xi_n : n \in \mathbb{Z}\} \subseteq \mathcal{C}_0^{\infty}(\mathcal{O})$ as in Remark 2.48(ii). In order to prove the assertion we are going to show the estimates

$$\sum_{(i,j,k)\in\Lambda_j^0}\left(\rho_{j,k}^{\gamma-\nu}|u|_{B_{p,p}^{\gamma}(Q_{j,k}^{\circ})}\right)^p \leq C \sum_{n\in\mathbb{N}_0} 2^{-(j-n)(\gamma-\nu)p}|\xi_{j-n}u|_{B_{p,p}^{\gamma}(\mathbb{R}^d)}^p, \tag{4.36}$$

and

$$|\xi_{j-n}u|_{B_{p,p}^{\gamma}(\mathbb{R}^d)}^p \leq C\, 2^{-(j-n)(d-\gamma p)}\left\|\xi_{j-n}\big(2^{-(j-n)}\cdot\big)u\big(2^{-(j-n)}\cdot\big)\right\|_{H_p^{\gamma}(\mathbb{R}^d)}^p, \tag{4.37}$$

where the constant C does not depend on j and n. This will prove the assertion since, assuming that (4.36) and (4.37) are true, their combination gives

$$\sum_{(i,j,k)\in\Lambda_j^0}\left(\rho_{j,k}^{\gamma-\nu}|u|_{B_{p,p}^{\gamma}(Q_{j,k}^{\circ})}\right)^p \leq C \sum_{n\in\mathbb{N}_0} 2^{-(j-n)(d-\nu p)}\left\|\xi_{j-n}\big(2^{-(j-n)}\cdot\big)u\big(2^{-(j-n)}\cdot\big)\right\|_{H_p^{\gamma}(\mathbb{R}^d)}^p$$

$$\leq C \sum_{n\in\mathbb{Z}} 2^{n(d-\nu p)}\left\|\xi_{-n}\big(2^n\cdot\big)u\big(2^n\cdot\big)\right\|_{H_p^{\gamma}(\mathbb{R}^d)}^p,$$

which by Remark 2.48(ii) and Lemma 2.47 yields

$$\sum_{(i,j,k)\in\Lambda_j^0}\left(\rho_{j,k}^{\gamma-\nu}|u|_{B_{p,p}^{\gamma}(Q_{j,k}^{\circ})}\right)^p \leq C\,\|u\|_{H_{p,d-\nu p}^{\gamma}(\mathcal{O})}^p.$$

Let us first verify inequality (4.37). To this end, let r be the smallest integer strictly greater than γ. Recall that $\Delta_h^r[f]$ denotes the the r-th difference of a function $f : \mathbb{R}^d \to \mathbb{R}$ with step $h \in \mathbb{R}^d$, compare Subsection 2.3.4. Writing out the Besov semi-norm and applying the transformation formula for integrals we see that

$$|\xi_{j-n}u|_{B_{p,p}^\gamma(\mathbb{R}^d)}^p = \int_0^\infty t^{-\gamma p} \sup_{|h|<t} \left\| \Delta_h^r[\xi_{j-n}u] \right\|_{L_p(\mathbb{R}^d)}^p \frac{\mathrm{d}t}{t}$$

$$= 2^{-(j-n)d} \int_0^\infty t^{-\gamma p} \sup_{|h|<t} \left\{ \int_{\mathbb{R}^d} \left| \Delta_h^r[\xi_{j-n}u]\big(2^{-(j-n)}x\big) \right|^p \mathrm{d}x \right\} \frac{\mathrm{d}t}{t}$$

Since the equality
$$\Delta_h^r[f](cx) = \Delta_{h/c}^r[f(c \cdot)](x), \qquad x \in \mathbb{R}^d,$$

holds for any function $f : \mathbb{R}^d \to \mathbb{R}$ and $c > 0$, we obtain

$$|\xi_{j-n}u|_{B_{p,p}^\gamma(\mathbb{R}^d)}^p = 2^{-(j-n)d} \int_0^\infty t^{-\gamma p} \sup_{|h|<2^{j-n}t} \left\{ \int_{\mathbb{R}^d} \left| \Delta_h^r\big[\xi_{j-n}\big(2^{-(j-n)} \cdot \big) u\big(2^{-(j-n)} \cdot \big)\big](x) \right|^p \mathrm{d}x \right\} \frac{\mathrm{d}t}{t}.$$

A further application of the transformation formula for integrals yields

$$|\xi_{j-n}u|_{B_{p,p}^\gamma(\mathbb{R}^d)}^p = 2^{-(j-n)d} 2^{(j-n)\gamma p} \int_0^\infty t^{-\gamma p} \sup_{|h|<t} \left\| \Delta_h^r\big[\xi_{j-n}\big(2^{-(j-n)} \cdot \big) u\big(2^{-(j-n)} \cdot \big)\big] \right\|_{L_p(\mathbb{R}^d)}^p \frac{\mathrm{d}t}{t}$$

$$= 2^{-(j-n)(d-\gamma p)} \big|\xi_{j-n}\big(2^{-(j-n)} \cdot \big) u\big(2^{-(j-n)} \cdot \big)\big|_{B_{p,p}^\gamma(\mathbb{R}^d)}^p,$$

which implies (4.37) since the space $H_p^\gamma(\mathbb{R}^d)$ of Bessel potentials is continuously embedded in the Besov space $B_{p,p}^\gamma(\mathbb{R}^d)$, see [116, Theorem 2.3.2(d) combined with Theorem 2.3.3(a)].

It remains to prove inequality (4.36). Recall that the index i referring to the different types of wavelets on a cube $Q_{j,k}$ ranges from 1 to $2^d - 1$. Since Λ_j^0 consists of those indices $(i,j,k) \in \Lambda_j$ with $2^{-j} \leq \rho_{j,k}$, we have

$$\sum_{(i,j,k)\in\Lambda_j^0} \left(\rho_{j,k}^{\gamma-\nu} |u|_{B_{p,p}^\gamma(Q_{j,k})} \right)^p = (2^d - 1) \sum_{k\in\Lambda_j^\star} \left(\rho_{j,k}^{\gamma-\nu} |u|_{B_{p,p}^\gamma(Q_{j,k})} \right)^p, \tag{4.38}$$

where we used the notation

$$\Lambda_j^\star := \left\{ k \in \mathbb{Z}^d : (i,j,k) \in \Lambda_j^0 \right\}.$$

Now we get the required estimate in three steps.

Step 1. We first show that the cubes supporting the wavelets fit into the stripes where the cut-off functions (ξ_n) are identical to one. That is, we claim that the proper choice of k_1, see (4.35), leads to the fact that, for any $k \in \Lambda_j^\star$, there exists a non-negative integer $n \in \mathbb{N}_0$ such that

$$Q_{j,k} \subseteq S_{j-n} := \left\{ x \in \mathcal{O} : 2^{-(j-n)} 2^{-k_1} \leq \rho(x) \leq 2^{-(j-n)} 2^{k_1} \right\}.$$

To prove this, we first note that, since $k_1 \geq 1$,

$$\bigcup_{k\in\Lambda_j^\star} Q_{j,k} \subseteq \bigcup_{n\in\mathbb{N}_0} S_{j-n}. \tag{4.39}$$

Fix $k \in \Lambda_j^*$. Because of (4.39), we can define n^* to be the smallest non-negative integer such that $Q_{j,k} \cap S_{j-n^*} \neq \emptyset$, i.e.,

$$n^* := \inf \left\{ n \in \mathbb{N}_0 : Q_{j,k} \cap S_{j-n} \neq \emptyset \right\}.$$

Then, there are two possibilities: On the one hand, $Q_{j,k}$ might be contained completely in S_{j-n^*}, i.e., $Q_{j,k} \subseteq S_{j-n^*}$. Then we are done. On the other hand, it might happen that $Q_{j,k}$ is not completely contained in the stripe S_{j-n^*}. In this case, we claim that $Q_{j,k} \subseteq S_{j-(n^*+1)}$, i.e.,

$$\rho(x) \in \left[2^{-j+n^*+1}2^{-k_1}, 2^{-j+n^*+1}2^{k_1}\right] \text{ for all } x \in Q_{j,k}.$$

Let us therefore fix $x \in Q_{j,k}$. Then, since the length of the diagonal of $Q_{j,k}$ is $2^{-j}2M\sqrt{d}$, we have

$$\rho(x) \leq \rho_{j,k} + 2^{-j}2M\sqrt{d}.$$

Also, $\rho_{j,k} \leq 2^{-j+n^*}2^{k_1}$ since $Q_{j,k} \cap S_{j-n^*} \neq \emptyset$. Hence,

$$\rho(x) \leq 2^{-j+n^*}2^{k_1} + 2^{-j}2M\sqrt{d}.$$

Since $2M\sqrt{d} \leq 2^{k_1}$, we conclude that

$$\rho(x) \leq 2^{-j+n^*+1}2^{k_1}\left(\frac{1}{2} + \frac{2M\sqrt{d}}{2^{n^*+1}2^{k_1}}\right) \leq 2^{-j+n^*+1}2^{k_1}\left(\frac{1}{2} + \frac{1}{2^{n^*+1}}\right) \leq 2^{-j+n^*+1}2^{k_1}. \qquad (4.40)$$

It remains to show that $\rho(x) \geq 2^{-j+n^*+1}2^{-k_1}$. We argue as follows: Since $Q_{j,k}$ is not completely contained in S_{j-n^*}, there exists a point $x_0 \in Q_{j,k}$ such that $\rho(x_0) > 2^{-j+n^*}2^{k_1}$. Therefore, since the length of the diagonal of $Q_{j,k}$ is $2^{-j}2M\sqrt{d}$ and since (4.35) holds, we have

$$\rho(x) > 2^{-j+n^*}2^{k_1} - 2M\sqrt{d}2^{-j} = 2^{-j+n^*+1}2^{-k_1}\left(\frac{2^{2k_1}}{2} - \frac{2M\sqrt{d}2^{k_1}}{2^{n^*+1}}\right) \geq 2^{-j+n^*+1}2^{-k_1}. \qquad (4.41)$$

Thus, since (4.40) and (4.41) hold for arbitrary $x \in Q_{j,k}$, we have shown that $Q_{j,k} \subseteq S_{j-(n^*+1)}$.

Step 2. Let us fix $k \in \Lambda_j^*$ and estimate the Besov semi-norm of the restriction of u to the corresponding cube $Q_{j,k}^\circ$. To this end, we use the results from [68], where the modulus of smoothness $\omega^r(t, f, G)_p$, $t \in (0, \infty)$, of a function f defined on a domain G is compared with the Peetre K-functional

$$K_r(t, f, G)_p := \inf_{g \in W_p^r(G)} \left\{\|f - g\|_{L_p(G)} + t\,|g|_{W_p^r(G)}\right\}, \quad t \in (0, \infty), \quad f \in L_p(G).$$

In particular, it is shown therein that for all $t \in (0, \infty)$, and $f \in L_p(G)$ for some $p \geq 1$,

$$\omega^r(t, f, G)_p \leq \max\left\{2^r, d^{r/2}\right\}K_r(t, f, G)_p$$

holds for $r \in \mathbb{N}$, see [68, Lemma 1]. Using this, we obtain the following estiamte:

$$
\begin{aligned}
|u|_{B_{p,p}^\gamma(Q_{j,k}^\circ)}^p &= \int_0^\infty t^{-\gamma p}\omega^r(t, u, Q_{j,k}^\circ)_p^p \frac{dt}{t} \\
&\leq C\int_0^\infty t^{-\gamma p}K_r(t^r, u, Q_{j,k}^\circ)_p^p \frac{dt}{t} \\
&= C\int_0^\infty t^{-\gamma p} \inf_{g \in W_p^r(Q_{j,k}^\circ)} \left\{\|u - g\|_{L_p(Q_{j,k})} + t^r|g|_{W_p^r(Q_{j,k}^\circ)}\right\}^p \frac{dt}{t} \\
&\leq C\int_0^\infty t^{-\gamma p} \inf_{g \in W_p^r(Q_{j,k}^\circ)} \left\{\|u - g\|_{L_p(Q_{j,k})}^p + t^{rp}|g|_{W_p^r(Q_{j,k}^\circ)}^p\right\} \frac{dt}{t},
\end{aligned}
$$

where the constant C depends only on r, d and p. (Recall that r is the smallest integer strictly greater than γ.)

Step 3. Now we collect the fruits of our work and approximate the sum on the right hand side of (4.38). As in Part One of the proof of Theorem 4.7 from the previous section, we use the partition $R_{j,m}$, $m \in \{1, \ldots, (2M)^d\}$, of the set $\{Q_{j,k} : k \in \mathbb{Z}^d\}$ defined in (4.24). Furthermore, we write

$$R_{j,m}^\star := \{k \in \Lambda_j^\star : Q_{j,k} \in R_{j,m}\}, \qquad m \in \{1, \ldots, (2M)^d\}, \tag{4.42}$$

and

$$S_{j,n}^\star := \{k \in \Lambda_j^\star : Q_{j,k} \in S_{j-n}\}, \qquad n \in \mathbb{N}_0.$$

Form Step 1 we can deduce that

$$\Lambda_j^\star = \bigcup_{n \in \mathbb{N}_0} S_{j,n}^\star.$$

Thus, since (4.25) holds, we have

$$\Lambda_j^\star = \Lambda_j^\star \cap \bigcup_{m=1}^{(2M)^d} R_{j,m}^\star = \bigcup_{n \in \mathbb{N}_0} \bigcup_{m=1}^{(2M)^d} S_{j,n}^\star \cap R_{j,m}^\star.$$

Therefore,

$$\sum_{k \in \Lambda_j^\star} \left(\rho_{j,k}^{\gamma-\nu} |u|_{B_{p,p}^\gamma(Q_{j,k}^\circ)} \right)^p \le \sum_{n \in \mathbb{N}_0} \sum_{m=1}^{(2M)^d} \sum_{k \in S_{j,n}^\star \cap R_{j,m}^\star} \left(\rho_{j,k}^{(\gamma-\nu)p} |u|_{B_{p,p}^\gamma(Q_{j,k}^\circ)}^p \right). \tag{4.43}$$

Let us fix $n \in \mathbb{N}_0$ such that $S_{j,n}^\star \ne \emptyset$ as well as $m \in \{1, \ldots, (2M)^d\}$. Then, $\rho_{j,k} \le 2^{k_1} 2^{-(j-n)}$ for $k \in S_{j,n}^\star$, so that using the estimate from Step 2 we obtain

$$\sum_{k \in S_{j,n}^\star \cap R_{j,m}^\star} \left(\rho_{j,k}^{(\gamma-\nu)p} |u|_{B_{p,p}^\gamma(Q_{j,k}^\circ)}^p \right)$$

$$\le C \sum_{k \in S_{j,n}^\star \cap R_{j,m}^\star} \left(2^{-(j-n)(\gamma-\nu)p} \int_0^\infty t^{-\gamma p} \inf_{g \in W_p^r(Q_{j,k}^\circ)} \left\{ \|u - g\|_{L_p(Q_{j,k})}^p + t^{rp} |g|_{W_p^r(Q_{j,k}^\circ)}^p \right\} \frac{dt}{t} \right)$$

$$\le C 2^{-(j-n)(\gamma-\nu)p} \sum_{k \in S_{j,n}^\star \cap R_{j,m}^\star} \int_0^\infty t^{-\gamma p} \inf_{g \in W_p^r(\mathcal{O})} \left\{ \|u - g\|_{L_p(Q_{j,k})}^p + t^{rp} |g|_{W_p^r(Q_{j,k}^\circ)}^p \right\} \frac{dt}{t}.$$

Furthermore, since $\xi_{j-n} = 1$ on $Q_{j,k}$ for any $k \in S_{j,n}^\star$,

$$\sum_{k \in S_{j,n}^\star \cap R_{j,m}^\star} \left(\rho_{j,k}^{(\gamma-\nu)p} |u|_{B_{p,p}^\gamma(Q_{j,k}^\circ)}^p \right)$$

$$\le C 2^{-(j-n)(\gamma-\nu)p} \sum_{k \in S_{j,n}^\star \cap R_{j,m}^\star} \left(\int_0^\infty t^{-\gamma p} \inf_{g \in W_p^r(\mathcal{O})} \left\{ \|\xi_{j-n} u - g\|_{L_p(Q_{j,k})}^p + t^{rp} |g|_{W_p^r(Q_{j,k}^\circ)}^p \right\} \frac{dt}{t} \right)$$

$$\le C 2^{-(j-n)(\gamma-\nu)p} \int_0^\infty t^{-\gamma p} \inf_{g \in W_p^r(\mathcal{O})} \left\{ \sum_{k \in S_{j,n}^\star \cap R_{j,m}^\star} \left(\|\xi_{j-n} u - g\|_{L_p(Q_{j,k})}^p + t^{rp} |g|_{W_p^r(Q_{j,k}^\circ)}^p \right) \right\} \frac{dt}{t}.$$

Since $Q^\circ_{j,k} \cap Q^\circ_{j,\ell} = \emptyset$ for $k, \ell \in R^\star_{j,m}$ if $k \neq \ell$, see (4.26) together with (4.42), we obtain

$$\sum_{k \in S^\star_{j,n} \cap R^\star_{j,m}} \left(\rho_{j,k}^{(\gamma-\nu)p} |u|^p_{B^\gamma_{p,p}(Q^\circ_{j,k})} \right)$$

$$\leq C2^{-(j-n)(\gamma-\nu)p} \int_0^\infty t^{-\gamma p} \inf_{g \in W^r_p(\mathcal{O})} \left\{ \|\xi_{j-n}u - g\|^p_{L_p(\mathcal{O})} + t^{rp}|g|^p_{W^r_p(\mathcal{O})} \right\} \frac{dt}{t}$$

$$\leq C2^{-(j-n)(\gamma-\nu)p} \int_0^\infty t^{-\gamma p} K_r(t^r, \xi_{j-n}u, \mathcal{O})_p^p \frac{dt}{t}.$$

Now we use another result from [68], which shows that the K-functional can be estimated by the modulus of smoothness. That is, [68, Theorem 1] yields the existence of a constant C, depending only on r, p and $\mathcal{O}$, such that

$$K_r(t^r, \xi_{j-n}u, \mathcal{O})_p \leq C \omega^r(t, \xi_{j-n}u, \mathcal{O})_p.$$

Putting everything together, we have shown that there exists a constant C which does not depend on j, n or m such that

$$\sum_{k \in S^\star_{j,n} \cap R^\star_{j,m}} \left(\rho_{j,k}^{(\gamma-\nu)p} |u|^p_{B^\gamma_{p,p}(Q^\circ_{j,k})} \right) \leq C2^{-(j-n)(\gamma-\nu)p} \int_0^\infty t^{-\gamma p} \omega^r(t, \xi_{j-n}u, \mathcal{O})_p^p \frac{dt}{t}$$

$$= C2^{-(j-n)(\gamma-\nu)p} |\xi_{j-n}u|^p_{B^\gamma_{p,p}(\mathcal{O})}$$

$$\leq C2^{-(j-n)(\gamma-\nu)p} |\xi_{j-n}u|^p_{B^\gamma_{p,p}(\mathbb{R}^d)}.$$

Inserting this estimate into (4.43) yields

$$\sum_{k \in \Lambda^\star_j} \left(\rho_{j,k}^{(\gamma-\nu)p} |u|^p_{B^\gamma_{p,p}(Q^\circ_{j,k})} \right) \leq C \sum_{n \in \mathbb{N}_0} 2^{-(j-n)(\gamma-\nu)p} |\xi_{j-n}u|^p_{B^\gamma_{p,p}(\mathbb{R}^d)},$$

which combined with (4.38) proves (4.36). $\qquad\square$

Chapter 5

Spatial Besov regularity of SPDEs on bounded Lipschitz domains

In this chapter, we are concerned with the spatial regularity of solutions to SPDEs on bounded Lipschitz domains $\mathcal{O} \subset \mathbb{R}^d$ in the non-linear approximation scale

$$B^\alpha_{\tau,\tau}(\mathcal{O}), \quad \frac{1}{\tau} = \frac{\alpha}{d} + \frac{1}{p}, \quad \alpha > 0, \tag{$*$}$$

with $p \geq 2$ fixed (i.e., topic (T1) in the introduction). We use the same setting and notation as introduced in Chapter 3.

The embedding of weighted Sobolev spaces into Besov spaces from the scale $(*)$ proven in the previous chapter (Theorem 4.7), shows that—to a certain extent—the analysis of the regularity of SPDEs in terms of the scale $(*)$ can be traced back to the analysis of such equations in terms of the spaces $\mathfrak{H}^{\gamma,q}_{p,\theta}(\mathcal{O},T)$. In particular, the following embeddings hold.

Theorem 5.1. *Let $\mathcal{O}$ be a bounded Lipschitz domain in $\mathbb{R}^d$. Fix $\gamma \in (0,\infty)$, $p,q \in [2,\infty)$, and $\theta \in \mathbb{R}$. Then*

$$\mathfrak{H}^{\gamma,q}_{p,\theta}(\mathcal{O},T) \hookrightarrow \mathbb{H}^{\gamma,q}_{p,\theta-p}(\mathcal{O},T) \hookrightarrow L_q(\Omega_T; B^\alpha_{\tau,\tau}(\mathcal{O})),$$

for all α and τ with

$$\frac{1}{\tau} = \frac{\alpha}{d} + \frac{1}{p} \quad \text{and} \quad 0 < \alpha < \min\left\{\gamma, \left(1 + \frac{d-\theta}{p}\right)\frac{d}{d-1}\right\}.$$

Proof. The first embedding follows from the definition of the stochastic parabolic weighted Sobolev spaces $\mathfrak{H}^{\gamma,q}_{p,\theta}(\mathcal{O},T)$, see Definition 3.3, and holds actually on arbitrary domains with non-empty boundary. Since

$$H^\gamma_{p,\theta-p}(\mathcal{O}) = H^\gamma_{p,d-\nu p}(\mathcal{O}), \quad \text{with} \quad \nu := 1 + \frac{d-\theta}{p},$$

the second embedding follows immediately from Theorem 4.7. $\qquad\square$

We use this result to prove spatial Besov regularity of the solutions to SPDEs in the scale $(*)$ of Besov spaces. We divide this chapter into two sections. We start with the linear equations introduced in Section 3.2. As outlined therein, in this setting, the L_p-theory from [75], already provides existence (and uniqueness) of solutions in the classes $\mathfrak{H}^\gamma_{p,\theta}(\mathcal{O},T) = \mathfrak{H}^{\gamma,p}_{p,\theta}(\mathcal{O},T)$, $p \in [2,\infty)$, $\gamma,\theta \in \mathbb{R}$. Thus, we can apply Theorem 5.1 directly and obtain spatial regularity results in the right scale, see Theorem 5.2. Afterwards, in Section 5.2, we generalize our results to a class of semi-linear SPDEs: The linear part will be of the same form as in [75], whereas the non-linearities fulfil certain Lipschitz conditions. Since in this case, existence of solutions has not been established yet, we first have to extend the main existence result of the aforementioned L_p-theory to this class of equations. This will be done in Theorem 5.13. Afterwards, we can apply Theorem 5.1 and obtain spatial regularity in the scale $(*)$ of Besov spaces, see Theorem 5.15.

The examples and remarks presented in Section 5.1 have been partially worked out in collaboration with F. Lindner, S. Dahlke, S. Kinzel, T. Raasch, K. Ritter, and R.L. Schilling [25].

5.1 Linear equations

In this section we use the scale $(*)$ with fixed $p \geq 2$ to analyse the spatial regularity of the solutions $u \in \mathfrak{H}^\gamma_{p,\theta}(\mathcal{O}, T)$ of the linear SPDEs of the form (3.1) studied in [75], see Section 3.2. Since we already have an existence and uniqueness result for this type of equations in $\mathfrak{H}^\gamma_{p,\theta}(\mathcal{O}, T)$, see Theorem 3.13, we can immediately extract an assertion about the spatial regularity of the solution in the scale $(*)$ by applying Theorem 5.1. After stating and proving this result, we present several examples and make some additional remarks. In particular, we enlighten the fact that, on bounded Lipschitz domains, the spatial smoothness of the solution in the non-linear approximation scale $(*)$ of Besov spaces is generically higher than its spatial Sobolev regularity. The relevance of this characteristic from the point of view of approximation theory and numerical analysis has been pointed out in Section 1.1.

We begin with the main result on the spatial regularity in the scale $(*)$ of the solutions to linear SPDEs. It is an improvement of [25, Theorems 3.1 and B.3], see also Remark 5.3 below.

Theorem 5.2. *Let $\mathcal{O}$ be a bounded Lipschitz domain in $\mathbb{R}^d$. Given $\gamma \in (-2, \infty)$, let a^{ij}, b^i, c, σ^{ik} and μ^k, $i, j \in \{1, \ldots, d\}$, $k \in \mathbb{N}$, satisfy Assumption 3.1 with suitable constants δ_0 and K. Furthermore, assume that $u \in \mathfrak{H}^{\gamma+2}_{p,\theta}(\mathcal{O}, T)$ is the unique solution of Eq. (3.1) with $f \in \mathbb{H}^\gamma_{p,\theta+p}(\mathcal{O}, T)$, $g \in \mathbb{H}^{\gamma+1}_{p,\theta}(\mathcal{O}, T; \ell_2)$ and $u_0 \in U^{\gamma+2}_{p,\theta}(\mathcal{O})$, where*

(i) $p \in [2, \infty)$ *and* $\theta \in (d + p - 2 - \kappa_0, d + p - 2 + \kappa_0)$

or, alternatively,

(ii) $p \in [2, p_0)$ *and* $\theta \in (d - \kappa_1, d + \kappa_1)$,

with $\kappa_0, \kappa_1 \in (0, 1)$ and $p_0 > 2$ as in Theorem 3.13. Then,

$$u \in L_p(\Omega_T; B^\alpha_{\tau,\tau}(\mathcal{O})), \quad \frac{1}{\tau} = \frac{\alpha}{d} + \frac{1}{p}, \quad \text{for all} \quad 0 < \alpha < \min\left\{\gamma + 2, \left(1 + \frac{d - \theta}{p}\right)\frac{d}{d - 1}\right\}. \quad (5.1)$$

Moreover, for any α and τ fulfilling (5.1), there exists a constant C, which does not depend on u, f, g and u_0 such that

$$\|u\|^p_{L_p(\Omega_T; B^\alpha_{\tau,\tau}(\mathcal{O}))} \leq C\left(\|f\|^p_{\mathbb{H}^\gamma_{p,\theta+p}(\mathcal{O},T)} + \|g\|^p_{\mathbb{H}^{\gamma+1}_{p,\theta}(\mathcal{O},T;\ell_2)} + \|u_0\|^p_{U^{\gamma+2}_{p,\theta}(\mathcal{O})}\right).$$

Proof. The assertion is an immediate consequence of Theorem 5.1 and the existence and uniqueness statements from Theorem 3.13. $\square$

Remark 5.3. A result similar to Theorem 4.4 has been proven in [25, Theorem 3.1, see also Theorem B.3]. However, there are three major improvements in Theorem 5.2 compared to [25, Theorems 3.1 and B.3][1]. Firstly, we have no restriction on $\gamma + 2 \in (0, \infty)$, whereas in [25] only

---

[1]We remark that the assumptions made in [25, Theorem B.3] are stronger than actually needed. In the notation used therein, the assumptions [K1]–[K5] only need to be fulfilled for $\gamma - 2$ instead of γ.

integer $\gamma + 2 \in \mathbb{N}$ are considered. Secondly, we obtain L_p-integrability in time of the $B_{\tau,\tau}^\alpha(\mathcal{O})$-valued process for arbitrary $p \geq 2$ fulfilling the assumptions (i) or (ii) from Theorem 3.13. With the techniques used in [25] just L_τ-integrability in time can be established. Thirdly, we do not need the extra assumption $u \in L_p([0,T] \times \Omega; B_{p,p}^s(\mathcal{O}))$ for some $s > 0$. Due to Corollary 4.2, it suffices that $u \in \mathbb{H}_{p,\theta-p}^{\gamma+2}(\mathcal{O},T)$.

Next, we give some examples of applications of Theorem 5.2 and interpret our result from the point of view of the question whether adaptivity pays, cf. our motivation for studying topic (T1) from Section 1.1. We are mainly interested in the Hilbert space case $p = 2$ since, as already pointed out in Section 1.1, it provides a natural setting for numerical discretization techniques like adaptive wavelet methods, see also the expositions in [105, 125] for more details. We begin with an application of Theorem 5.2 for particular parameters $\gamma, \theta \in \mathbb{R}$ and $p = 2$.

Example 5.4. Assume that we have given coefficients a^{ij}, b^i, c, σ^{ik}, and μ^k, with $i, j \in \{1, \ldots, d\}$ and $k \in \mathbb{N}$, fulfilling Assumption 3.1 with $\gamma = 0$. Furthermore, fix arbitrary $f \in \mathbb{H}_{2,d+2}^0(\mathcal{O},T)$, $g \in \mathbb{H}_{2,d}^1(\mathcal{O},T;\ell_2)$ and $u_0 \in U_{2,d}^2 = L_2(\Omega, \mathcal{F}_0, \mathbb{P}; H_{2,d}^1(\mathcal{O}))$. Then, by an application of Theorem 3.13 with $\gamma = 0$, $p = 2$ and $\theta = d$, Eq. (3.1) has a unique solution $u \in \mathfrak{H}_{2,d}^2(\mathcal{O},T)$. Due to Theorem 5.2,

$$u \in L_2(\Omega_T; B_{\tau,\tau}^\alpha(\mathcal{O})), \qquad \frac{1}{\tau} = \frac{\alpha}{d} + \frac{1}{2}, \qquad \text{for all} \quad 0 < \alpha < \frac{d}{d-1}. \tag{5.2}$$

In the two-dimensional case, this means that

$$u \in L_2(\Omega_T; B_{\tau,\tau}^\alpha(\mathcal{O})), \qquad \frac{1}{\tau} = \frac{\alpha}{2} + \frac{1}{2}, \qquad \text{for all} \quad 0 < \alpha < 2.$$

Note that if we assume slightly more regularity on the coefficients, the initial condition u_0 and the free terms f and g, we can included the border case $\alpha = d/(d-1)$ in (5.2). To this end, assume that the coefficients a^{ij}, b^i, c, σ^{ik}, and μ^k, with $i, j \in \{1, \ldots, d\}$ and $k \in \mathbb{N}$, fulfil Assumption 3.1 for some arbitrary positive $\gamma > 0$. Furthermore, fix an arbitrary $\varepsilon > 0$ and assume that $f \in \mathbb{H}_{2,d-\varepsilon+2}^\gamma(\mathcal{O},T)$, $g \in \mathbb{H}_{2,d-\varepsilon}^{\gamma+1}(\mathcal{O},T;\ell_2)$ and $u_0 \in U_{2,d-\varepsilon}^{\gamma+2} = L_2(\Omega, \mathcal{F}_0, \mathbb{P}; H_{2,d-\varepsilon}^{\gamma+2}(\mathcal{O}))$. Then, there exists an $\varepsilon_1 \in (0, \kappa_1)$ with $\kappa_1 > 0$ from Theorem 3.13(ii), such that Eq. (3.1) has a unique solution $u \in \mathfrak{H}_{2,d-\varepsilon_1}^{\gamma+2}(\mathcal{O},T)$. Due to Theorem 5.2,

$$u \in L_2(\Omega_T; B_{\tau,\tau}^\alpha(\mathcal{O})), \qquad \frac{1}{\tau} = \frac{\alpha}{d} + \frac{1}{2}, \qquad \text{for all} \quad 0 < \alpha < \min\left\{\gamma+2, \left(1 + \frac{\varepsilon_1}{2}\right)\frac{d}{d-1}\right\},$$

and therefore, since γ and ε_1 are strictly positive,

$$u \in L_2(\Omega_T; B_{\tau,\tau}^\alpha(\mathcal{O})), \qquad \frac{1}{\tau} = \frac{\alpha}{d} + \frac{1}{2}, \qquad \text{for all} \quad 0 < \alpha \leq \frac{d}{d-1},$$

which in the two-dimensional case yields

$$u \in L_2(\Omega_T; B_{\tau,\tau}^\alpha(\mathcal{O})), \qquad \frac{1}{\tau} = \frac{\alpha}{d} + \frac{1}{2}, \qquad \text{for all} \quad 0 < \alpha \leq 2.$$

The example above shows that equations of the type (3.1) on general bounded Lipschitz domains have spatial Besov regularity in the scale (*) up to order $\alpha = 2$. In order to answer the question whether this is enough for justifying the development of spatially adaptive wavelet methods, we have to compare this result with the spatial Sobolev regularity of the solution under consideration. We give now a concrete example of an SPDE of the type (3.1) with solution $u \in \mathfrak{H}_{2,d}^2(\mathcal{O},T)$ whose spatial Besov regularity in the scale (*) is strictly higher than its spatial Sobolev regularity.

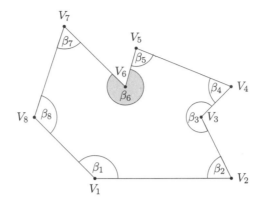

Figure 5.1: Polygon in $\mathbb{R}^2$ with $\beta_{\max} = \beta_6 = 5\pi/4$.

Example 5.5. We consider an equation of the type (3.1) on a polygonal domain $\mathcal{O} \subset \mathbb{R}^2$ and show that, under natural conditions on the data of the equation, if the underlying domain is not convex, the spatial Besov regularity of the solution in the scale $(*)$ is strictly higher than its spatial Sobolev smoothness. In particular, this shows that, generically, solutions to linear SPDEs on bounded Lipschitz domains behave as described in (1.11), so that the use of spatially adaptive methods is recommended. For more details on the link between regularity theory and the convergence rates of numerical methods we refer to Section 1.1.

Let $\mathcal{O} \subset \mathbb{R}^2$ be a simply connected bounded domain in $\mathbb{R}^2$ with a polygonal boundary $\partial\mathcal{O}$ such that $\mathcal{O}$ lies on one side of $\partial\mathcal{O}$. It can be described by a finite set $\{V_n : n = 1, \ldots, N\}$ of vertices of the boundary numbered, e.g., according to their order in $\partial\mathcal{O}$ in counter-clockwise orientation. For $n \in \{1, \ldots, N\}$, we write $\beta_n \in (0, 2\pi)$ for the interior angle at the vertex V_n and denote by $\beta_{\max}$ the maximal interior angle of $\mathcal{O}$, i.e.,

$$\beta_{\max} := \max\{\beta_n : n = 1, \ldots, N\}.$$

An example of such a domain with $\beta_{\max} = 5\pi/4$ is shown in Figure 5.1. Assume that we have an initial condition $u_0 \in U_{2,2}^2(\mathcal{O})$ additionally satisfying

$$u_0 \in L_2(\Omega, \mathcal{F}_0, \mathbb{P}; \mathring{W}_2^1(\mathcal{O})) \cap L_q(\Omega, \mathcal{F}_0, \mathbb{P}; L_2(\mathcal{O}))$$

for some $q > 2$. Furthermore, let $f \in L_2(\Omega_T; L_2(\mathcal{O})) \hookrightarrow \mathbb{H}_{2,4}^0(\mathcal{O}, T)$ and let $g \in H_{2,2}^1(\mathcal{O}; \ell_2)$. Typically, we make slight abuse of notation and write g also for the constant stochastic process $g \in \mathbb{H}_{2,d}^1(\mathcal{O}, T; \ell_2)$ with $g(\omega, t) := g$ for all $(\omega, t) \in \Omega_T$. Then, due to Theorem 3.13, the stochastic heat equation

$$\left.\begin{aligned} du &= (\Delta u + f)\, dt + g^k\, dw_t^k \quad \text{on } \Omega_T \times \mathcal{O}, \\ u(0) &= u_0 \quad \text{on } \Omega \times \mathcal{O}, \end{aligned}\right\} \tag{5.3}$$

has a unique solution $u \in \mathfrak{H}_{2,2}^2(\mathcal{O}, T)$.

We want to compare the spatial Besov regularity of the solution to Eq. (5.3) in the scale $(*)$ with its spatial Sobolev regularity. Regarding the spatial regularity in the non-linear approximation scale $(*)$, an application of Theorem 5.2 yields

$$u \in L_2(\Omega_T; B_{\tau,\tau}^\alpha(\mathcal{O})), \qquad \frac{1}{\tau} = \frac{\alpha}{2} + \frac{1}{2} = \frac{\alpha+1}{2}, \qquad \text{for all} \quad 0 < \alpha < 2. \tag{5.4}$$

Concerning the spatial Sobolev regularity of the solution, by our analysis so far, we can only guarantee that

$$u \in L_2(\Omega_T; \mathring{W}_2^1(\mathcal{O})),$$

which is a consequence of Proposition 4.1. Together with (5.4), this suggests that the Besov regularity of the solution to Eq. (5.3) in the scale (∗) is generically higher than its spatial Sobolev regularity in the following sense: There exist polygonal domains $\mathcal{O} \subset \mathbb{R}^2$ and free terms f and g fulfilling the assumptions from above, such that

$$\tilde{s}_{\max}^{\mathrm{Sob}}(u) < 2, \tag{5.5}$$

with $\tilde{s}_{\max}^{\mathrm{Sob}}(u)$ as introduced in (1.12). We can confirm this statement by exploiting the recent results from [92]. Therefore, let us denote by $\Delta_{2,w}^D : D(\Delta_{2,w}^D) \subseteq L_2(\mathcal{O}) \to L_2(\mathcal{O})$ the *weak Dirichlet-Laplacian* on $L_2(\mathcal{O})$, i.e.,

$$D(\Delta_{2,w}^D) := \{ u \in \mathring{W}_2^1(\mathcal{O}) \,:\, \Delta u \in L_2(\mathcal{O}) \},$$
$$\Delta_{2,w}^D u := \Delta u, \qquad u \in D(\Delta_{2,w}^D).$$

From Proposition 3.18 we already know that our solution $u \in \mathfrak{H}_{2,2}^2(\mathcal{O}, T)$ is also the unique weak solution (in the sense of Da Prato and Zabczyk [32]) of the $L_2(\mathcal{O})$-valued ordinary SDE

$$\left. \begin{aligned} \mathrm{d}u(t) - \Delta_{2,w}^D u(t)\,\mathrm{d}t &= f(t)\,\mathrm{d}t + \mathrm{d}W^Q(t), \qquad t \in [0,T], \\ u(0) &= u_0, \end{aligned} \right\} \tag{5.6}$$

driven by the $H_{2,2}^1(\mathcal{O})$ valued Q-Wiener process $(W^Q(t))_{t\in[0,T]} := \big(\sum_{k\in\mathbb{N}} g^k w_t^k \big)_{t\in[0,T]}$ with covariance operator $Q := \sum_{k\in\mathbb{N}} \langle g^k, \cdot \rangle_{H_{2,2}^1(\mathcal{O})} g^k \in \mathcal{L}_1(H_{2,2}^1(\mathcal{O}))$. Moreover, due to Theorem 3.8(ii),

$$\sup_{t\in[0,T]} \mathbb{E}\Big[\|u(t)\|_{L_2(\mathcal{O})}^2 \Big] \leq \mathbb{E}\Big[\sup_{t\in[0,T]} \|u(t)\|_{L_2(\mathcal{O})}^2 \Big] \leq C \, \|u\|_{\mathfrak{H}_{2,2}^2(\mathcal{O},T)}^2 < \infty,$$

and by [32, Theorem 5.4], for all $t \in [0,T]$,

$$u(t) = S_2(t)u_0 + \int_0^t S_2(t-s)f(s)\,\mathrm{d}s + \int_0^t S_2(t-s)\,\mathrm{d}W^Q(s) \qquad \mathbb{P}\text{-a.s.},$$

where $\{S_2(t)\}_{t\geq 0}$ denotes the contraction semigroup on $L_2(\mathcal{O})$ generated by $(\Delta_{2,w}^D, D(\Delta_{2,w}^D))$. Thus, u is the unique (up to modifications) *mild solution* of Eq. (5.6) which is studied in [92], see also [91, Chapter 4]. Therein, techniques from [57, 58] have been adapted to the stochastic setting, and it has been shown that this solution can be divided into a spatially regular and a spatially irregular part, regularity being measured by means of Sobolev spaces. In particular, if we assume that the range of the covariance operator Q is dense in $H_{2,2}^1(\mathcal{O}) \hookrightarrow L_2(\mathcal{O})$, it follows from [92, Example 3.6] that

$$u \notin L_2(\Omega_T; W_2^s(\mathcal{O})) \quad \text{for any} \quad s > 1 + \frac{\pi}{\beta_{\max}}. \tag{5.7}$$

Thus, if $\mathcal{O}$ is not convex, we have

$$\tilde{s}_{\max}^{\mathrm{Sob}}(u) \leq 1 + \frac{\pi}{\beta_{\max}} < 2,$$

with $\tilde{s}_{\max}^{\mathrm{Sub}}(u)$ as defined in (1.12). Together with (5.4), this shows that the solution to Eq. (5.3) generically behaves as described in (1.11). Therefore, the development of suitable spatially adaptive numerical methods is completely justified.

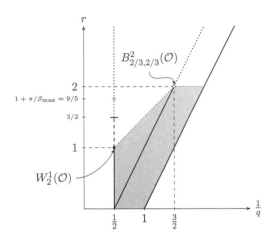

Figure 5.2:　Spatial Besov regularity in the scale $B^\alpha_{\tau,\tau}(\mathcal{O})$, $1/\tau = (\alpha + 1)/2$, versus spatial Sobolev regularity of the solution of Eq. (5.3), illustrated in a DeVore/Triebel diagram.

Figure 5.2 shows a DeVore/Triebel diagram illustrating the situation described above (see Remark 2.63 for details on the visualisation of Besov spaces using this type of diagrams). The fact that (5.4) holds, is represented by the solid segment $\{(1/\tau, \alpha) : 1/\tau = (\alpha + 1)/2, 0 \le \alpha < 2\}$ of the $L_2(\mathcal{O})$-non-linear approximation line and the annulus at $(3/2, 2)$, which stands for the Besov space $B^2_{2/3,2/3}(\mathcal{O})$. The point at $(1/2, 1)$ shows that $u \in L_2(\Omega_T; W^1_2(\mathcal{O}))$. In this situation, by Theorem 2.61 and standard interpolation results, see, e.g. [117, Corollary 1.111], $u \in L_2(\Omega_T; B^r_{q,q}(\mathcal{O}))$ for all $(1/q, r)$ in the interior of the polygon with vertices at the points $(1/2, 0)$, $(1/2, 1)$, $(3/2, 2)$, $(2, 2)$, and $(1, 0)$. This is indicated by the shaded area. The border at $(1/2, 3/2)$ illustrates the following consequence of (5.7): For any $\varepsilon > 0$, there exists a polygonal domain $\mathcal{O} \subset \mathbb{R}^2$, such that $u \notin L_2(\Omega_T; W^{3/2+\varepsilon}_2(\mathcal{O}))$. The concrete border for the example in Figure 5.1 is indicated by the annulus at $(1/2, 1 + \pi/\beta_{\max}) = (1/2, 9/5)$, which stands for the Sobolev space $W^{9/5}_2(\mathcal{O})$.

In the following example we are concerned with equations of the form (3.1) driven by a specific type of noise.

Example 5.6. We consider an equation of the type (3.1) driven by a time-dependent version of the stochastic wavelet expansion introduced in [1] in the context of Bayesian non-parametric regression and generalized in [14] and [24]. This noise model is formulated in terms of a wavelet basis expansion on the domain $\mathcal{O} \subset \mathbb{R}^d$ with random coefficients of prescribed sparsity and thus tailor-made for applying adaptive techniques with regard to the numerical approximation of the corresponding SPDEs. Via the choice of certain parameters specifying the distributions of the wavelet coefficients it also allows for an explicit control of the spatial Besov regularity of the noise. We first describe the general noise model and then deduce a further example for the application of Theorem 5.2.

Let $\{\psi_\lambda : \lambda \in \nabla\}$ be a multiscale Riesz basis of $L_2(\mathcal{O})$ consisting of scaling functions at a fixed scale level $j_0 \in \mathbb{Z}$ and of wavelets at level j_0 and all finer levels. We follow [27] and use the same notation as in Section 1.1. Information like scale level, spatial location and type of the wavelets or scaling functions are encoded in the indices $\lambda \in \nabla$. In particular, we write

$\nabla = \bigcup_{j \geq j_0-1} \nabla_j$, where for $j \geq j_0$ the set $\nabla_j \subset \nabla$ contains the indices of all wavelets ψ_λ at scale level j and where $\nabla_{j_0-1} \subset \nabla$ is the index set referring to the scaling functions at scale level j_0 which we denote by ψ_λ, $\lambda \in \nabla_{j_0-1}$, for the sake of notational simplicity; $|\lambda| := j$ for all $\lambda \in \nabla_j$. We make the following assumptions concerning our basis. Firstly, the cardinalities of the index sets ∇_j, $j \geq j_0 - 1$, satisfy

$$C^{-1}2^{jd} \leq |\nabla_j| \leq C2^{jd}, \qquad j \geq j_0 - 1, \tag{5.8}$$

with a constant C which does not depend on j. Secondly, we assume that the basis admits norm equivalences similar to those described in Theorem 4.4. That is, there exists an $s_0 \in \mathbb{N}$ (depending on the smoothness of the scaling functions ψ_λ, $\lambda \in \nabla_{j_0-1}$, and on the degree of polynomial exactness of their linear span), such that, given $p, q > 0$, $\max\{0, d(1/p - 1)\} < s < s_0$, and a real valued distribution $v \in \mathcal{D}'(\mathcal{O})$, we have $v \in B^s_{p,q}(\mathcal{O})$ if, and only if, v can be represented as $v = \sum_{\lambda \in \nabla} c_\lambda \psi_\lambda$, $(c_\lambda)_{\lambda \in \nabla} \subset \mathbb{R}$ (convergence in $\mathcal{D}'(\mathcal{O})$), such that

$$\left(\sum_{j=j_0-1}^{\infty} 2^{jq\left(s+d\left(\frac{1}{2}-\frac{1}{p}\right)\right)} \left(\sum_{\lambda \in \nabla_j} |c_\lambda|^p \right)^{\frac{q}{p}} \right)^{\frac{1}{q}} < \infty. \tag{5.9}$$

Furthermore, $\|v\|_{B^s_{p,q}(\mathcal{O})}$ is equivalent to the (quasi-)norm (5.9). Concrete constructions of bases satisfying these assumptions can be found e.g. in [42–44] or [19, 20], see also [27, Section 2.12 together with Section 3.9] for a detailed discussion. Concerning the family of independent standard Brownian motions $(w^k_t)_{t \in [0,T]}$, $k \in \mathbb{N}$, in (3.1), we modify our notation and write $(w^\lambda_t)_{t \in [0,T]}$, $\lambda \in \nabla$, instead. The description of the noise model involves parameters $a_1 \geq 0$, $a_2 \in [0, 1]$, $b \in \mathbb{R}$, with $a_1 + a_2 > 1$. For every $j \geq j_0 - 1$ we set $\varsigma_j := (j - (j_0 - 2))^{\frac{bd}{2}} 2^{-\frac{a_1(j-(j_0-1))d}{2}}$ and let Y_λ, $\lambda \in \nabla_j$, be $\{0, 1\}$-valued Bernoulli distributed random variables on $(\Omega, \mathcal{F}_0, \mathbb{P})$ with parameter $p_j = 2^{-a_2(j-(j_0-1))d}$, such that the random variables and processes Y_λ, $(w^\lambda_t)_{t \in [0,T]}$, $\lambda \in \nabla$, are stochastically independent. The noise in our equation will be described by the $L_2(\mathcal{O})$-valued stochastic process $(M_t)_{t \in [0,T]}$ defined by

$$M_t := \sum_{j=j_0-1}^{\infty} \sum_{\lambda \in \nabla_j} \varsigma_j Y_\lambda \psi_\lambda \cdot w^\lambda_t, \qquad t \in [0, T]. \tag{5.10}$$

Using (5.9), (5.8) and $a_1 + a_2 > 1$, it is easy to check that the infinite sum converges in $L_2(\Omega_T; L_2(\mathcal{O}))$ as well as in $L_2(\Omega; \mathcal{C}([0, T]; L_2(\mathcal{O})))$. Moreover, by the choice of the parameters a_1, a_2 and b one has an explicit control of the convergence of the infinite sum in (5.10) in the (quasi-)Banach spaces $L_{p_2}(\Omega_T; B^s_{p_1,q}(\mathcal{O}))$, $s < s_0$, $p_1, q > 0$, $p_2 \leq q$. (Compare [24] which can easily be adapted to our setting.)

For simplicity, let us consider the two-dimensional case, i.e., $d = 2$. Assume that we have a given $f \in \mathbb{H}^0_{2,2}(\mathcal{O}, T)$, an initial condition $u_0 \in U^2_{2,2}(\mathcal{O})$, and coefficients a^{ij}, b^i and c, with $i, j \in \{1, \ldots, d\}$, fulfilling Assumption 3.1 with $\sigma = 0$ and $\mu = 0$. We consider the equation

$$\left. \begin{aligned} du &= \left(a^{ij}u_{x^i x^j} + b^i u_{x^i} + cu + f\right) dt + \varsigma_{|\lambda|} Y_\lambda \psi_\lambda \, dw^\lambda_t \quad \text{on } \Omega_T \times \mathcal{O}, \\ u(0) &= u_0 \quad \text{on } \Omega \times \mathcal{O}, \end{aligned} \right\} \tag{5.11}$$

where we sum over all $\lambda \in \nabla$ instead of $k \in \mathbb{N}$. That is, we understand this equation similar to equations of the type (3.1), where the role of $k \in \mathbb{N}$ in the required definitions is taken by $\lambda \in \nabla$. In this setting, let $g := (g^\lambda)_{\lambda \in \nabla} := (\varsigma_{|\lambda|} Y_\lambda \psi_\lambda)_{\lambda \in \nabla}$. Since $a_1 + a_2 > 1$ and $\|g\|_{\mathbb{H}^0_{2,2}(\mathcal{O},T;\ell_2)} = \sqrt{2/T}\|M\|_{L_2(\Omega_T;L_2(\mathcal{O}))}$ we have $g \in \mathbb{H}^0_{2,2}(\mathcal{O}, T; \ell_2)$. Let us impose a bit more smoothness on g

and assume that $a_1 + a_2 > 2$. This is sufficient to ensure that $g \in \mathbb{H}^1_{2,2}(\mathcal{O}, T; \ell_2)$, since

$$
\begin{aligned}
\|g\|^2_{\mathbb{H}^1_{2,2}(\mathcal{O}, T; \ell_2)} &= \mathbb{E}\left[\int_0^T \|g\|^2_{H^1_{2,2}(\mathcal{O}; \ell_2)} \, dt \right] \\
&= \mathbb{E}\left[\int_0^T \sum_{\lambda \in \nabla} \|g^\lambda(t, \cdot)\|^2_{H^1_{2,2}(\mathcal{O})} \, dt \right] \\
&= T \, \mathbb{E}\left[\sum_{j=j_0-1}^\infty \sum_{\lambda \in \nabla_j} \varsigma_j^2 Y_\lambda^2 \|\psi_\lambda\|^2_{H^1_{2,2}(\mathcal{O})} \right],
\end{aligned}
$$

so that by (2.28),

$$
\begin{aligned}
\|g\|^2_{\mathbb{H}^1_{2,2}(\mathcal{O}, T; \ell_2)} &\leq C \sum_{j=j_0-1}^\infty \sum_{\lambda \in \nabla_j} \varsigma_j^2 p_j \sum_{|\alpha| \leq 1} \|\rho^{|\alpha|} D^\alpha \psi_\lambda\|^2_{L_2(\mathcal{O})} \\
&\leq C \sum_{j=j_0-1}^\infty \sum_{\lambda \in \nabla_j} \varsigma_j^2 p_j \|\psi_\lambda\|^2_{W^1_2(\mathcal{O})}.
\end{aligned}
$$

Since $W^1_2(\mathcal{O}) = B^1_{2,2}(\mathcal{O})$, see Theorem 2.60(ii), we can use the equivalence (5.9) with $v = \psi_\lambda$ followed by (5.8) with $d = 2$ to obtain

$$
\begin{aligned}
\|g\|^2_{\mathbb{H}^1_{2,2}(\mathcal{O}, T; \ell_2)} &\leq C \sum_{j=j_0-1}^\infty \sum_{\lambda \in \nabla_j} \varsigma_j^2 p_j 2^{2j} \\
&= C \sum_{j=j_0-1}^\infty |\nabla_j| (j - (j_0 - 2))^{2b} 2^{-2a_1(j-(j_0-1))} 2^{-2a_2(j-(j_0-1))} 2^{2j} \\
&\leq C \sum_{j=j_0-1}^\infty (j - (j_0 - 2))^{2b} 2^{-2j(a_1+a_2-2)}.
\end{aligned}
$$

Thus $g \in \mathbb{H}^1_{2,2}(\mathcal{O}, T; \ell_2)$ and for any $\varphi \in \mathcal{C}_0^\infty(\mathcal{O})$,

$$
\sum_{\lambda \in \nabla} \int_0^{\cdot} (g^\lambda, \varphi) \, dw_t^\lambda = (M_\cdot, \varphi) \qquad \mathbb{P}\text{-a.s.}
$$

in $\mathcal{C}([0, T]; \mathbb{R})$, see also Proposition 3.6 and the definition of stochastic integrals from Subsection 2.2.3 for details. As in the examples above, by Theorem 3.13, there exists a unique solution of Eq. (5.11) in the class $\mathfrak{H}^2_{2,2}(\mathcal{O}, T)$. As shown in Examples 5.5, in general, the solution process is not in $L_2(\Omega_T; W^s_2(\mathcal{O}))$ for all $s < 2$, but, by Theorem 5.2, it belongs to every space $L_2(\Omega_T; B^\alpha_{\tau, \tau}(\mathcal{O}))$ with $\alpha < 2$ and $\tau = 2/(\alpha + 1)$.

We make the following note regarding adaptive versus uniform methods in Sobolev spaces.

Remark 5.7. As already mentioned in the introduction, in different deterministic settings, there exist adaptive wavelet-based schemes realising the convergence rate of the best m-term approximation error in the energy norm. This norm is determined by the equation and is usually equivalent to an $L_2(\mathcal{O})$-Sobolev norm and not to the $L_2(\mathcal{O})$ norm itself. Thus, the question arises whether our regularity results underpin the use of adaptivity also in the case that the error is measured in a suitable Sobolev norm. Again this question can be decided after a rigorous regularity analysis of the target function, since the results on the link between regularity theory and the convergence rate of approximation methods discussed in Section 1.1 can be generalised to the case where the error is measured in a Sobolev spaces $W^r_2(\mathcal{O})$ with $r > 0$ instead of $L_p(\mathcal{O})$.

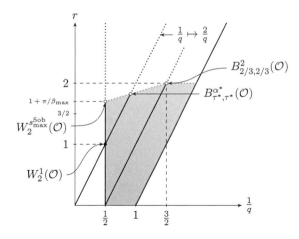

Figure 5.3: Spatial Besov regularity in the scale $B^\alpha_{\tau,\tau}(\mathcal{O})$, $1/\tau = \alpha/2$, versus spatial Sobolev regularity of the solution to Eq. (5.3), illustrated in a DeVore/Triebel diagram.

Let us denote by $\{\eta_\lambda : \lambda \in \nabla\}$ a wavelet basis of $W^r_2(\mathcal{O})$ for some $r > 0$. Such a basis can be obtained by rescaling a wavelet basis $\{\psi_\lambda : \lambda \in \nabla\}$ of $L_2(\mathcal{O})$ as the one used in Example 5.6 and by using the norm equivalence (5.9), see, e.g., [27] or [41]. For the error of the best m-term wavelet approximation error in this Sobolev norm, it is well-known that

$$u \in B^\alpha_{\tau,\tau}(\mathcal{O}), \quad \frac{1}{\tau} = \frac{\alpha - r}{d} + \frac{1}{2} \qquad \text{implies} \qquad \sigma_{m,W^r_2(\mathcal{O})}(u) \le C\, m^{-(\alpha-r)/d}, \qquad (5.12)$$

where

$$\sigma_{m,W^r_2(\mathcal{O})}(u) := \inf\left\{ \|u - u_m\|_{W^r_2(\mathcal{O})} : u_m \in \widetilde{\Sigma}_{m,W^r_2(\mathcal{O})} \right\}$$

with

$$\widetilde{\Sigma}_{m,W^r_2(\mathcal{O})} := \left\{ \sum_{\lambda \in \Lambda} c_\lambda \eta_\lambda : \Lambda \subset \nabla, |\Lambda| = m, c_\lambda \in \mathbb{R}, \lambda \in \Lambda \right\},$$

see, e.g., [125, Corollary 3.2] and the references therein, in particular, [27]. Therefore, similar to the $L_2(\mathcal{O})$-setting, on the one hand, the decay rate of the best m-term wavelet approximation error in $E = W^r_2(\mathcal{O})$ depends on the Besov regularity of the target function. On the other hand, the convergence rate of uniform numerical methods is determined by the Sobolev regularity of the solution to be approximated. It is well-known that, under fairly natural conditions, if u_m, $m \in \mathbb{N}$, is a uniform approximation scheme of u, then,

$$\|u - u_m\|_{W^r_2(\mathcal{O})} \le C\, m^{-(s-r)/d} \|u\|_{W^s_2(\mathcal{O})}, \qquad m \in \mathbb{N};$$

see, e.g., [37], [46] or [61] for details. If we consider uniform wavelet approximation, the following converse assertion also holds: If $u \notin W^s_2(\mathcal{O})$, then the convergence rate of the uniform method in $W^r_2(\mathcal{O})$ is limited by $(s-r)/d$, see, e.g., [125, Proposition 3.2] and the references therein. This means that, if the error is measured in $W^r_2(\mathcal{O})$, adaptivity pays if the spatial smoothness of the solution in the Besov spaces from (5.12) is strictly higher than its spatial Sobolev regularity.

Let us consider the setting from Example 5.5 and discuss the relationship between the spatial Sobolev and Besov regularity in view of approximation in $W_2^1(\mathcal{O})$, i.e., $r = 1$. We use a DeVore/Triebel diagram to visualise our explanations, see Figure 5.3. Due to (5.7),

$$\tilde{s}_{\max}^{\text{Sob}}(u) \leq 1 + \frac{\pi}{\beta_{\max}}, \tag{5.13}$$

with $\tilde{s}_{\max}^{\text{Sob}}(u)$ as defined in (1.12). Thus,

$$\sup\left\{(s-1)/2 : u \in L_2(\Omega_T; W_2^s(\mathcal{O}))\right\} \leq \frac{\pi}{2\beta_{\max}}. \tag{5.14}$$

Let us assume that the spatial Sobolev regularity of the solution u reaches its maximum, i.e., that $u \in L_2(\Omega_T; W_2^{1+\pi/\beta_{\max}}(\mathcal{O}))$, cf. (5.7). Then, due to (5.4), by Theorem 2.61 and standard interpolation results, see, e.g. [117, Corollary 1.111],

$$u \in L_2(\Omega_T; B_{\tau,\tau}^\alpha(\mathcal{O})), \quad \frac{1}{\tau} = \frac{\alpha-1}{2} + \frac{1}{2} = \frac{\alpha}{2} \quad \text{for all} \quad 0 < \alpha < \alpha^* := \frac{\beta_{\max} + 3\pi}{\beta_{\max} + \pi}.$$

This is illustrated in Figure 5.3 by the solid segment of the line $1/q \mapsto 2/q$ delimited by the origin and the annulus at $(1/\tau^*, \alpha^*) = (\alpha^*/2, \alpha^*)$. Thus, the decay rate of the best m-term wavelet approximation error in $W_2^1(\mathcal{O})$ with respect to the space coordinates goes up to $\pi/(\beta_{\max} + \pi)$, which is greater than $\pi/(2\beta_{\max})$ whenever $\beta_{\max} > \pi$, i.e., whenever the polygonal domain $\mathcal{O}$ is not convex. Therefore, also in this setting, the implementation of adaptive wavelet methods is justified.

In all the other examples from above we consider general bounded Lipschitz domains. In this case, we do not have an explicit bound for the spatial Sobolev regularity of the solution. Thus, we can only assume the limit case $\beta_{\max} = 2\pi$. Inserting this into the calculations from above, we can say that, in the worst case,

$$\sup\left\{(s-1)/2 : u \in L_2(\Omega_T; W_2^s(\mathcal{O}))\right\} \leq \frac{1}{4}. \tag{5.15}$$

Simultaneously,

$$u \in L_2(\Omega_T; B_{\tau,\tau}^\alpha(\mathcal{O})), \quad \frac{1}{\tau} = \frac{\alpha-1}{2} + \frac{1}{2} = \frac{\alpha}{2} \quad \text{for all} \quad 0 < \alpha < \frac{5}{3}.$$

Since $(5/3 - 1)/2 = 2/6 > 1/4$, the development of optimal adaptive algorithms with respect to the space coordinates, where the error is measured in $W_2^1(\mathcal{O})$, is recommended. We illustrate this limiting case in Figure 5.4 by using again a DeVore/Triebel diagram.

We conclude this section with an example showing that, in contrast to what is known to hold for deterministic equations, adaptive wavelet methods for SPDEs may pay even if the underlying domain is smooth.

Example 5.8. Let $\mathcal{O}$ be a bounded $\mathcal{C}_u^1$-domain (and, therefore, a bounded Lipschitz domain) in $\mathbb{R}^d$. Furthermore, let a^{ij}, b^i, c, and μ^k, with $i, j \in \{1, \ldots, d\}$ and $k \in \mathbb{N}$, be given coefficients satisfying Assumption 3.1 with $\gamma = 0$, $\sigma = 0$ and suitable constants δ_0 and K. Fix $p \in [2, \infty)$ and let $f \in \mathbb{H}_{p,d-1+p}^1(\mathcal{O}, T)$, $g \in \mathbb{H}_{p,d-1}^2(\mathcal{O}, T; \ell_2)$ and $u_0 \in U_{p,d-1}^3(\mathcal{O})$. Then, by [72, Theorem 2.9] there exists a unique solution u of Eq. (3.1), which is in the class $\mathfrak{H}_{p,d-\varepsilon}^3(\mathcal{O}, T)$ for any $\varepsilon > 0$; see also Remark 3.14(ii). Due to Theorem 5.1 this yields

$$u \in L_p(\Omega_T; B_{\tau,\tau}^\alpha(\mathcal{O})), \quad \frac{1}{\tau} = \frac{\alpha}{d} + \frac{1}{p}, \quad \text{for all} \quad 0 < \alpha < \left(1 + \frac{\varepsilon}{p}\right)\frac{d}{d-1}, \quad \varepsilon \in (0, 1).$$

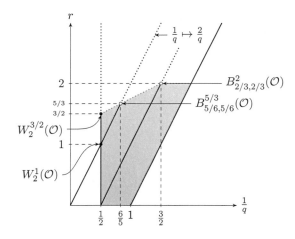

Figure 5.4: Spatial Besov regularity in the scale $B^\alpha_{\tau,\tau}(\mathcal{O})$, $1/\tau = \alpha/2$, versus spatial Sobolev regularity of the solution to equations of type (3.1), illustrated in a DeVore/Triebel diagram.

Thus, in the two-dimensional case, if $p = 2$, we have

$$u \in L_p(\Omega_T; B^\alpha_{\tau,\tau}(\mathcal{O})), \quad \frac{1}{\tau} = \frac{\alpha+1}{2}, \quad \text{for all} \quad 0 < \alpha < 3.$$

What about the spatial Sobolev regularity of this solution? It is known from [78, Example 1.2] that if we consider $\mathbb{R}_+$ instead of $\mathcal{O}$, there exists a non-random g, continuously differentiable on $[0, \infty) \times [0, \infty)$ such that the second partial derivatives with respect to the space coordinates of the solution to the heat equation

$$\mathrm{d}u = \Delta u \, \mathrm{d}t + g^k \, \mathrm{d}w^k_t, \qquad u\big|_{\partial\mathbb{R}_+} = 0, \qquad u(0) = 0,$$

on $\mathbb{R}_+$, do *not* lie in $L_2(\Omega_T; L_2(\mathbb{R}_+))$. This is due to the incompatibility of the noise with the zero Dirichlet boundary condition. Exploiting the compatibility results from [55], it is reasonable to expect that we can construct similar examples on smooth bounded domains, with maximal spatial Sobolev regularity strictly less than the spatial Besov regularity in the non-linear approximation scale (∗). If this is indeed the case, it shows that in the stochastic setting, adaptive methods are a serious alternative to uniform methods even if the underlying domain is smooth. It is worth noting that this would be completely different from what is known to hold in the deterministic setting, where adaptivity does not pay on smooth domains.

5.2 Semi-linear equations

In this section we continue our analysis of the spatial regularity of SPDEs in the scale (∗) of Besov spaces. We generalize the results from the previous section to a class of semi-linear equations on bounded Lipschitz domains $\mathcal{O} \subset \mathbb{R}$. However, for semi-linear SPDEs, existence of solutions in the classes $\mathfrak{H}^\gamma_{p,\theta}(\mathcal{O}, T)$, $\gamma \in \mathbb{R}$, is yet to be proven. Therefore, before we can apply Theorem 5.1 in order to obtain spatial regularity in the non-linear approximation scale (∗), we have to extend the main existence (and uniqueness) result of the L_p-theory from [75], cf. Theorem 3.13, to semi-linear SPDEs. The equations considered in this section are of the form

$$\begin{aligned}
\mathrm{d}u = \left(a^{ij}u_{x^i x^j} + b^i u_{x^i} + cu + f + L(u)\right)\mathrm{d}t \\
+ \left(\sigma^{ik}u_{x^i} + \mu^k u + g^k + (\Lambda(u))^k\right)\mathrm{d}w_t^k \quad \text{on } \Omega_T \times \mathcal{O}, \\
u(0) = u_0 \quad \text{on } \Omega \times \mathcal{O}.
\end{aligned} \qquad (5.16)$$

While the linear parts of Eq. (5.16) are supposed to satisfy Assumption 3.1 as in the previous section, we impose the following conditions on the non-linearities L and Λ.

Assumption 5.9. The functions

$$L : \mathfrak{H}_{p,\theta}^{\gamma+2}(\mathcal{O},T) \to \mathbb{H}_{p,\theta+p}^{\gamma}(\mathcal{O},T) \quad \text{and} \quad \Lambda : \mathfrak{H}_{p,\theta}^{\gamma+2}(\mathcal{O},T) \to \mathbb{H}_{p,\theta}^{\gamma+1}(\mathcal{O},T;\ell_2)$$

fulfil the following conditions:

(i) For all $u, v \in \mathfrak{H}_{p,\theta}^{\gamma+2}(\mathcal{O},T)$ and $t \in [0,T]$

$$\begin{aligned}
\|L(u) - L(v)\|_{\mathbb{H}_{p,\theta+p}^{\gamma}(\mathcal{O},t)}^p + \|\Lambda(u) - \Lambda(v)\|_{\mathbb{H}_{p,\theta}^{\gamma+1}(\mathcal{O},t;\ell_2)}^p \\
\leq \varepsilon \|u - v\|_{\mathbb{H}_{p,\theta-p}^{\gamma+2}(\mathcal{O},t)}^p + K_1 \|u - v\|_{\mathbb{H}_{p,\theta}^{\gamma+1}(\mathcal{O},t)}^p
\end{aligned} \qquad (5.17)$$

with $\varepsilon > 0$ and $K_1 \in [0,\infty)$ independent of u, v and $t \in [0,T]$.

(ii) The non-linearities vanish at the origin, i.e., $L(0) = 0$ and $\Lambda(0) = 0$.

We use the following solutions concept, which is a straight-forward generalization of the solution concept presented in Definition 3.10.

Definition 5.10. Given $\gamma \in \mathbb{R}$, let a^{ij}, b^i, c, σ^{ik} and μ^k, $i,j \in \{1,\dots,d\}$, $k \in \mathbb{N}$, fulfil Assumption 3.1. Furthermore, let Assumption 5.9(i) be satisfied for given $\theta \in \mathbb{R}$ and $p \in [2,\infty)$. A stochastic process $u \in \mathbb{H}_{p,\theta-p}^{\gamma}(\mathcal{O},T)$ is called a *solution* of Eq. (5.16) *in the class* $\mathfrak{H}_{p,\theta}^{\gamma}(\mathcal{O},T)$ if, and only if, $u \in \mathfrak{H}_{p,\theta}^{\gamma}(\mathcal{O},T)$ with

$$u(0,\cdot) = u_0, \quad \mathbb{D}u = a^{ij}u_{x^i x^j} + b^i u_{x^i} + cu + f + L(u), \quad \text{and} \quad \mathbb{S}u = \left(\sigma^{ik}u_{x^i} + \mu^k u + g^k + (\Lambda(u))^k\right)_{k \in \mathbb{N}}$$

in the sense of Definition 3.3.

Remark 5.11. In this thesis, if we call an element $u \in \mathfrak{H}_{p,\theta}^{\gamma}(G,T)$ a *solution* of Eq. (5.16), we mean that u is a solution of Eq. (5.16) in the class $\mathfrak{H}_{p,\theta}^{\gamma}(G,T)$.

Remark 5.12. As already mentioned, throughout this thesis, for a better readability, we omit the notation of the sums $\sum_{i,j}$ and $\sum_k$ when writing down equations and use the so-called summation convention on the repeated indices i,j,k. Thus, the expression

$$\mathrm{d}u = \left(a^{ij}u_{x^i x^j} + b^i u_{x^i} + cu + f + L(u)\right)\mathrm{d}t + \left(\sigma^{ik}u_{x^i} + \mu^k u + g^k + (\Lambda(u))^k\right)\mathrm{d}w_t^k$$

is short-hand for

$$\mathrm{d}u = \left(\sum_{i,j=1}^d a^{ij}u_{x^i x^j} + \sum_{i=1}^d b^i u_{x^i} + cu + f + L(u)\right)\mathrm{d}t + \left(\sum_{i=1}^d \sigma^{ik}u_{x^i} + \mu^k u + g^k + (\Lambda(u))^k\right)\mathrm{d}w_t^k$$

in the sense of Definition 3.3.

We first state our main result on the existence and uniqueness of solutions of Eq. (5.16) in weighted Sobolev spaces.

Theorem 5.13. *Let $\mathcal{O}$ be a bounded Lipschitz domain in $\mathbb{R}^d$, and $\gamma \in \mathbb{R}$. For $i, j \in \{1, \ldots, d\}$ and $k \in \mathbb{N}$, let a^{ij}, b^i, c, σ^{ik}, and μ^k be given coefficients satisfying Assumption 3.1 with suitable constants δ_0 and K.*

(i) *For $p \in [2, \infty)$, there exists a constant $\tilde{\kappa}_0 \in (0, 1)$, depending only on d, p, δ_0, K and $\mathcal{O}$, such that for any $\theta \in (d + p - 2 - \tilde{\kappa}_0, d + p - 2 + \tilde{\kappa}_0)$, $f \in \mathbb{H}_{p,\theta+p}^{\gamma}(\mathcal{O}, T)$, $g \in \mathbb{H}_{p,\theta}^{\gamma+1}(\mathcal{O}, T; \ell_2)$ and $u_0 \in U_{p,\theta}^{\gamma+2}(\mathcal{O})$, the following holds: There exists an $\varepsilon_0 > 0$ such that, if L and Λ fulfil Assumption 5.9 with $\varepsilon < \varepsilon_0$ and $K_1 \in [0, \infty)$, Eq. (5.16) has a unique solution u in the class $\mathfrak{H}_{p,\theta}^{\gamma+2}(\mathcal{O}, T)$. For this solution, the a priori estimate*

$$\|u\|_{\mathfrak{H}_{p,\theta}^{\gamma+2}(\mathcal{O},T)}^p \leq C \left(\|f\|_{\mathbb{H}_{p,\theta+p}^{\gamma}(\mathcal{O},T)}^p + \|g\|_{\mathbb{H}_{p,\theta}^{\gamma+1}(\mathcal{O},T;\ell_2)}^p + \|u_0\|_{U_{p,\theta}^{\gamma+2}(\mathcal{O})}^p \right), \qquad (5.18)$$

holds with a constant C which does not depend on u, f, g and u_0.

(ii) *There exists $p_0 > 2$, such that, if $p \in [2, p_0)$, then there exists a constant $\tilde{\kappa}_1 \in (0, 1)$, depending only on d, p, δ_0, K and $\mathcal{O}$, such that for any $\theta \in (d - \tilde{\kappa}_1, d + \tilde{\kappa}_1)$, $f \in \mathbb{H}_{p,\theta+p}^{\gamma}(\mathcal{O}, T)$, $g \in \mathbb{H}_{p,\theta}^{\gamma+1}(\mathcal{O}, T; \ell_2)$ and $u_0 \in U_{p,\theta}^{\gamma+2}(\mathcal{O})$, the following holds: There exists an $\varepsilon_0 > 0$ such that, if L and Λ fulfil Assumption 5.9 with $\varepsilon < \varepsilon_0$ and $K_1 \in [0, \infty)$, Eq. (5.16) has a unique solution u in the class $\mathfrak{H}_{p,\theta}^{\gamma+2}(\mathcal{O}, T)$. For this solution, the a priori estimate (5.18) holds.*

Before we present a proof of this result (starting on page 101), we make some remarks on the assumptions therein. Furthermore, we state and prove the consequences concerning the spatial Besov regularity in the scale $(*)$ of solutions to Eq. (5.11), and prove an auxiliary theorem, which we will need for proving Theorem 5.13.

Remark 5.14. **(i)** The constants $\tilde{\kappa}_0$ and $\tilde{\kappa}_1$ coincide with the constants κ_0 and κ_1, respectively, from Theorem 3.13. Moreover, $p_0 > 2$ is the same p_0 as in Theorem 3.13(ii).

(ii) The statement of Theorem 5.13(i) with $K_1 = 0$ has been already proven in [22], see Theorem 4.1 together with Remark 4.2 therein. Note that the assumptions made in [22, Theorem 4.1] are stronger than actually needed, since the assumptions (K1)–(K5) therein only need to be fulfilled with $\gamma - 2$ instead of γ.

(iii) The assumption that the non-linearities vanish at the origin is made only for convenience. To see this, let all assumptions of Theorem 5.13 be fulfilled except Assumption 5.9(ii), i.e., we allow $L(0) \neq 0$ and $\Lambda(0) \neq 0$. Then, Eq. (5.16) can be rewritten as

$$\begin{aligned} \mathrm{d}u &= \left(a^{ij}u_{x^ix^j} + b^i u_{x^i} + cu + f + L(0) + L_1(u)\right) \mathrm{d}t \\ &\quad + \left(\sigma^{ik}u_{x^i} + \mu^k u + g^k + (\Lambda(0))^k + (\Lambda_1(u))^k\right) \mathrm{d}w_t^k \quad \text{on } \Omega_T \times \mathcal{O}, \\ u(0) &= u_0 \quad \text{on } \Omega \times \mathcal{O}. \end{aligned} \right\} \qquad (5.19)$$

with $L_1(u) := L(u) - L(0)$ and $\Lambda_1(u) := \Lambda(u) - \Lambda(0)$ for $u \in \mathfrak{H}_{p,\theta}^{\gamma+2}(\mathcal{O})$. Then, Assumption 5.9 is fulfilled with L_1 and Λ_1 instead of L and Λ, respectively. Thus, since $L(0) \in \mathbb{H}_{p,\theta+p}^{\gamma}(\mathcal{O}, T)$ and $\Lambda(0) \in \mathbb{H}_{p,\theta}^{\gamma+1}(\mathcal{O}, T; \ell_2)$, applying Theorem 5.13 yields the existence of a unique solution $u \in \mathfrak{H}_{p,\theta}^{\gamma+2}(\mathcal{O}, T)$ of Eq. (5.16), which fulfils the estimate

$$\|u\|_{\mathfrak{H}_{p,\theta}^{\gamma+2}(\mathcal{O},T)}^p \leq C \left(\|f + L(0)\|_{\mathbb{H}_{p,\theta+p}^{\gamma}(\mathcal{O},T)}^p + \|g + \Lambda(0)\|_{\mathbb{H}_{p,\theta}^{\gamma+1}(\mathcal{O},T;\ell_2)}^p + \|u_0\|_{U_{p,\theta}^{\gamma+2}(\mathcal{O})}^p \right),$$

with a constant C which does not depend on u, f, g, and u_0.

(iv) Let $\varepsilon_0 > 0$. Assume that there exists an $\eta \in (0,1)$ such that the two functions

$$L : \mathfrak{H}_{p,\theta}^{\gamma+2}(\mathcal{O},T) \to \mathbb{H}_{p,\theta+p}^{\gamma}(\mathcal{O},T) \quad \text{and} \quad \Lambda : \mathfrak{H}_{p,\theta}^{\gamma+2}(\mathcal{O},T) \to \mathbb{H}_{p,\theta}^{\gamma+1}(\mathcal{O},T;\ell_2)$$

fulfil the estimate

$$\begin{aligned}
\|L(u) - L(v)\|_{\mathbb{H}_{p,\theta+p}^{\gamma}(\mathcal{O},t)}^p &+ \|\Lambda(u) - \Lambda(v)\|_{\mathbb{H}_{p,\theta}^{\gamma+1}(\mathcal{O},t;\ell_2)}^p \\
&\leq \tilde{\varepsilon}\|u - v\|_{\mathbb{H}_{p,\theta-p}^{\gamma+2}(\mathcal{O},t)}^p + \widetilde{K}_1\|u - v\|_{\mathbb{H}_{p,\theta+p-2\eta p}^{\gamma+2\eta}(\mathcal{O},t)}^p,
\end{aligned} \tag{5.20}$$

for some $\tilde{\varepsilon} \in (0,\varepsilon_0)$ and $\widetilde{K}_1 \in [0,\infty)$ independent of $u, v \in \mathfrak{H}_{p,\theta}^{\gamma+2}(\mathcal{O},T)$ and $t \in [0,T]$. Then, by Lemma 2.45(v), the estimate

$$\begin{aligned}
\|L(u) - L(v)\|_{\mathbb{H}_{p,\theta+p}^{\gamma}(\mathcal{O},t)}^p &+ \|\Lambda(u) - \Lambda(v)\|_{\mathbb{H}_{p,\theta}^{\gamma+1}(\mathcal{O},t;\ell_2)}^p \\
&\leq \tilde{\varepsilon}\|u - v\|_{\mathbb{H}_{p,\theta-p}^{\gamma+2}(\mathcal{O},t)}^p + \widetilde{K}_1\|u - v\|_{\mathbb{H}_{p,\theta+p}^{\gamma}(\mathcal{O},t)}^p,
\end{aligned}$$

holds with potentially different $\tilde{\varepsilon} \in (0,\varepsilon_0)$ and $\widetilde{K}_1 \in [0,\infty)$, which again do not depend on $u, v \in \mathfrak{H}_{p,\theta}^{\gamma+2}(\mathcal{O},T)$ and $t \in [0,T]$. Also, the reverse direction holds, since $\mathcal{O} \subset \mathbb{R}^d$ is assumed to be bounded and therefore $H_{p,d+p-2\eta p}^{\gamma+2\eta}(\mathcal{O}) \hookrightarrow H_{p,d+p}^{\gamma}(\mathcal{O})$ for every $\eta \in (0,1)$, see also Lemma 2.45(vii). Note that (5.20) with $\eta = 1/2$ is exactly (5.17) from Assumption 5.9(i).

Next, we present our main result concerning the spatial regularity of semi-linear equations in the non-linear approximation scale $(*)$ of Besov spaces. It is an extension and an improvement of [22, Theorem 4.4].

Theorem 5.15. *Given the setting from Theorem 5.13, let* $u \in \mathfrak{H}_{p,\theta}^{\gamma+2}(\mathcal{O},T)$ *be the unique solution of Eq. (5.16). Then,*

$$u \in L_p(\Omega_T; B_{\tau,\tau}^\alpha), \quad \frac{1}{\tau} = \frac{\alpha}{d} + \frac{1}{p}, \quad \text{for all} \quad 0 < \alpha < \min\left\{\gamma + 2, \left(1 + \frac{d-\theta}{p}\right)\frac{d}{d-1}\right\}. \tag{5.21}$$

Moreover, for any α and τ fulfilling (5.21), there exists a constant C, which does not depend on u, f, g and u_0 such that

$$\|u\|_{L_p(\Omega_T; B_{\tau,\tau}^\alpha(\mathcal{O}))}^p \leq C\left(\|f\|_{\mathbb{H}_{p,\theta+p}^\gamma(\mathcal{O},T)}^p + \|g\|_{\mathbb{H}_{p,\theta}^{\gamma+1}(\mathcal{O},T;\ell_2)}^p + \|u_0\|_{U_{p,\theta}^{\gamma+2}(\mathcal{O})}^p\right).$$

Proof. This is an immediate consequence of Theorem 5.13 and Theorem 5.1. $\qquad\square$

Now we state and prove an auxiliary result, which we will use later on in order to prove Theorem 5.13. It shows how fixed point arguments can be used to prove existence and uniqueness of a solution to the semi-linear equation (5.16), provided this result is already established for the corresponding linear equation. One needs Assumption 5.9 for the non-linearities and suitable a priori estimates for the linear equations with vanishing initial value. Similar ideas have been already used in the context of SPDEs on the whole space $\mathbb{R}^d$ by N.V. Krylov, see [80, Theorem 6.4].

Lemma 5.16. *Given the setting from Theorem 3.13, assume that the solution* $u \in \mathfrak{H}_{p,\theta}^{\gamma+2}(\mathcal{O},T)$ *of Eq. (3.1) with* $f \in \mathbb{H}_{p,\theta+p}^\gamma(\mathcal{O},T)$, $g \in \mathbb{H}_{p,\theta}^{\gamma+1}(\mathcal{O},T;\ell_2)$ *and* $u_0 = 0$, *fulfils the estimate*

$$\|u\|_{\mathfrak{H}_{p,\theta}^{\gamma+2}(\mathcal{O},t)}^p \leq C_0\left(\|f\|_{\mathbb{H}_{p,\theta+p}^\gamma(\mathcal{O},t)}^p + \|g\|_{\mathbb{H}_{p,\theta}^{\gamma+1}(\mathcal{O},t;\ell_2)}^p\right) \tag{5.22}$$

for all $t \in [0, T]$ with a constant $C_0 \in (0, \infty)$ independent of $t \in [0, T]$, u, f and g. Then there exists an $\varepsilon_0 > 0$ (depending on T in general), such that, if Assumption 5.9 is fulfilled with $\varepsilon < \varepsilon_0$ and $K_1 \in [0, \infty)$, the following holds: For any $f \in \mathbb{H}_{p,\theta+p}^{\gamma}(\mathcal{O}, T)$, $g \in \mathbb{H}_{p,\theta}^{\gamma+1}(\mathcal{O}, T; \ell_2)$ and $u_0 \in U_{p,\theta}^{\gamma+2}(\mathcal{O})$, there exists a unique solution u^* of Eq. (5.16) in the class $\mathfrak{H}_{p,\theta}^{\gamma+2}(\mathcal{O}, T)$. Moreover, there exists a constant $C \in (0, \infty)$ which does not depend on u^*, f, g and u_0, such that

$$\|u^*\|_{\mathfrak{H}_{p,\theta}^{\gamma+2}(\mathcal{O},T)}^p \leq C \Big(\|f\|_{\mathbb{H}_{p,\theta+p}^{\gamma}(\mathcal{O},T)}^p + \|g\|_{\mathbb{H}_{p,\theta}^{\gamma+1}(\mathcal{O},T;\ell_2)}^p + \|u_0\|_{U_{p,\theta}^{\gamma+2}(\mathcal{O})}^p \Big). \tag{5.23}$$

Proof. For $i, j \in \{1, \ldots, d\}$ and $k \in \mathbb{N}$, let a^{ij}, b^i, c, σ^{ik}, and μ^k, fulfil Assumption 3.1 for some $\gamma \in \mathbb{R}$. Furthermore, let p and θ be as in Theorem 3.13(i) or (ii) and fix $f \in \mathbb{H}_{p,\theta+p}^{\gamma}(\mathcal{O}, T)$, $g \in \mathbb{H}_{p,\theta}^{\gamma+1}(\mathcal{O}, T; \ell_2)$ and $u_0 \in U_{p,\theta}^{\gamma+2}(\mathcal{O})$. Then, the operator

$$\mathcal{N} : \mathfrak{H}_{p,\theta}^{\gamma+2}(\mathcal{O}, T) \to \mathfrak{H}_{p,\theta}^{\gamma+2}(\mathcal{O}, T)$$
$$u \mapsto \mathcal{N}(u),$$

where $\mathcal{N}(u)$ is the unique solution in the class $\mathfrak{H}_{p,\theta}^{\gamma+2}(\mathcal{O}, T)$ of the linear equation

$$\left.\begin{aligned} \mathrm{d}v = \big(a^{ij}v_{x^ix^j} + b^i v_{x^i} + cv + f + L(u)\big)\,\mathrm{d}t \\ + \big(\sigma^{ik}v_{x^i} + \mu^k v + g^k + (\Lambda(u))^k\big)\,\mathrm{d}w_t^k \quad \text{on } \Omega_T \times \mathcal{O}, \\ v(0) = u_0 \quad \text{on } \Omega \times \mathcal{O}, \end{aligned}\right\}$$

is well-defined by Theorem 3.13. Fix arbitrary $u, v \in \mathfrak{H}_{p,\theta}^{\gamma+2}(\mathcal{O}, T)$. Then $\mathcal{N}(u) - \mathcal{N}(v)$ is the unique solution in the class $\mathfrak{H}_{p,\theta}^{\gamma+2}(\mathcal{O}, T)$ of the equation

$$\left.\begin{aligned} \mathrm{d}\tilde{v} = \big(a^{ij}\tilde{v}_{x^ix^j} + b^i \tilde{v}_{x^i} + c\tilde{v} + L(u) - L(v)\big)\,\mathrm{d}t \\ + \big(\sigma^{ik}\tilde{v}_{x^i} + \mu^k \tilde{v} + (\Lambda(u))^k - (\Lambda(v))^k\big)\,\mathrm{d}w_t^k \quad \text{on } \Omega_T \times \mathcal{O}, \\ \tilde{v}(0) = 0 \quad \text{on } \Omega \times \mathcal{O}. \end{aligned}\right\}$$

By (5.22),

$$\big\|\mathcal{N}(u) - \mathcal{N}(v)\big\|_{\mathfrak{H}_{p,\theta}^{\gamma+2}(\mathcal{O},t)}^p \leq C_0 \Big(\|L(u) - L(v)\|_{\mathbb{H}_{p,\theta+p}^{\gamma}(\mathcal{O},t)}^p + \|\Lambda(u) - \Lambda(v)\|_{\mathbb{H}_{p,\theta}^{\gamma+1}(\mathcal{O},t;\ell_2)}^p \Big)$$

for all $t \in [0, T]$. If Assumption 5.9 is fulfilled with some $\varepsilon > 0$ and $K_1 \in [0, \infty)$, this leads to

$$\big\|\mathcal{N}(u) - \mathcal{N}(v)\big\|_{\mathfrak{H}_{p,\theta}^{\gamma+2}(\mathcal{O},t)}^p \leq C_0\,\varepsilon\,\|u - v\|_{\mathbb{H}_{p,\theta-p}^{\gamma+2}(\mathcal{O},t)}^p + C_0\,K_1\,\|u - v\|_{\mathbb{H}_{p,\theta}^{\gamma+1}(\mathcal{O},t)}^p. \tag{5.24}$$

Let us first assume that $K_1 = 0$. In this case we are done: If we choose $\varepsilon > 0$ small enough, e.g., $\varepsilon < \varepsilon_0 := 1/C_0$, the operator $\mathcal{N}$ turns out to be a contraction on the Banach space $\mathfrak{H}_{p,\theta}^{\gamma+2}(\mathcal{O}, T)$. Therefore, by the Banach fixed point theorem, $\mathcal{N}$ has a unique fixed point. Regarding the fact that any solution of Eq. (5.16) in the class $\mathfrak{H}_{p,\theta}^{\gamma+2}(\mathcal{O}, T)$ is a fixed point of $\mathcal{N}$ and vice versa, we have just proven the fact that, given the setting from Theorem 3.13, if Assumption 5.9(i) holds with $\varepsilon > 0$ small enough and $K_1 = 0$, Eq. (5.16) has a unique solution u^* in the class $\mathfrak{H}_{p,\theta}^{\gamma+2}(\mathcal{O}, T)$. We can also obtain Estimate (5.23) arguing as follows: Start the fixed point iteration with $u^{(0)} := 0 \in \mathfrak{H}_{p,\theta+p}^{\gamma+2}(\mathcal{O}, T)$ and set $u^{(j+1)} := \mathcal{N}(u^{(j)})$ for $j \geq 0$. Then $(u^{(j)})_{j \in \mathbb{N}}$ converges to the unique solution u^* in $\mathfrak{H}_{p,\theta}^{\gamma+2}(\mathcal{O}, T)$. Furthermore, since Assumption 5.9(ii) is fulfilled and estimate (3.9) holds, we have

$$\big\|\mathcal{N}(u^{(0)})\big\|_{\mathfrak{H}_{p,\theta}^{\gamma+2}(\mathcal{O},T)} \leq C^{1/p} \Big(\|f\|_{\mathbb{H}_{p,\theta+p}^{\gamma}(\mathcal{O},T)}^p + \|g\|_{\mathbb{H}_{p,\theta}^{\gamma+1}(\mathcal{O},T;\ell_2)}^p + \|u_0\|_{U_{p,\theta}^{\gamma+2}(\mathcal{O})}^p \Big)^{1/p}. \tag{5.25}$$

Hence, using the a priori estimate from the Banach fixed point theorem, as it can be found in [66, Theorem 3.1.2], we obtain

$$\|u^*\|^p_{\mathfrak{H}^{\gamma+2}_{p,\theta}(\mathcal{O},T)} \leq \left(\frac{1}{1 - C_0^{1/p}\varepsilon^{1/p}}\right)^p C\left(\|f\|^p_{\mathbb{H}^{\gamma}_{p,\theta+p}(\mathcal{O},T)} + \|g\|^p_{\mathbb{H}^{\gamma+1}_{p,\theta}(\mathcal{O},T;\ell_2)} + \|u_0\|^p_{U^{\gamma+2}_{p,\theta}(\mathcal{O})}\right).$$

For $K_1 \in (0,\infty)$ we still have to work in order to prove that the operator $\mathcal{N}$ has a unique fixed point under suitable assumptions on ε. To this end, we will prove that, given the setting from Theorem 3.13, if Assumption 5.9(i) is fulfilled with $\varepsilon > 0$ small enough, we can find an $m \in \mathbb{N}$, depending (among others) on ε, K_1 and T, such that $\mathcal{N}^m$ becomes a contraction on $\mathfrak{H}^{\gamma+2}_{p,\theta}(\mathcal{O},T)$. Due to the Banach fixed point theorem, this leads to the existence of a unique fixed point of $\mathcal{N}^m$, which automatically implies that $\mathcal{N}$ has a unique fixed point, and, therefore, Eq. (5.16) has a unique solution u^* in the class $\mathfrak{H}^{\gamma+2}_{p,\theta}(\mathcal{O},T)$, cf. Remark 5.17 below. By Theorem 3.8(ii), there exists a constant C_1, depending on T in general, such that for all $t \in [0,T]$

$$\|u-v\|^p_{\mathbb{H}^{\gamma+1}_{p,\theta}(\mathcal{O},t)} \leq C_1 \int_0^t \|u-v\|^p_{\mathfrak{H}^{\gamma+2}_{p,\theta}(\mathcal{O},s)}\,\mathrm{d}s.$$

Using this, we obtain from (5.24), that

$$\|\mathcal{N}(u) - \mathcal{N}(v)\|^p_{\mathfrak{H}^{\gamma+2}_{p,\theta}(\mathcal{O},t)} \leq C_0\,\varepsilon\,\|u-v\|^p_{\mathbb{H}^{\gamma+2}_{p,\theta-p}(\mathcal{O},t)} + C_0\,K_1\,C_1 \int_0^t \|u-v\|^p_{\mathfrak{H}^{\gamma+2}_{p,\theta}(\mathcal{O},s)}\,\mathrm{d}s,$$

for all $t \in [0,T]$. As a consequence, we can prove by induction, that for any $m \in \mathbb{N}$ the following estimate holds for all $t \in [0,T]$:

$$\|\mathcal{N}^m(u) - \mathcal{N}^m(v)\|^p_{\mathfrak{H}^{\gamma+2}_{p,\theta}(\mathcal{O},t)} \leq C_0^m\,\varepsilon^m\,\|u-v\|^p_{\mathfrak{H}^{\gamma+2}_{p,\theta}(\mathcal{O},t)}$$
$$+ \sum_{k=1}^m \binom{m}{k} C_0^m\,\varepsilon^{m-k}\,K_1^k\,C_1^k \int_0^t \frac{(t-s)^{k-1}}{(k-1)!}\|u-v\|^p_{\mathfrak{H}^{\gamma+2}_{p,\theta}(\mathcal{O},s)}\,\mathrm{d}s.$$

For $t = T$ and each $k \in \{1,\ldots,m\}$ we can approximate the integrals on the right-hand side by $\|u-v\|^p_{\mathfrak{H}^{\gamma+2}_{p,\theta}(\mathcal{O},T)} T^k/(k-1)!$. Consequently,

$$\|\mathcal{N}^m(u) - \mathcal{N}^m(v)\|^p_{\mathfrak{H}^{\gamma+2}_{p,\theta}(\mathcal{O},T)}$$
$$\leq C_0^m\,\varepsilon^m\,\|u-v\|^p_{\mathfrak{H}^{\gamma+2}_{p,\theta}(\mathcal{O},T)}$$
$$+ C_0^m\,\varepsilon^m \sum_{k=1}^m \binom{m}{k}\left(\frac{K_1 C_1 T}{\varepsilon}\right)^k \frac{1}{(k-1)!}\|u-v\|^p_{\mathfrak{H}^{\gamma+2}_{p,\theta}(\mathcal{O},T)}$$
$$\leq C_0^m\,\varepsilon^m\left(1 + 2^m \max_{1 \leq k \leq m}\left\{\left(\frac{K_1 C_1 T}{\varepsilon}\right)^k \frac{1}{(k-1)!}\right\}\right)\|u-v\|^p_{\mathfrak{H}^{\gamma+2}_{p,\theta}(\mathcal{O},T)}.$$

Now assume that Assumption 5.9(i) holds with $\varepsilon > 0$ small enough, e.g., $\varepsilon \leq 1/(8C_0)$, and $K_1 \in (0,\infty)$. Then,

$$\|\mathcal{N}^m(u) - \mathcal{N}^m(v)\|^p_{\mathfrak{H}^{\gamma+2}_{p,\theta}(\mathcal{O},T)} \leq \left(\frac{1}{8^m} + \frac{1}{4^m} \max_{k \in \mathbb{N}}\left\{\left(\frac{K_1 C_1 T}{\varepsilon}\right)^k \frac{1}{(k-1)!}\right\}\right)\|u-v\|^p_{\mathfrak{H}^{\gamma+2}_{p,\theta}(\mathcal{O},T)}.$$

Since for any fixed $x \in (0,\infty)$ the function $k \mapsto x^k/(k-1)!$ is decreasing for sufficiently large k, we have

$$\max_{k \in \mathbb{N}}\left\{\left(\frac{K_1 C_1 T}{\varepsilon}\right)^k \frac{1}{(k-1)!}\right\} = C_3 < \infty,$$

and, consequently,

$$\left\|\mathcal{N}^m(u) - \mathcal{N}^m(v)\right\|^p_{\mathfrak{H}^{\gamma+2}_{p,\theta}(\mathcal{O},T)} \leq \frac{1}{4^m}\left(1+C_3\right)\left\|u-v\right\|^p_{\mathfrak{H}^{\gamma+2}_{p,\theta}(\mathcal{O},T)}, \qquad m \in \mathbb{N}. \tag{5.26}$$

Thus, there exists an $m \in \mathbb{N}$, such that $\mathcal{N}^m$ is a contraction on the Banach space $\mathfrak{H}^{\gamma+2}_{p,\theta}(\mathcal{O},T)$. Consequently, by the Banach fixed point theorem, $\mathcal{N}^m$ (and therefore $\mathcal{N}$) has a unique fixed point, and, therefore, Eq. (5.16) has a unique solution u^* in the class $\mathfrak{H}^{\gamma+2}_{p,\theta}(\mathcal{O},T)$. In order to prove Estimate (5.23), we argue as follows: Take the sequence $(u^{(j)})_{j\in\mathbb{N}_0}$ defined above. Then for any $j \in \mathbb{N}_0$, since $u^{(0)} = 0$,

$$\left\|u^{(j+1)}\right\|_{\mathfrak{H}^{\gamma+2}_{p,\theta}(\mathcal{O},T)} \leq \sum_{i=0}^{j}\left\|u^{(i+1)} - u^{(i)}\right\|_{\mathfrak{H}^{\gamma+2}_{p,\theta}(\mathcal{O},T)} = \sum_{i=0}^{j}\left\|\mathcal{N}^{(i+1)}\left(u^{(0)}\right) - \mathcal{N}^{(i)}\left(u^{(0)}\right)\right\|_{\mathfrak{H}^{\gamma+2}_{p,\theta}(\mathcal{O},T)}.$$

Using (5.26) and $u^{(0)} = 0$, we obtain

$$\left\|u^{(j+1)}\right\|_{\mathfrak{H}^{\gamma+2}_{p,\theta}(\mathcal{O},T)} \leq \sum_{i=0}^{j}\frac{1}{4^i}\left(1+C_3\right)\left\|\mathcal{N}\left(u^{(0)}\right)\right\|_{\mathfrak{H}^{\gamma+2}_{p,\theta}(\mathcal{O},T)},$$

and by (5.25),

$$\left\|u^{(j+1)}\right\|^p_{\mathfrak{H}^{\gamma+2}_{p,\theta}(\mathcal{O},T)} \leq \left(\frac{1+C_3}{1-1/4}\right)^p C\left(\left\|f\right\|^p_{\mathbb{H}^{\gamma}_{p,\theta+p}(\mathcal{O},T)} + \left\|g\right\|^p_{\mathbb{H}^{\gamma+1}_{p,\theta}(\mathcal{O},T;\ell_2)} + \left\|u_0\right\|^p_{U^{\gamma+2}_{p,\theta}(\mathcal{O})}\right).$$

Since by the Banach fixed point theorem, there exists a subsequence of $(u^{(j)})_{j\in\mathbb{N}_0}$ converging to u^* in $\mathfrak{H}^{\gamma+2}_{p,\theta}(\mathcal{O},T)$, this finishes the proof. $\qquad\square$

Remark 5.17. In the proof of Lemma 5.16 above, we use the following fact:

If the m-th power $\mathcal{N}^m$, $m \in \mathbb{N}$, of a mapping $\mathcal{N} : E \to E$ on a Banach space $(E, \|\cdot\|_E)$ has a unique fixed point, then, so does $\mathcal{N}$.

This can be seen as follows: Let u be the unique fixed point of $\mathcal{N}^m$ for some $m \in \mathbb{N}$. In particular,

$$\mathcal{N}^m u = u,$$

and, therefore, due to the associativity of the composition of functions,

$$\mathcal{N}^m \mathcal{N} u = \mathcal{N} u.$$

Consequently, $\mathcal{N}u$ is a fixed point of $\mathcal{N}^m$, and, due to the uniqueness assumption, $\mathcal{N}u = u$. The uniqueness of the fixed point of $\mathcal{N}$ follows immediately from the uniqueness of the fixed point of $\mathcal{N}^m$.

After these preparations, we are able to prove Theorem 5.13 in the following way.

Proof of Theorem 5.13. We prove that in the different situations of Theorem 3.13, the solution $u \in \mathfrak{H}^{\gamma+2}_{p,\theta}(\mathcal{O},T)$ of Eq. (3.1) with $u_0 = 0$ fulfils the a priori estimate (5.22) for all $t \in [0,T]$ with a constant C_0 which does not depend on $t \in [0,T]$. Since we have proven Lemma 5.16, this automatically implies our assertion with $\tilde{\kappa}_0 = \kappa_0$ and $\tilde{\kappa}_1 = \kappa_1$ from Theorem 3.13. We prove the a priori estimate in four different situations.

Case 1. Assume that a^{ij} and σ^{ik} do not depend on $x \in \mathcal{O}$ and fulfil Assumption 3.1 with $\gamma \geq 0$ and $b^i = c = \mu^k = 0$ for all $i \in \{1,\dots,d\}$ and $k \in \mathbb{N}$. Furthermore assume that

$\theta \in (d + p - 2 - \kappa_0, d + p - 2 + \kappa_0)$ with $\kappa_0 \in (0,1)$ as in Theorem 3.13(i). In this case we can argue as follows: Given $f \in \mathbb{H}^{\gamma}_{p,\theta+p}(\mathcal{O},T)$ and $g \in \mathbb{H}^{\gamma+1}_{p,\theta}(\mathcal{O},T)$, the solution of the equation

$$\left. \begin{aligned} du &= \left(a^{ij} u_{x^i x^j} + f \right) dt + \left(\sigma^{ik} u_{x^i} + g^k \right) dw_t^k \quad \text{on } \Omega_T \times \mathcal{O}, \\ u(0) &= 0 \quad \text{on } \Omega \times \mathcal{O}. \end{aligned} \right\} \tag{5.27}$$

fulfils the estimate

$$\|u\|^p_{\mathfrak{H}^{\gamma+2}_{p,\theta}(\mathcal{O},T)} \le C \left(\|f\|^p_{\mathbb{H}^{\gamma}_{p,\theta+p}(\mathcal{O},T)} + \|g\|^p_{\mathbb{H}^{\gamma+1}_{p,\theta}(\mathcal{O},T;\ell_2)} \right) \tag{5.28}$$

with a constant C which does not depend on T. This has been proven in [75], see especially Corollary 3.6 therein. Therefore, since the restriction $u|_{\Omega \times [0,t]}$ is the unique solution in the class $\mathfrak{H}^{\gamma+2}_{p,\theta}(\mathcal{O},t)$ of the equation

$$\left. \begin{aligned} dv &= \left(a^{ij} v_{x^i x^j} + f|_{\Omega \times [0,t]} \right) ds + \left(\sigma^{ik} v_{x^i} + g^k|_{\Omega \times [0,t]} \right) dw_s^k \quad \text{on } \Omega \times [0,t] \times \mathcal{O}, \\ v(0) &= 0 \quad \text{on } \Omega \times \mathcal{O}. \end{aligned} \right\}$$

for any $t \in [0,T]$, Estimate (5.22) is fulfilled with C independent of $t \in [0,T]$.

Case 2. Consider the situation from Case 1 and relax the restriction $\gamma \ge 0$ allowing γ to be negative. In order to prove that Estimate (5.22) holds also in this situation with a constant independent of $t \in [0,T]$, we will prove that (5.28) holds with a constant C independent of T. We follow the lines of Case 2 in the proof of [73, Theorem 4.7]. Let us assume that $\gamma \in [-1,0]$. (The case $\gamma < 1$ can be proven analogously by iterating the proof for $\gamma \in [-1,0]$.) For $\nu \in [0,\infty)$, let

$$\mathcal{R} : \mathbb{H}^{\nu}_{p,\theta+p}(\mathcal{O},T) \times \mathbb{H}^{\nu+1}_{p,\theta}(\mathcal{O},T;\ell_2) \to \mathfrak{H}^{\nu+2}_{p,\theta}(\mathcal{O},T)$$
$$(f,g) \mapsto \mathcal{R}(f,g)$$

be the solution operator for Eq. (5.27), i.e., $\mathcal{R}(f,g) \in \mathfrak{H}^{\nu+2}_{p,\theta}(\mathcal{O},T)$ denotes the unique solution of Eq. (5.27), given $f \in \mathbb{H}^{\nu}_{p,\theta+p}(\mathcal{O},T)$ and $g \in \mathbb{H}^{\nu+1}_{p,\theta}(\mathcal{O},T;\ell_2)$. Notice that by the uniqueness of the solution, this operator does not depend on $\nu \in [0,\infty)$. Furthermore, by the a priori estimate (5.28), it is a bounded operator with operator norm independent of T, which we will denote by $\|\mathcal{R}\|_\nu$. Fix $(f,g) \in \mathbb{H}^{\gamma}_{p,\theta+p}(\mathcal{O},T) \times \mathbb{H}^{\gamma+1}_{p,\theta}(\mathcal{O},T;\ell_2)$. Furthermore, let ψ be an infinitely differentiable function on $\mathcal{O}$ fulfilling (2.25) and choose $c_0 > 0$, such that the operator $\mathfrak{L} := \psi^2 \Delta - c_0$ is an isomorphism between $H^{\gamma+2}_{p,\theta+p}(\mathcal{O})$ and $H^{\gamma}_{p,\theta+p}(\mathcal{O})$ and between $H^{\gamma+3}_{p,\theta}(\mathcal{O};\ell_2)$ and $H^{\gamma+1}_{p,\theta}(\mathcal{O};\ell_2)$, respectively; this is possible due to Lemma 2.45(vi) and Lemma 2.55(iii), respectively. Set

$$(\tilde{f},\tilde{g}) := \mathfrak{L}^{-1}(f,g).$$

Then,

$$f = \psi D^r (\psi D^r \tilde{f}) - \psi \psi_{x^r} D^r \tilde{f} - c_0 \tilde{f} \quad \text{and} \quad g = \psi D^r (\psi D^r \tilde{g}) - \psi \psi_{x^r} D^r \tilde{g} - c_0 \tilde{g}.$$

For $r = 1, \dots, d$, denote

$$u^r := \mathcal{R}(\psi D^r \tilde{f}, \psi D^r \tilde{g}), \quad u^0 := \mathcal{R}(-\psi \psi_{x^r} D^r \tilde{f} - c_0 \tilde{f}, -\psi \psi_{x^r} D^r \tilde{g} - c_0 \tilde{g}),$$

and set

$$v := u^0 + \sum_{r=1}^{d} \psi D^r u^r.$$

Note that, by Theorem 3.13 and Lemma 2.45(iii) and (iv) together with Lemma 2.55(i) and (ii), $u^0, u^r \in \mathfrak{H}_{p,\theta}^{\gamma+3}(\mathcal{O},T)$, $r = 1, \ldots, d$, and $v \in \mathbb{H}_{p,\theta-p}^{\gamma+2}(\mathcal{O},T)$ are well-defined. A short calculation shows that for all $\varphi \in \mathcal{C}_0^\infty(\mathcal{O})$, with probability one, the equality

$$(v(t,\cdot),\varphi) = \int_0^t (a^{ij}(s)v_{x^i x^j}(s,\cdot) + f_0(s,\cdot) + f(s,\cdot),\varphi)\,\mathrm{d}s$$
$$+ \sum_{k=1}^d \int_0^t (\sigma^{ik}(s)v_{x^i}(s,\cdot) + g_0^k(s,\cdot) + g^k(s,\cdot),\varphi)\,\mathrm{d}w_s^k$$

holds for all $t \in [0,T]$ with

$$f_0 := -a^{ij}\big(\psi_{x^i x^j}D^r u^r + \psi_{x^i}D^r u_{x^j}^r + \psi_{x^j}D^r u_{x^i}^r\big) \quad \text{and} \quad g_0 := (-\sigma^{ik}\psi_{x^i}D^r u^r)_{k\in\mathbb{N}}.$$

Using the decay properties (2.25) of ψ and its derivatives, Assumption 3.1 as well as Lemma 2.45, we can deduce that there exists a constant C, which does not depend on T, such that

$$\|f_0\|_{\mathbb{H}_{p,\theta+p}^{\gamma+1}(\mathcal{O},T)} + \|g_0\|_{\mathbb{H}_{p,\theta}^{\gamma+2}(\mathcal{O},T;\ell_2)} \leq C \sum_{r=1}^d \|u^r\|_{\mathbb{H}_{p,\theta-p}^{\gamma+3}(\mathcal{O},T)}. \tag{5.29}$$

Thus, by Lemma 2.45(iii) and Lemma 2.55(i), $v \in \mathfrak{H}_{p,\theta}^{\gamma+2}(\mathcal{O},T)$ and solves the equation

$$\left.\begin{array}{l} \mathrm{d}v = \big(a^{ij}v_{x^i x^j} + f_0 + f\big)\,\mathrm{d}t + \big(\sigma^{ik}v_{x^i} + g_0^k + g^k\big)\,\mathrm{d}w_t^k \quad \text{on } \Omega_T \times \mathcal{O}, \\ v(0) = 0 \quad \text{on } \Omega \times \mathcal{O}. \end{array}\right\} \tag{5.30}$$

Set $\tilde{v} := \mathcal{R}(f_0, g_0) \in \mathfrak{H}_{p,\theta}^{\gamma+3}(\mathcal{O},T)$. Then, obviously $u := v - \tilde{v} \in \mathfrak{H}_{p,\theta}^{\gamma+2}(\mathcal{O},T)$ solves Eq. (5.27). Moreover,

$$\|u\|_{\mathfrak{H}_{p,\theta}^{\gamma+2}(\mathcal{O},T)} \leq \|v\|_{\mathfrak{H}_{p,\theta}^{\gamma+2}(\mathcal{O},T)} + \|\tilde{v}\|_{\mathfrak{H}_{p,\theta}^{\gamma+2}(\mathcal{O},T)}. \tag{5.31}$$

We first prove that there exists a constant C, which does not depend on T, such that

$$\|\tilde{v}\|_{\mathfrak{H}_{p,\theta}^{\gamma+2}(\mathcal{O},T)} \leq C\left(\|f\|_{\mathbb{H}_{p,\theta+p}^{\gamma}(\mathcal{O},T)} + \|g\|_{\mathbb{H}_{p,\theta}^{\gamma+1}(\mathcal{O},T;\ell_2)}\right). \tag{5.32}$$

We argue as follows: Since $\mathcal{R}$ is a bounded operator,

$$\|\tilde{v}\|_{\mathfrak{H}_{p,\theta}^{\gamma+2}(\mathcal{O},T)} \leq \|\tilde{v}\|_{\mathfrak{H}_{p,\theta}^{\gamma+3}(\mathcal{O},T)} < \|\mathcal{R}\|_{\gamma+1}\left(\|f_0\|_{\mathbb{H}_{p,\theta+p}^{\gamma+1}(\mathcal{O},T)} + \|g_0\|_{\mathbb{H}_{p,\theta}^{\gamma+2}(\mathcal{O},T;\ell_2)}\right).$$

The same argument, together with Lemma 2.45(iii) and Lemma 2.55(i), shows that

$$\sum_{r=1}^d \|u^r\|_{\mathbb{H}_{p,\theta-p}^{\gamma+3}(\mathcal{O},T)} \leq C_2 \|\mathcal{R}\|_{\gamma+1}\left(\|\tilde{f}\|_{\mathbb{H}_{p,\theta+p}^{\gamma+2}(\mathcal{O},T)} + \|\tilde{g}\|_{\mathbb{H}_{p,\theta}^{\gamma+3}(\mathcal{O},T;\ell_2)}\right)$$

with a constant C_2 independent of T. Consequently,

$$\sum_{r=1}^d \|u^r\|_{\mathbb{H}_{p,\theta-p}^{\gamma+3}(\mathcal{O},T)} \leq C_2 \|\mathcal{R}\|_{\gamma+1} \|\mathfrak{L}^{-1}\|\left(\|f\|_{\mathbb{H}_{p,\theta+p}^{\gamma}(\mathcal{O},T)} + \|g\|_{\mathbb{H}_{p,\theta}^{\gamma+1}(\mathcal{O},T;\ell_2)}\right),$$

where

$$\|\mathfrak{L}^{-1}\| := \max\left\{\|\mathfrak{L}^{-1}\|_{\mathcal{L}(H_{p,\theta}^{\gamma+1}(\mathcal{O};\ell_2),H_{p,\theta}^{\gamma+3}(\mathcal{O};\ell_2))}, \|\mathfrak{L}^{-1}\|_{\mathcal{L}(H_{p,\theta+p}^{\gamma}(\mathcal{O}),H_{p,\theta+p}^{\gamma+2}(\mathcal{O}))}\right\}.$$

Finally, using estimate (5.29), we obtain (5.32) with a constant C independent of T. We move on and estimate $\|v\|_{\mathfrak{H}^{\gamma+2}_{p,\theta}(\mathcal{O},T)}$. Since $v \in \mathfrak{H}^{\gamma+2}_{p,\theta}(\mathcal{O},T)$ solves Eq. (5.30),

$$\|v\|_{\mathfrak{H}^{\gamma+2}_{p,\theta}(\mathcal{O},T)} \leq \|v\|_{\mathbb{H}^{\gamma+2}_{p,\theta-p}(\mathcal{O},T)} + \|a^{ij} v_{x^i x^j}\|_{\mathbb{H}^{\gamma}_{p,\theta+p}(\mathcal{O},T)} + \|f_0\|_{\mathbb{H}^{\gamma}_{p,\theta+p}(\mathcal{O},T)} + \|f\|_{\mathbb{H}^{\gamma}_{p,\theta+p}(\mathcal{O},T)}$$
$$+ \|\sigma^{ik} u_{x^i}\|_{\mathbb{H}^{\gamma+1}_{p,\theta}(\mathcal{O},T;\ell_2)} + \|g_0\|_{\mathbb{H}^{\gamma+1}_{p,\theta}(\mathcal{O},T;\ell_2)} + \|g\|_{\mathbb{H}^{\gamma+1}_{p,\theta}(\mathcal{O},T;\ell_2)}.$$

Thus, we can argue as before when estimating $\|\tilde{v}\|_{\mathfrak{H}^{\gamma+2}_{p,\theta}(\mathcal{O},T)}$ and obtain

$$\|v\|_{\mathfrak{H}^{\gamma+2}_{p,\theta}(\mathcal{O},T)} \leq C \left(\|f\|_{\mathbb{H}^{\gamma}_{p,\theta+p}(\mathcal{O},T)} + \|g\|_{\mathbb{H}^{\gamma+1}_{p,\theta+p}(\mathcal{O},T;\ell_2)} \right)$$

with a constant C which does not depend on T. This, together with (5.31) and (5.32), proves that (5.28) holds with a constant C independent of T.

Case 3. Assume that a^{ij} and σ^{ik} do not depend on $x \in \mathcal{O}$ and fulfil Assumption 3.1 with $\gamma \in \mathbb{R}$ and $b^i = c = \mu^k = 0$ for all $i \in \{1,\ldots,d\}$ and $k \in \mathbb{N}$. Furthermore assume that $\theta \in (d+\kappa_1, d+\kappa_1)$ with $\kappa_1 \in (0,1)$ and $p \in [2,p_0)$ as in Theorem 3.13(ii). In this situation, the assertion for $\gamma \geq 0$ can be proven by following the lines of [75, Section 5]. Essentially, this strategy makes use of the fact that the complex interpolation method is an exact interpolation method and that for two compatible couples (A_0, B_0) and (A_1, B_1) of Banach spaces,

$$[A_0 \times A_1, B_0 \times B_1]_\eta = [A_0, B_0]_\eta \times [A_1, B_1]_\eta,$$

with equivalent norms ($\eta \in (0,1)$). Additionally, in order to apply this strategy in the case of bounded Lipschitz domains, Lemma 2.45(v) and Lemma 2.55(v), concerning complex interpolation of weighted Sobolev spaces and of their generalizations $H^{\gamma}_{p,\theta}(\mathcal{O};\ell_2)$, are required. Using the argumentation line from Case 2 above, we can obtain the assertion also for $\gamma < 0$.

Case 4. Finally, we consider the general case. We assume that $\theta \in (d+p-2-\kappa_0, 2+p-2+\kappa_0)$ and $p \in [2,\infty)$ or, alternatively, that $\theta \in (d-\kappa_1, d+\kappa_1)$ and $p \in [2,p_0)$ as in the different situations of Theorem 3.13(i) and (ii), respectively. Following the lines of the proof of [75, Theorem 3.7] (see also Section 5 in [73]) and using what we have proved in the first three cases, we can show that there exists a constant C_3 independent of $t \in [0,T]$ such that

$$\|u\|^p_{\mathfrak{H}^{\gamma+2}_{p,\theta}(\mathcal{O},t)} \leq C_3 \left(\|u\|^p_{\mathbb{H}^{\gamma+1}_{p,\theta}(\mathcal{O},t)} + \|f\|^p_{\mathbb{H}^{\gamma}_{p,\theta+p}(\mathcal{O},t)} + \|g\|^p_{\mathbb{H}^{\gamma+1}_{p,\theta}(\mathcal{O},t;\ell_2)} \right)$$

for all $t \in [0,T]$. By Theorem 3.8 this leads to

$$\|u\|^p_{\mathfrak{H}^{\gamma+2}_{p,\theta}(\mathcal{O},t)} \leq C_4 \int_0^t \|u\|^p_{\mathfrak{H}^{\gamma+2}_{p,\theta}(\mathcal{O},s)} \, ds + C_3 \left(\|f\|^p_{\mathbb{H}^{\gamma}_{p,\theta+p}(\mathcal{O},t)} + \|g\|^p_{\mathbb{H}^{\gamma+1}_{p,\theta}(\mathcal{O},t;\ell_2)} \right)$$

for all $t \in [0,T]$ with a constant C_4 independent of $t \in [0,T]$. Using Gronwall's lemma (see, e.g., [7, Corollary (6.2)]) this proves that for all $t \in [0,T]$,

$$\|u\|^p_{\mathfrak{H}^{\gamma+2}_{p,\theta}(\mathcal{O},t)} \leq C_3 \, e^{tC_4} \left(\|f\|^p_{\mathbb{H}^{\gamma}_{p,\theta+p}(\mathcal{O},t)} + \|g\|^p_{\mathbb{H}^{\gamma+1}_{p,\theta}(\mathcal{O},t;\ell_2)} \right).$$

Thus, Estimate (5.22) is fulfilled with $C_0 := C_3 e^{tC_4}$, which does not depend on $t \in [0,T]$. $\square$

We conclude this section with two examples. The first one is put in a setting similar to the one presented in [121, Section 7]. However, we are able to treat the case of general bounded Lipschitz domains, whereas [121, Section 7] is restricted to bounded domains with $\mathcal{C}^2$ boundary. As pointed out in Remark 4.3, we are only concerned with equations fulfilling zero Dirichlet boundary conditions. Using the notation from [121], this means that $\Gamma_0 = \partial\mathcal{O}$ and therefore $\Gamma_1 = \emptyset$.

Example 5.18. For $i, j \in \{1, \ldots, d\}$ and $k \in \mathbb{N}$, let a^{ij}, b^i, c, σ^{ik}, and μ^k be given coefficients fulfilling Assumption 3.1 with $\gamma = 0$. Let

$$f_L : \Omega_T \times H^1_{2,d}(\mathcal{O}) \to H^0_{2,d+2}(\mathcal{O}) \qquad \text{and} \qquad g^k_\Lambda : \Omega_T \times H^1_{2,d}(\mathcal{O}) \to H^1_{2,d}(\mathcal{O}), \quad k \in \mathbb{N},$$

be strongly $\mathcal{P}_T \otimes \mathcal{B}(H^1_{2,d}(\mathcal{O}))$-measurable mappings. Assume that

$$f_L(\omega, t, 0) = g^k(\omega, t, 0) = 0 \quad \text{for all} \quad (\omega, t) \in \Omega_T, \quad k \in \mathbb{N},$$

and that there exist $C_L \in L_\infty(\Omega_T; \mathbb{R})$ and $C_\Lambda = (C^k_\Lambda)_{k \in \mathbb{N}} \in L_\infty(\Omega_T; \ell_2)$ such that for all $u, v \in H^1_{2,d}(\mathcal{O})$,

$$\|f_L(\omega, t, u) - f_L(\omega, t, v)\|_{H^0_{2,d+2}(\mathcal{O})} \leq C_L(\omega, t) \|u - v\|_{H^1_{2,d}(\mathcal{O})}$$

and

$$\|g^k_\Lambda(\omega, t, u) - g^k_\Lambda(\omega, t, v)\|_{H^1_{2,d}(\mathcal{O})} \leq C^k_\Lambda(\omega, t) \|u - v\|_{H^1_{2,d}(\mathcal{O})}, \quad k \in \mathbb{N}.$$

Then, the functions

$$L : \mathfrak{H}^2_{2,d}(\mathcal{O}, T) \to \mathbb{H}^0_{2,d+2}(\mathcal{O}, T)$$
$$u \mapsto L(u) := \left((\omega, t) \mapsto f_L(\omega, t, u(\omega, t, \cdot)) \right)$$

and

$$\Lambda : \mathfrak{H}^2_{2,d}(\mathcal{O}, T) \to \mathbb{H}^1_{2,d}(\mathcal{O}, T; \ell_2)$$
$$u \mapsto \Lambda(u) := \left(\left((\omega, t) \mapsto g^k_\Lambda(\omega, t, u(\omega, t, \cdot)) \right)_{k \in \mathbb{N}} \right)$$

are well-defined and fulfil Assumption 5.9 with

$$\varepsilon = 0 \qquad \text{and} \qquad K_1 = \max \left\{ \|C_L\|^2_{L_\infty(\Omega_T; \mathbb{R})}, \|C_\Lambda\|^2_{L_\infty(\Omega_T; \ell_2)} \right\} < \infty.$$

Therefore, by Theorem 5.13, Eq. (5.16) with L and Λ as defined above has a unique solution $u \in \mathfrak{H}^2_{2,d}(\mathcal{O}, T)$. Furthermore, due to Theorem 5.15,

$$u \in L_2(\Omega_T; B^\alpha_{\tau,\tau}(\mathcal{O})), \qquad \frac{1}{\tau} = \frac{\alpha}{d} + \frac{1}{2}, \qquad \text{for all} \quad 0 < \alpha < \frac{d}{d-1}.$$

In the two-dimensional case, this yields

$$u \in L_2(\Omega_T; B^\alpha_{\tau,\tau}(\mathcal{O})), \qquad \frac{1}{\tau} = \frac{\alpha}{2} + \frac{1}{2}, \qquad \text{for all} \quad 0 < \alpha < 2.$$

Also in this case, we expect that, at least on non-smooth and non-convex domains, $\tilde{s}^{\text{Sob}}_{\max}(u) < 2$, cf. Example 5.5.

The following example is inspired from [59, Section 6.1]. Therein space time discretization schemes for SPDEs are discussed.

Example 5.19. Let again a^{ij}, b^i, c, σ^{ik}, and μ^k with $i, j \in \{1, \ldots, d\}$ and $k \in \mathbb{N}$ be given coefficients fulfilling Assumption 3.1 with $\gamma = 0$. Furthermore, let

$$F : \Omega_T \times \mathcal{O} \times \mathbb{R}^d \times \mathbb{R} \to \mathbb{R}$$

be a strongly $\mathcal{P}_T \otimes \mathcal{B}(\mathcal{O}) \otimes \mathcal{B}(\mathbb{R}^d) \otimes \mathcal{B}(\mathbb{R})$-measurable function satisfying the following conditions:

[F1] There exists a constant C_F, which does not depend on $(\omega, t, x) \in \Omega_T \times \mathcal{O}$, $p_1, p_2 \in \mathbb{R}^d$, and $r_1, r_2 \in \mathbb{R}$ such that

$$\left| F(\omega, t, x, p_1, r_1) - F(\omega, t, x, p_2, r_2) \right| \leq C_F \left(|p_1 - p_2| + \rho(x)^{-1} |r_1 - r_2| \right);$$

[F2] For all $(\omega, t, x) \in \Omega_T \times \mathcal{O}$:

$$F(\omega, t, x, 0, 0) = 0.$$

Then, for any $u \in \mathbb{H}^1_{2,d}(\mathcal{O}, T)$, the function

$$\Omega_T \ni (\omega, t) \mapsto F(\omega, t, \cdot, u_x(\omega, t, \cdot), u(\omega, t, \cdot)) \in H^0_{2,d+2}(\mathcal{O})$$

is well-defined, strongly $\mathcal{P}_T$-measurable and for $u, v \in \mathbb{H}^1_{2,d}(\mathcal{O}, T)$ and arbitrary $t \in [0, T]$,

$$\int_\Omega \int_0^t \|F(\omega, s, \cdot, u_x(\omega, s, \cdot), u(\omega, s, \cdot)) - F(\omega, s, \cdot, v_x(\omega, s, \cdot), v(\omega, s, \cdot))\|^2_{H^0_{2,d+2}(\mathcal{O})} \, \mathbb{P} \otimes \lambda^1(d(\omega, s))$$

$$\leq C \int_\Omega \int_0^t \int_\mathcal{O} \left| F(\omega, s, x, u_x(\omega, s, x), u(\omega, s, x)) \right.$$

$$\left. - F(\omega, s, x, v_x(\omega, s, x), v(\omega, s, \cdot)) \right|^2 \rho(x)^2 \, dx \, \mathbb{P} \otimes \lambda^1(d(\omega, s))$$

$$\leq C \int_\Omega \int_0^t \int_\mathcal{O} 4 C_F^2 \left(|u_x(\omega, s, x) - v_x(\omega, s, x)|^2 \rho(x)^2 \right.$$

$$\left. + |u(\omega, s, x) - v(\omega, s, x)|^2 \right) dx \, \mathbb{P} \otimes \lambda^1(d(\omega, s))$$

$$\leq C \|u - v\|^2_{\mathbb{H}^1_{2,d}(\mathcal{O}, T)},$$

where in the last step we have used the norm equivalence (2.28). Therefore, there exists a constant $K_1 \in [0, \infty)$ such that Assumption 5.9 is fulfilled with

$$L : \mathfrak{H}^2_{2,d+2}(\mathcal{O}, T) \to \mathbb{H}^0_{2,d+2}(\mathcal{O}, T)$$

$$u \mapsto L(u) := \left((\omega, t) \mapsto F(\omega, t, \cdot, u_x(\omega, t, \cdot), u(\omega, t, \cdot)) \right)$$

and $\varepsilon = 0$ ($\Lambda = 0$). Thus, by Theorem 5.13, there exists a solution $u \in \mathfrak{H}^2_{2,d}(\mathcal{O}, T)$ of Eq. (5.16) with L as defined above and $\Lambda = 0$. Due to Theorem 5.15, this solution also fulfils

$$u \in L_2(\Omega_T; B^\alpha_{\tau,\tau}(\mathcal{O})), \qquad \frac{1}{\tau} = \frac{\alpha}{d} + \frac{1}{2}, \qquad \text{for all} \quad 0 < \alpha < \frac{d}{d-1}.$$

In the two-dimensional case, this means that

$$u \in L_2(\Omega_T; B^\alpha_{\tau,\tau}(\mathcal{O})), \qquad \frac{1}{\tau} = \frac{\alpha}{2} + \frac{1}{2}, \qquad \text{for all} \quad 0 < \alpha < 2.$$

In the light of Example 5.5, in general, we expect that $\tilde{s}^{\text{Sob}}_{\text{max}}(u) < 2$; at least on non-smooth and non-convex domains.

Chapter 6

Space time regularity of the inhomogeneous heat equation with additive noise

In this chapter we are concerned with the Hölder regularity of the paths of the solution to the *inhomogeneous stochastic heat equation with additive noise*

$$\left.\begin{aligned} du &= \left(\Delta u + f\right) dt + g^k \, dw_t^k \quad \text{on } \Omega_T \times \mathcal{O}, \\ u(0) &= 0 \quad \text{on } \Omega \times \mathcal{O}, \end{aligned}\right\} \tag{6.1}$$

considered as a stochastic process taking values in Besov spaces from the scale (∗). Eq. (6.1) is understood in the sense of Definition 3.10 with $a^{ij} = \delta_{i,j}$, $i, j \in \{1, \ldots, d\}$. It will be sometimes referred to as the *stochastic heat equation*.

As we have already seen in Chapter 3, see Theorem 3.13, it is known that for $\gamma \in \mathbb{R}$, certain $p \in [2, \infty)$ and corresponding $\theta \in \mathbb{R}$, Eq. (6.1) has a unique solution u within the class $\mathfrak{H}_{p,\theta}^{\gamma+2}(\mathcal{O}, T)$, provided $f \in \mathbb{H}_{p,\theta+p}^\gamma(\mathcal{O}, T)$ and $g \in \mathbb{H}_{p,\theta}^{\gamma+1}(\mathcal{O}, T; \ell_2)$. Applying Theorem 3.8 we obtain

$$\mathbb{E}\big[u\big]_{\mathcal{C}^{\tilde{\beta}/2 - 1/p}([0,T]; H_{p,\theta-(1-\beta)p}^{\gamma+2-\beta}(\mathcal{O}))}^p \leq C \, T^{(\beta-\tilde{\beta})p/2} \|u\|_{\mathfrak{H}_{p,\theta}^\gamma(\mathcal{O}, T)}^p < \infty,$$

with the restriction

$$\frac{2}{p} < \tilde{\beta} < \beta \leq 1. \tag{6.2}$$

Thus, a simple application of the embedding (4.13) already yields a first result concerning the Hölder regularity of the paths of u, seen as a stochastic process with values in the Besov spaces from the scale (∗). That is,

$$\mathbb{E}\big[u\big]_{\mathcal{C}^{\tilde{\beta}/2 - 1/p}([0,T]; B_{\tau,\tau}^\alpha(\mathcal{O}))}^p \leq C \, T^{(\beta-\tilde{\beta})p/2} \|u\|_{\mathfrak{H}_{p,\theta}^{\gamma+2}(\mathcal{O}, T)}^p,$$

and therefore

$$\mathbb{P}\Big(\big[u\big]_{\mathcal{C}^{\tilde{\beta}/2 - 1/p}([0,T]; B_{\tau,\tau}^\alpha(\mathcal{O}))} < \infty\Big) = 1, \tag{6.3}$$

for all α and τ with

$$\frac{1}{\tau} = \frac{\alpha}{d} + \frac{1}{p} \quad \text{and} \quad 0 < \alpha < \min\left\{\gamma + 2 - \beta, \left(1 + \frac{d-\theta}{p} - \beta\right)\frac{d}{d-1}\right\}.$$

However, this result turns out to be not at all satisfactory: The range of parameters in Theorem 3.13 is restricted to

- $p \in [2, \infty)$ and $\theta \in (d + p - 2 - \kappa_0, d + p - 2 + \kappa_0)$

or, alternatively,

- $p \in [2, p_0)$ and $\theta \in (d - \kappa_1, d + \kappa_1)$,

with $\kappa_0, \kappa_1 \in (0,1)$ depending on d, p and $\mathcal{O}$, and $p_0 \in (2, 4]$, see also Remark 3.14(iii) and (i). Since we do not have any lower bound for κ_0, in the first case we can only guarantee that for arbitrary $p \in [2, \infty)$ our solution exists in $\mathfrak{H}_{p,\theta_0}^{\gamma+2}(\mathcal{O}, T)$ with $\theta_0 := d + p - 2$ (provided f and g are smooth enough). But, in this case

$$\min \left\{ \gamma + 2 - \beta, \left(1 + \frac{d - \theta_0}{p} - \beta \right) \frac{d}{d-1} \right\} < 0,$$

so that the calculations above are useless. In the second case, due to the same arguments, we can only guarantee that, for sufficiently smooth f and g, the solution is in $\mathfrak{H}_{p,d}^{\gamma+2}(\mathcal{O}, T)$ for $p \in [2, p_0)$. Moreover, if we consider general bounded Lipschitz domains, we have to assume that $p_0 \leq 4$, compare Remark 3.14(i). Thus, by the calculations above, if $p \in (2, p_0)$, this solution fulfils (6.3) for all α and τ with

$$\frac{1}{\tau} = \frac{\alpha}{d} + \frac{1}{p} \quad \text{and} \quad 0 < \alpha < \min \left\{ \gamma + 2 - \beta, (1 - \beta) \frac{d}{d-1} \right\} < 1, \tag{6.4}$$

since (6.2) has to be fulfilled. This is indeed a first result. However, it still has two drawbacks. Firstly, it does not allow to consider the Hilbert space case $p = 2$. Secondly, in the view of the convergence rate of the best m-term wavelet approximation error, $\alpha > 1$ would be desirable, cf. Section 1.1.

In order to overcome these difficulties, we apply the following strategy. We start with the analysis of the Hölder regularity of the paths of elements of $\mathfrak{H}_{p,\theta}^{\gamma,q}(\mathcal{O}, T)$ for the case $q > p$. We are motivated by the fact that in [83, Theorem 4.1] it has been proved that for $q \geq p \geq 2$,

$$\mathbb{E}[u]_{C^{\tilde{\beta}/2 - 1/q}([0,T];H_{p,\theta-(1-\beta)p}^{\gamma+2-\beta}(\mathbb{R}_+^d))}^q \leq C T^{(\beta - \tilde{\beta})q/2} \|u\|_{\mathfrak{H}_{p,\theta}^{\gamma,q}(\mathbb{R}_+^d, T)}^q,$$

provided

$$\frac{2}{q} < \tilde{\beta} < \beta \leq 1.$$

Note that a generalization of this result to the case of bounded Lipschitz domains $\mathcal{O} \subset \mathbb{R}^d$ instead of $\mathbb{R}_+^d$ would allow us to choose simultaneously, e.g., $p = 2$ and β close to zero, such that α in (6.4) might become greater than 1. We prove this generalization in Section 6.1. After applying Embedding (4.13), this leads to Hölder regularity results for elements of $\mathfrak{H}_{p,\theta}^{\gamma,q}(\mathcal{O}, T)$, considered as stochastic processes with values in the scale (∗) of Besov spaces. From the point of non-linear approximation theory, the permitted range of p and α is much more satisfactory than in the case $q = p$. In Section 6.2 we prove one aspect of the 'suitability' of the spaces $\mathfrak{H}_{p,\theta}^{\gamma,q}(\mathcal{O}, T)$ for the regularity analysis of SPDEs: We show that, if we have a solution $u \in \mathfrak{H}_{p,\theta}^{\gamma,q}(\mathcal{O}, T)$ with low regularity $\gamma \geq 0$, but f and g have high $L_q(L_p)$-regularity then we can lift up the regularity of the solution. Finally, in Section 6.3 we prove that under suitable assumptions on the $L_q(L_p)$-regularity of f and g, the stochastic heat equation has a solution in the space $\mathfrak{H}_{p,\theta}^{\gamma,q}(\mathcal{O}, T)$. Then we can apply the results from Section 6.1 and obtain space time regularity results for the solution to the stochastic heat equation.

Slightly different versions of the results and proofs presented in this chapter have been partially worked out in collaboration with K.-H. Kim, K. Lee and F. Lindner [26].

6.1 Space time regularity of elements from $\mathfrak{H}_{p,\theta}^{\gamma,q}(\mathcal{O},T)$

In this section we analyse the Hölder regularity of the paths of elements of $\mathfrak{H}_{p,\theta}^{\gamma,q}(\mathcal{O},T)$. We are mainly interested in the case where the summability parameters p and q in space and time, respectively, do *not* coincide. We start by presenting the two main results: the first one is a generalization of Theorem 3.8 to the case $q > p$; $u \in \mathfrak{H}_{p,\theta}^{\gamma,q}(\mathcal{O},T)$ is seen as a stochastic process taking values in the weighted Sobolev spaces $H_{p,\tilde{\theta}}^{\nu}(\mathcal{O})$, ν, $\tilde{\theta} \in \mathbb{R}$. The second one is concerned with the Hölder regularity of $u \in \mathfrak{H}_{p,\theta}^{\gamma,q}(\mathcal{O},T)$, seen as a stochastic process taking values in the Besov spaces from the scale $(*)$.

Theorem 6.1. *Let $\mathcal{O}$ be a bounded Lipschitz domain in $\mathbb{R}^d$. Let $2 \leq p \leq q < \infty$, $\gamma \in \mathbb{N}$, $\theta \in \mathbb{R}$, and $u \in \mathfrak{H}_{p,\theta}^{\gamma,q}(\mathcal{O},T)$. Moreover, let*

$$\frac{2}{q} < \tilde{\beta} < \beta \leq 1.$$

Then there exists a constant C, which does not depend on T and u, such that

$$\mathbb{E}[u]_{\mathcal{C}^{\tilde{\beta}/2-1/q}([0,T];H_{p,\theta-(1-\beta)p}^{\gamma-\beta}(\mathcal{O}))}^{q}$$
$$\leq CT^{(\beta-\tilde{\beta})\frac{q}{2}} \left(\|u\|_{\mathbb{H}_{p,\theta-p}^{\gamma,q}(\mathcal{O},T)}^{q} + \|\mathbb{D}u\|_{\mathbb{H}_{p,\theta+p}^{\gamma-2,q}(\mathcal{O},T)}^{q} + \|\mathbb{S}u\|_{\mathbb{H}_{p,\theta}^{\gamma-1,q}(\mathcal{O},T;\ell_2)}^{q} \right) \tag{6.5}$$
$$\leq CT^{(\beta-\tilde{\beta})\frac{q}{2}} \|u\|_{\mathfrak{H}_{p,\theta}^{\gamma,q}(\mathcal{O},T)}^{q},$$

and

$$\mathbb{E}\|u\|_{\mathcal{C}^{\tilde{\beta}/2-1/q}([0,T];H_{p,\theta-(1-\beta)p}^{\gamma-\beta}(\mathcal{O}))}^{q}$$
$$\leq CT^{(\beta-\tilde{\beta})\frac{q}{2}} \left(\mathbb{E}\|u(0)\|_{H_{p,\theta-(1-\beta)p}^{\gamma-\beta}(\mathcal{O})}^{q} + \right.$$
$$\left. \|u\|_{\mathbb{H}_{p,\theta-p}^{\gamma,q}(\mathcal{O},T)}^{q} + \|\mathbb{D}u\|_{\mathbb{H}_{p,\theta+p}^{\gamma-2,q}(\mathcal{O},T)}^{q} + \|\mathbb{S}u\|_{\mathbb{H}_{p,\theta}^{\gamma-1,q}(\mathcal{O},T;\ell_2)}^{q} \right) \tag{6.6}$$
$$\leq CT^{(\beta-\tilde{\beta})\frac{q}{2}} \|u\|_{\mathfrak{H}_{p,\theta}^{\gamma,q}(\mathcal{O},T)}^{q}.$$

Before we prove this theorem, we present our second main result, which follows immediately form Theorem 6.1 by applying Embedding (4.13).

Theorem 6.2. *Let $\mathcal{O}$ be a bounded Lipschitz domain in $\mathbb{R}^d$. Let $2 \leq p \leq q < \infty$, $\gamma \in \mathbb{N}$, $\theta \in \mathbb{R}$, and $u \in \mathfrak{H}_{p,\theta}^{\gamma,q}(\mathcal{O},T)$. Moreover, let*

$$\frac{2}{q} < \tilde{\beta} < \min\left\{1, 1 + \frac{d-\theta}{p}\right\}.$$

Then, for all α and τ with

$$\frac{1}{\tau} = \frac{\alpha}{d} + \frac{1}{p} \quad \text{and} \quad 0 < \alpha < \min\left\{\gamma - \tilde{\beta}, \left(1 + \frac{d-\theta}{p} - \tilde{\beta}\right)\frac{d}{d-1}\right\},$$

we have

$$\mathbb{E}[u]_{\mathcal{C}^{\tilde{\beta}/2-1/q}([0,T];B_{\tau,\tau}^{\alpha}(\mathcal{O}))}^{q} \leq C(T)\left(\|u\|_{\mathbb{H}_{p,\theta-p}^{\gamma,q}(\mathcal{O},T)}^{q} + \|\mathbb{D}u\|_{\mathbb{H}_{p,\theta+p}^{\gamma-2,q}(\mathcal{O},T)}^{q} + \|\mathbb{S}u\|_{\mathbb{H}_{p,\theta}^{\gamma-1,q}(\mathcal{O},T;\ell_2)}^{q}\right), \tag{6.7}$$

and

$$\mathbb{E}\|u\|_{\mathcal{C}^{\tilde{\beta}/2-1/q}([0,T];B_{\tau,\tau}^{\alpha}(\mathcal{O}))}^{q} \leq C(T)\|u\|_{\mathfrak{H}_{p,\theta}^{\gamma,q}(\mathcal{O},T)}^{q}. \tag{6.8}$$

The constants $C(T)$ in (6.7) and (6.8) are given by $C(T) = C\sup_{\beta \in [\tilde{\beta},1]}\{T^{(\beta-\tilde{\beta})q/2}\}$, with C from (6.5) and (6.6) respectively.

Proof. The assertion is an immediate consequence of Theorem 6.1 and Theorem 4.7. $\qquad\square$

Now we turn our attention to the proof of Theorem 6.1. For the case that the summability parameters in time and space coincide, i.e., $q = p$, the assertion is covered by Theorem 3.8. A proof of Theorem 3.8 can be found in [75, Theorem 2.9]. It is straightforward and relies on [83, Corollary 4.12], which is the analogue of Theorem 6.1 on the whole space $\mathbb{R}^d$. However, we are explicitly interested in the case $q > p$ since it allows for a wider range of parameters $\tilde{\beta}$ and β, and therefore leads to better regularity results. Unfortunately, the proof technique used in [75, Proposition 2.9] does not work any more in this case. Therefore, we use a different approach: We make use of [83, Proposition 4.1], which covers the assertion of Theorem 6.1 with $\mathbb{R}^d_+ := \{(x^1, x') \in \mathbb{R}^d : x^1 > 0\}$ instead of $\mathcal{O}$, and the Lipschitz character of $\partial\mathcal{O}$ to derive Theorem 6.1 via a boundary flattening argument. To this end, we need the following two lemmas, which we prove first. We start with a transformation rule for weighted Sobolev spaces, where the transformation and its inverse are assumed to be Lipschitz. Remember that $\rho_G(x)$ denotes the distance of a point $x \in G$ to the boundary ∂G of a domain $G \subset \mathbb{R}^d$.

Lemma 6.3. *Let $G^{(1)}, G^{(2)}$ be two domains in $\mathbb{R}^d$ with non-empty boundaries, and let $\phi : G^{(1)} \to G^{(2)}$ be a bijective map, such that ϕ and ϕ^{-1} are Lipschitz continuous. Furthermore, assume that there exists a constant $C \in (0, \infty)$, such that*

$$\frac{1}{C}\rho_{G^{(1)}}(\phi^{-1}(y)) \leq \rho_{G^{(2)}}(y) \leq C\rho_{G^{(1)}}(\phi^{-1}(y)) \text{ for all } y \in G^{(2)},$$

and that the (a.e. existing) Jacobians $J\phi$ and $J\phi^{-1}$ fulfil

$$\left| \operatorname{Det} J\phi \right| = 1 \text{ and } \left| \operatorname{Det} J\phi^{-1} \right| = 1 \quad (\text{a.e.}).$$

Then, for any $\gamma \in [-1, 1]$, $p \in (1, \infty)$, and $\theta \in \mathbb{R}$, there exists a constant $C = C(d, \gamma, p, \theta, \phi) \in (0, \infty)$, which does not depend on u, such that

$$\frac{1}{C}\|u\|_{H^\gamma_{p,\theta}(G^{(1)})} \leq \left\|u \circ \phi^{-1}\right\|_{H^\gamma_{p,\theta}(G^{(2)})} \leq C\|u\|_{H^\gamma_{p,\theta}(G^{(1)})}$$

in the sense that, if one of the norms exists, so does the other one and the above inequality holds.

Remark 6.4. (i) A Lipschitz continuous function $\phi : G^{(1)} \to G^{(2)}$ with Lipschitz continuous inverse, as in the assumptions of Lemma 6.3, is usually called *bi-Lipschitz*.

(ii) The Jacobians $J\phi$ and $J\phi^{-1}$ in Lemma 6.3 exist λ^d-a.e. on $G^{(1)}$ and $G^{(2)}$, respectively, due to *Rademacher's theorem*: "Let $\mathcal{U} \subseteq \mathbb{R}^d$ be an open set and let $m \in \mathbb{N}$. A Lipschitz continuous function $f : \mathcal{U} \to \mathbb{R}^m$ is λ^d-a.e. (totally) differentiable (in the classical sense)." A proof can be found e.g. in [54, Section 3.1].

(iii) The meaning of $u \circ \phi^{-1}$ for $u \in H^\gamma_{p,\theta}(G^{(1)})$ with $\gamma \geq 0$ and ϕ as in Lemma 6.3 is naturally given as the composition of the two functions. However, for negative $\gamma \in [-1, 0)$ this definition is not suitable anymore, since in this case $u \in H^\gamma_{p,\theta}(G^{(1)})$ is not necessarily a function, but only a distribution. We will define $u \circ \phi^{-1}$ in this case during the proof of Lemma 6.3 in such a way that, in particular, the identity

$$(u \circ \phi^{-1}, \varphi) = (u, \varphi \circ \phi), \qquad \varphi \in \mathcal{C}^\infty_0(G^{(2)}), \tag{6.9}$$

holds.

Proof of Lemma 6.3. We consider consecutively the cases $\gamma = 0, 1, -1$. For fractional $\gamma \in (-1,1)$, the statement follows then by using interpolation arguments and Lemma 2.45(v). Furthermore, we restrict ourselves to the proof of the right inequality in the assertion of the Lemma, i.e., that there exists a constant $C = C(d, \gamma, p, \theta, \phi) \in (0, \infty)$, such that for any $u \in H_{p,\theta}^{\gamma}(G^{(1)})$ the following inequality holds:

$$\left\| u \circ \phi^{-1} \right\|_{H_{p,\theta}^{\gamma}(G^{(2)})} \leq C \|u\|_{H_{p,\theta}^{\gamma}(G^{(1)})}.$$

The left inequality can be proven analogously. For $\gamma = 0$, the assertion follows immediately by using the assumptions of the Lemma and the change of variables formula for bi-Lipschitz transformations, see, e.g., [62, Theorem 3]. Let us go on and look at the case $\gamma = 1$. Because of the density of the test functions $\mathcal{C}_0^{\infty}(G^{(1)})$ in $H_{p,\theta}^1(G^{(1)})$, see Lemma 2.45(ii), it suffices to prove the asserted inequality for $u \in \mathcal{C}_0^{\infty}(G^{(1)})$. In this case, because of the assumed Lipschitz-continuity of ϕ^{-1}, the classical partial derivatives of $u \circ \phi^{-1}$ exist a.e., compare Remark 6.4(ii), and

$$\left| \frac{\partial}{\partial y^j} \left(u \circ \phi^{-1} \right) \right| = \left| \sum_{i=1}^{d} \left(\frac{\partial}{\partial x^i} u \right) \circ \phi^{-1} \frac{\partial}{\partial y^j} (\phi^{-1})^i \right| \leq C \sum_{i=1}^{d} \left| \left(\frac{\partial}{\partial x^i} u \right) \circ \phi^{-1} \right| \qquad \text{(a.e.)},$$

since the absolute values of the derivatives $\frac{\partial}{\partial y^j}(\phi^{-1})^i$, $j = 1, \ldots, d$, of the coordinates $(\phi^{-1})^i$, $i = 1, \ldots, d$, are bounded by the Lipschitz constant of ϕ^{-1}. Thus, applying again the change of variables formula for bi-Lipschitz transformations and the assumed equivalence of $\rho_{G^{(1)}}$ and $\rho_{G^{(2)}} \circ \phi$ on $G^{(1)}$, we can use the norm equivalence (2.28) together with the fact that, since $u \in \mathcal{C}_0^{\infty}(G^{(1)})$, the classical derivatives $\frac{\partial}{\partial x^i} u$ coincide with the generalized derivatives u_{x^i}, for $i \in \{1, \ldots, d\}$, and estimate

$$\int_{G^{(2)}} \left| \left(u \circ \phi^{-1} \right)(y) \right|^p \rho_{G^{(2)}}(y)^{\theta - d} \, dy + \sum_{j=1}^{d} \int_{G^{(2)}} \left| \frac{\partial}{\partial y^j} \left(u \circ \phi^{-1} \right)(y) \right|^p \rho_{G^{(2)}}(y)^{p+\theta-d} dy$$

$$\leq C \left(\int_{G^{(2)}} \left| \left(u \circ \phi^{-1} \right)(y) \right|^p \rho_{G^{(2)}}(y)^{\theta - d} \, dy + \int_{G^{(2)}} \sum_{i=1}^{d} \left| \left(\frac{\partial}{\partial x^i} u \right) (\phi^{-1}(y)) \right|^p \rho_{G^{(2)}}(y)^{p+\theta-d} dy \right)$$

$$\leq C \left(\int_{G^{(1)}} |u(x)|^p \rho_{G^{(1)}}(x)^{\theta - d} \, dx + \sum_{i=1}^{d} \int_{G^{(1)}} \left| \frac{\partial}{\partial x^i} u(x) \right|^p \rho_{G^{(1)}}(x)^{p+\theta-d} dx \right)$$

$$\leq C \|u\|_{H_{p,\theta}^1(G^{(1)})}^p.$$

By the norm equivalence (2.28), these calculations yield

$$\left\| u \circ \phi^{-1} \right\|_{H_{p,\theta}^1(G^{(2)})} \leq C \|u\|_{H_{p,\theta}^1(G^{(1)})},$$

if we can guarantee that for any $j \in \{1, \ldots, d\}$, the a.e. existing classical partial derivative $\frac{\partial}{\partial y^j}(u \circ \phi^{-1})$ is a version of the corresponding generalized derivative $(u \circ \phi^{-1})_{y^j}$. This can be deduced as follows: By the above calculations, $u \circ \phi^{-1}$ and $\frac{\partial}{\partial y^j}(u \circ \phi^{-1})$ are locally integrable functions on $G^{(2)}$. Furthermore, $u \circ \phi^{-1}$ is Lipschitz continuous. Thus, $u \circ \phi^{-1}$ has a Lipschitz continuous extension to $\mathbb{R}^d$, compare, e.g., Theorem 1 in [54, Section 3.1.1], which we also denote by $u \circ \phi^{-1}$. Moreover,

$$\mathbb{R} \ni y^j \mapsto u \circ \phi^{-1}(y^1, \ldots, y^{j-1}, y^j, y^{j+1}, \ldots, y^d)$$

is absolutely continuous on compact subsets of $\mathbb{R}$ for every $(y^1, \ldots, y^{j-1}, y^{j+1}, \ldots, y^d) \in \mathbb{R}^{d-1}$ (see [54, Section 4.9] for a precise definition of absolute continuity). Thus, as in the proof of Theorem 2 in [54, Section 4.9.2], we can do integration by parts and obtain for every $\varphi \in \mathcal{C}_0^\infty(G^{(2)})$,

$$\int_{G^{(2)}} \frac{\partial}{\partial y^j}(u \circ \phi^{-1})(y)\varphi(y)\,\mathrm{d}y = -\int_{G^{(2)}} (u \circ \phi^{-1})(y)\frac{\partial}{\partial y^j}\varphi(y)\,\mathrm{d}y.$$

Therefore,

$$\frac{\partial}{\partial y^j}(u \circ \phi^{-1}) = (u \circ \phi^{-1})_{y^j} \qquad \text{(a.e.)},$$

and the assertion for $\gamma = 1$ follows. Finally let us consider the case $\gamma = -1$. Assume for a moment that $u \in \mathcal{C}_0^\infty(G^{(1)})$. By the change of variables formula for bi-Lipschitz transformations, we have

$$(u \circ \phi^{-1}, \varphi) = (u, \varphi \circ \phi), \qquad \varphi \in \mathcal{C}_0^\infty(G^{(2)});$$

see also (2.38) in Remark 2.49 for the extended meaning of $(\cdot, \cdot)$. Using Lemma 2.45(viii), i.e., the fact that

$$H_{p,\theta}^{-1}(G^{(i)}) \simeq \left(H_{p',\theta'}^1(G^{(i)})\right)^*, \quad \text{with} \quad \frac{1}{p} + \frac{1}{p'} = 1 \quad \text{and} \quad \frac{\theta}{p} + \frac{\theta'}{p'} = d \qquad (i = 1, 2), \quad (6.10)$$

see also Remark 2.49, we obtain

$$\left|(u, \varphi \circ \phi)\right| \leq C \|u\|_{H_{p,\theta}^{-1}(G^{(1)})} \|\varphi \circ \phi\|_{H_{p',\theta'}^1(G^{(1)})}, \qquad \varphi \in \mathcal{C}_0^\infty(G^{(2)}).$$

Thus, an application of the already proven assertion for $\gamma = 1$, yields

$$\left|(u, \varphi \circ \phi)\right| \leq C \|u\|_{H_{p,\theta}^{-1}(G^{(1)})} \|\varphi\|_{H_{p',\theta'}^1(G^{(2)})}, \qquad \varphi \in \mathcal{C}_0^\infty(G^{(2)}).$$

Hence, by the density of $\mathcal{C}_0^\infty(G^{(2)})$ in $H_{p',\theta'}^1(G^{(2)})$, cf. Lemma 2.45(ii), we obtain

$$\left\|(u \circ \phi^{-1}, \cdot)\right\|_{\left(H_{p',\theta'}^1(G^{(2)})\right)^*} \leq C \|u\|_{H_{p,\theta}^{-1}(G^{(1)})}.$$

Applying (6.10), this shows that

$$\|u \circ \phi^{-1}\|_{H_{p,\theta}^{-1}(G^{(2)})} \leq C \|u\|_{H_{p,\theta}^{-1}(G^{(1)})}, \qquad u \in \mathcal{C}_0^\infty(G^{(1)}). \qquad (6.11)$$

Let us consider the general case and assume that $u \in H_{p,\theta}^{-1}(G^{(1)})$. We fix a sequence $(u_n)_{n \in \mathbb{N}}$ approximating u in $H_{p,\theta}^{-1}(G^{(1)})$, which exists by another application of Lemma 2.45(ii). By (6.11), $(u_n \circ \phi^{-1})_{n \in \mathbb{N}}$ is a Cauchy sequence in the Banach space $H_{p,\theta}^{-1}(G^{(2)})$ and, therefore, converges in $H_{p,\theta}^{-1}(G^{(2)})$. We set

$$u \circ \phi^{-1} := \lim_{n \to \infty}\left(u_n \circ \phi^{-1}\right), \qquad \text{(convergence in } H_{p,\theta}^{-1}(G^{(2)})\text{)}.$$

Then, Equality (6.9) holds, and, by (6.11),

$$\|u \circ \phi^{-1}\|_{H_{p,\theta}^{-1}(G^{(2)})} \leq C \|u\|_{H_{p,\theta}^{-1}(G^{(1)})}, \qquad u \in H_{p,\theta}^{-1}(G^{(1)}). \qquad \square$$

We use Lemma 6.3 to prove the following rule for bi-Lipschitz transformations of elements of $\mathfrak{H}_{p,\theta}^{1,q}(\mathcal{O}, T)$.

Lemma 6.5. *Let $G^{(1)}, G^{(2)}$ be bounded domains in $\mathbb{R}^d$ and let $\phi : G^{(1)} \to G^{(2)}$ satisfy the assumptions of Lemma 6.3. Furthermore, let $u \in \mathfrak{H}_{p,\theta}^{1,q}(G^{(1)},T)$ with $2 \le p \le q < \infty$ and $\theta \in \mathbb{R}$. Then $u \circ \phi^{-1} \in \mathfrak{H}_{p,\theta}^{1,q}(G^{(2)},T)$ with deterministic part $\mathbb{D}(u \circ \phi^{-1}) = \mathbb{D}u \circ \phi^{-1}$ and stochastic part $\mathbb{S}(u \circ \phi^{-1}) = \mathbb{S}u \circ \phi^{-1}$. In particular, for any $\varphi \in \mathcal{C}_0^\infty(G^{(2)})$, with probability one, the equality*

$$
\bigl(u(t,\cdot) \circ \phi^{-1}, \varphi\bigr) = \bigl(u(0,\cdot) \circ \phi^{-1}, \varphi\bigr)
$$
$$
+ \int_0^t \bigl((\mathbb{D}u)(s,\cdot) \circ \phi^{-1}, \varphi\bigr)\,\mathrm{d}s + \sum_{k=1}^\infty \int_0^t \bigl((\mathbb{S}^k u)(s,\cdot) \circ \phi^{-1}, \varphi\bigr)\,\mathrm{d}w_s^k \qquad (6.12)
$$

holds for all $t \in [0,T]$.

Proof. We set $f := \mathbb{D}u$ and $g := \mathbb{S}u$. Since $u \in \mathfrak{H}_{p,\theta}^{1,q}(G^{(1)},T)$, Lemma 6.3 guarantees that $u \circ \phi^{-1} \in \mathbb{H}_{p,\theta-p}^{1,q}(G^{(2)},T)$, $f \circ \phi^{-1} \in \mathbb{H}_{p,\theta+p}^{-1,q}(G^{(2)},T)$, $g \circ \phi^{-1} \in \mathbb{H}_{p,\theta}^{0,q}(G^{(2)},T;\ell_2)$ and $u(0) \circ \phi^{-1} \in U_{p,\theta}^{1,q}(G^{(2)})$. Thus, all the terms in formula (6.12) are well-defined. In particular, since (6.9) holds, showing that for any $\varphi \in \mathcal{C}_0^\infty(G^{(2)})$, with probability one, the equality

$$
\bigl(u(t,\cdot), \varphi \circ \phi\bigr) = \bigl(u(0,\cdot), \varphi \circ \phi\bigr)
$$
$$
+ \int_0^t \bigl((\mathbb{D}u)(s,\cdot), \varphi \circ \phi\bigr)\,\mathrm{d}s + \sum_{k=1}^\infty \int_0^t \bigl((\mathbb{S}^k u)(s,\cdot), \varphi \circ \phi\bigr)\,\mathrm{d}w_s^k \qquad (6.13)
$$

holds for all $t \in [0,T]$, proves our assertion (with the right meaning of the brackets $(\cdot,\cdot)$, cf. Remark 2.49). We consider two different cases.

Case 1. Firstly, we assume that $p > 2$. Let us fix $\varphi \in \mathcal{C}_0^\infty(G^{(2)})$. By Lemma 6.3, $\varphi \circ \phi \in H_{\tilde{p},\tilde{\theta}-\tilde{p}}^1(G^{(1)})$ for any $\tilde{p} \in (1,\infty)$ and $\tilde{\theta} \in \mathbb{R}$, hence also for

$$
\tilde{p} := \frac{2p}{p-2}, \quad \text{i.e.,} \quad \tilde{p} \quad \text{fulfilling} \quad \frac{2}{p} + \frac{1}{\tilde{p}} = 1,
$$

and

$$
\tilde{\theta} := \theta'\left(1 + \frac{p}{p-2}\right) - d\frac{p}{p-2}, \quad \text{where} \quad \frac{\theta}{p} + \frac{\theta'}{p'} = d \quad \text{with} \quad \frac{1}{p} + \frac{1}{p'} = 1.
$$

Moreover, by Lemma 2.45(ii) we can choose a sequence $\tilde{\varphi}_n \subseteq \mathcal{C}_0^\infty(G^{(1)})$ approximating $\varphi \circ \phi$ in $H_{\tilde{p},\tilde{\theta}-\tilde{p}}^1(G^{(1)})$. We know that for all $n \in \mathbb{N}$, with probability one, the equality

$$
\bigl(u(t,\cdot), \tilde{\varphi}_n\bigr) = \bigl(u(0,\cdot), \tilde{\varphi}_n\bigr) + \int_0^t \bigl(f(s,\cdot), \tilde{\varphi}_n\bigr)\,\mathrm{d}s + \sum_{k=1}^\infty \int_0^t \bigl(g^k(s,\cdot), \tilde{\varphi}_n\bigr)\,\mathrm{d}w_s^k \qquad (6.14)
$$

holds for all $t \in [0,T]$. Thus, if we can show that each side of (6.14) converges in $L_2(\Omega; \mathcal{C}([0,T]))$ to the respective side of (6.13), the assertion follows. We write

$$
\tilde{v}_n := \tilde{\varphi}_n - \varphi \circ \phi \qquad \text{for } n \in \mathbb{N},
$$

and start with the right hand side. We estimate

$$
\mathbb{E}\left[\sup_{t \in [0,T]} \left|\bigl(u(0,\cdot), \tilde{v}_n\bigr) + \int_0^t \bigl(f(s,\cdot), \tilde{v}_n\bigr)\,\mathrm{d}s + \sum_{k=1}^\infty \int_0^t \bigl(g^k(s,\cdot), \tilde{v}_n\bigr)\,\mathrm{d}w_s^k\right|^2\right]
$$
$$
\le C\left(\mathbb{E}\left[\bigl|\bigl(u(0,\cdot), \tilde{v}_n\bigr)\bigr|^2\right] + \mathbb{E}\left[\sup_{t \in [0,T]} \left|\int_0^t \bigl(f(s,\cdot), \tilde{v}_n\bigr)\,\mathrm{d}s\right|^2\right] + \right. \qquad (6.15)
$$
$$
\left. \mathbb{E}\left[\sup_{t \in [0,T]} \left|\sum_{k=1}^\infty \int_0^t \bigl(g^k(s,\cdot), \tilde{v}_n\bigr)\,\mathrm{d}w_s^k\right|^2\right]\right)
$$
$$
=: C\bigl(I + I\!I + I\!I\!I\bigr),
$$

and prove that each of the terms on the right hand side converges to zero for $n \to \infty$. Before we do this, we show that the following embeddings hold:

$$H^1_{\tilde{p}, \tilde{\theta} - \tilde{p}}(G^{(1)}) \hookrightarrow L_{\tilde{p}, \tilde{\theta}}(G^{(1)}), \tag{6.16}$$

$$H^1_{\tilde{p}, \tilde{\theta} - \tilde{p}}(G^{(1)}) \hookrightarrow H^1_{p', \theta' - p'}(G^{(1)}), \tag{6.17}$$

$$H^1_{\tilde{p}, \tilde{\theta} - \tilde{p}}(G^{(1)}) \hookrightarrow L_{p', \theta'}(G^{(1)}). \tag{6.18}$$

The first one follows immediately from Lemma 2.45(vii). In order to prove the second embedding, we argue as follows: Using the fact that

$$\tilde{p} = p'\left(1 + \frac{p}{p-2}\right) \qquad \text{and} \qquad p' < \tilde{p},$$

together with Hölder's inequality, the boundedness of $G^{(1)}$, and the norm equivalence (2.28) yields

$$
\|\tilde{v}\|_{H^1_{p', \theta' - p'}(G^{(1)})} \leq C\left(\int_{G^{(1)}} |\tilde{v}(x)|^{p'} \rho(x)^{\theta' - p' - d}\right)^{\frac{1}{p'}} + C\sum_{|\alpha|=1}\left(\int_{G^{(1)}} |D^\alpha \tilde{v}(x)|^{p'} \rho(x)^{\theta' - d}\right)^{\frac{1}{p'}}
$$
$$
\leq C\left(\int_{G^{(1)}} |\tilde{v}(x)|^{\tilde{p}} \rho(x)^{\tilde{\theta} - \tilde{p} - d}\right)^{\frac{1}{\tilde{p}}} + C\sum_{|\alpha|=1}\left(\int_{G^{(1)}} |D^\alpha \tilde{v}(x)|^{\tilde{p}} \rho(x)^{\tilde{\theta} - d}\right)^{\frac{1}{\tilde{p}}}
$$
$$
\leq C\|\tilde{v}\|_{H^1_{\tilde{p}, \tilde{\theta} - \tilde{p}}(G^{(1)})},
$$

with a constant C independent of $\tilde{v} \in H^1_{\tilde{p}, \tilde{\theta} - \tilde{p}}(G^{(1)})$. The third embedding (6.18) follows with similar arguments. Let us return to (6.15). Since $(L_{p,\theta}(G^{(1)}))^* \simeq L_{p', \theta'}(G^{(1)})$, cf. Lemma 2.45(viii), using embedding (6.18) together with the fact that $H^{1-2/q}_{p, \theta - (1-2/q)p}(G^{(1)}) \hookrightarrow L_{p,\theta}(G^{(2)})$, we obtain

$$
I = \mathbb{E}\left[|(u(0,\cdot), \tilde{v}_n)|^2\right] \leq \mathbb{E}\left[\|u(0,\cdot)\|^2_{L_{p,\theta}(G^{(1)})}\right]\|\tilde{v}_n\|^2_{L_{p', \theta'}(G^{(1)})}
$$
$$
\leq C\|u(0,\cdot)\|^2_{U^{\gamma, q}_{p, \theta}(G^{(1)})}\|\tilde{v}_n\|^2_{H^1_{\tilde{p}, \tilde{\theta} - \tilde{p}}(G^{(1)})}. \tag{6.19}
$$

Furthermore, since $(H^{-1}_{p, \theta + p}(G^{(1)}))^* \simeq H^1_{p', \theta' - p'}(G^{(1)})$, cf. Lemma 2.45(viii), we can use embedding (6.17) together with Hölder's inequality and estimate the second term as follows:

$$
I\!\!I = \mathbb{E}\left[\sup_{t \in [0,T]}\left|\int_0^t (f(s,\cdot), \tilde{v}_n)\, ds\right|^2\right]
$$
$$
\leq C\,\mathbb{E}\left[\sup_{t \in [0,T]}\int_0^t \|f(s,\cdot)\|^2_{H^{-1}_{p, \theta + p}(G^{(1)})}\, ds\right]\|\tilde{v}_n\|^2_{H^1_{p', \theta' - p'}(G^{(1)})} \tag{6.20}
$$
$$
\leq C\|f\|^2_{\mathbb{H}^{-1, q}_{p, \theta + p}(G^{(1)}, T)}\|\tilde{v}_n\|^2_{H^1_{\tilde{p}, \tilde{\theta} - \tilde{p}}(G^{(1)})}.
$$

Finally, by Doob's inequality and Itô's isometry, together with Jensen's inequality and Fubini's theorem,

$$
\begin{aligned}
\mathit{III} &= \mathbb{E}\left[\sup_{t\in[0,T]}\left|\sum_{k=1}^{\infty}\int_0^t\left(g^k(s,\cdot),\tilde{v}_n\right)dw_s^k\right|^2\right] \\
&\leq C\,\mathbb{E}\left[\int_0^T\sum_{k=1}^{\infty}\left|\left(g^k(s,\cdot),\tilde{v}_n\right)\right|^2 ds\right] \\
&= C\,\mathbb{E}\left[\int_0^T\sum_{k=1}^{\infty}\left|\int_{G^{(1)}}g^k(s,x)\tilde{v}_n(x)\,dx\right|^2 ds\right] \\
&\leq C\,\mathbb{E}\left[\int_0^T\int_{G^{(1)}}\sum_{k=1}^{\infty}\left|g^k(s,x)\right|^2\left|\tilde{v}_n(x)\right|^2 dx\,ds\right].
\end{aligned}
$$

Thus, inserting $1 = \rho^{2(\theta-d)/p}\rho^{2(\theta'-d)/p'}$ and using Hölder's inequality twice, followed by an application of Embedding (6.16), yields

$$
\begin{aligned}
\mathit{III} &\leq C\,\mathbb{E}\left[\int_0^T\left(\int_{G^{(1)}}\left(\sum_{k=1}^{\infty}\left|g^k(s,x)\right|^2\right)^{\frac{p}{2}}\rho(x)^{\theta-d}\,dx\right)^{\frac{2}{p}}ds\right]\left(\int_{G^{(1)}}\left|\tilde{v}_n(x)\right|^{\tilde{p}}\rho(x)^{\tilde{\theta}-d}\,dx\right)^{\frac{1}{\tilde{p}}} \\
&\leq C\,\|g\|_{\mathbb{H}_{p,\theta}^{0,q}(G^{(1)},T;\ell_2)}^2\|\tilde{v}_n\|_{H_{\tilde{p},\tilde{\theta}-\tilde{p}}^1(G^{(1)})}^2.
\end{aligned}
\tag{6.21}
$$

The combination of the estimates (6.19), (6.20) and (6.21) with (6.15) yields the convergence of the right hand side of (6.14) to the right hand side of (6.13) in $L_2(\Omega;\mathcal{C}([0,T]))$. Let us now consider the corresponding left hand sides. An application of Theorem 3.8(ii) and the fact that $q\geq p$ lead to

$$
\begin{aligned}
\mathbb{E}\left[\sup_{t\in[0,T]}\left|\left(u(t,\cdot),\tilde{v}_n\right)\right|^2\right] &\leq \left(\mathbb{E}\left[\sup_{t\in[0,T]}\|u(t,\cdot)\|_{L_{p,\theta}(G^{(1)})}^p\right]\right)^{\frac{2}{p}}\|\tilde{v}_n\|_{L_{p',\theta'}(C^{(1)})}^2 \\
&\leq C\|u\|_{\mathfrak{H}_{p,\theta}^{1,p}(G^{(1)},T)}^2\|\tilde{v}_n\|_{L_{p',\theta'}(G^{(1)})}^2 \\
&\leq C\|u\|_{\mathfrak{H}_{p,\theta}^{1,q}(G^{(1)},T)}^2\|\tilde{v}_n\|_{L_{p',\theta'}(G^{(1)})}^2.
\end{aligned}
$$

Thus, by (6.18),

$$
\mathbb{E}\left[\sup_{t\in[0,T]}\left|\left(u(t,\cdot),\tilde{v}_n\right)\right|^2\right] \leq C\|u\|_{\mathfrak{H}_{p,\theta}^{1,q}(G^{(1)},T)}^2\|\tilde{v}_n\|_{H_{\tilde{p},\tilde{\theta}-\tilde{p}}^1(G^{(1)})}^2.
$$

Hence, also the left hand side of (6.14) converges to the left hand side of (6.13) in $L_2(\Omega;\mathcal{C}([0,T]))$ and the assertion is proved for $p > 2$.

Case 2. It remains to consider the case $p = 2$. Replacing $\tilde{p}$ by 2 and $\tilde{\theta}$ by $\theta' = 2d - \theta$ and arguing as in the first case using the inequality

$$
\mathbb{E}\left[\sup_{t\in[0,T]}\left|\sum_{k=1}^{\infty}\int_0^t\left(g^k(s,\cdot),\tilde{v}\right)dw_s^k\right|^2\right] \leq C\|g\|_{\mathbb{H}_{2,\theta}^{0,q}(G^{(1)},T;\ell_2)}^2\|\tilde{v}\|_{L_{2,\theta'}(G^{(1)})}^2
$$

for the estimate of III, proves the assertion also for $p = 2$. $\qquad\square$

Now we are ready to prove our main result in this section.

Proof of Theorem 6.1. As before, we simplify notation and write $f := \mathbb{D}u$ and $g := \mathbb{S}u$ throughout the proof. We will show that (6.5) is true by induction over $\gamma \in \mathbb{N}$; estimate (6.6) can be proven analogously.

We start with the case $\gamma = 1$. Fix $x_0 \in \partial\mathcal{O}$ and choose $r > 0$ small enough, e.g., $r := r_0(10K_0)^{-1}$ with r_0 and $K_0 > 1$ from Definition 2.1. Let us assume for a moment that the supports (in the sense of distributions) of u, f and g are contained in $B_r(x_0)$ for each t and ω. With μ_0 from Definition 2.1, we introduce the function

$$\phi: G^{(1)} := \mathcal{O} \cap B_{r_0}(x_0) \longrightarrow G^{(2)} := \phi(\mathcal{O} \cap B_{r_0}(x_0)) \subseteq \mathbb{R}_+^d$$
$$x = (x^1, x') \longmapsto \phi(x) := (x^1 - \mu_0(x'), x'),$$

which fulfils all the assumptions of Lemma 6.3. Note that, since r has been chosen sufficiently small, one has $\rho_{\mathcal{O}}(x) = \rho_{G^{(1)}}(x)$ for all $x \in \mathcal{O} \cap B_r(x_0)$, so that one can easily show that the equivalence

$$\|v\|_{H_{\tilde{p},\tilde{\theta}}^{\nu}(\mathcal{O})} \asymp \|v\|_{H_{\tilde{p},\tilde{\theta}}^{\nu}(G^{(1)})}, \qquad v \in \mathcal{D}'(\mathcal{O}), \ \mathrm{supp}\, v \subseteq B_r(x_0),$$

holds for all ν, $\tilde{\theta} \in \mathbb{R}$ and $\tilde{p} > 1$. Together with Lemma 6.3 we obtain for any $\nu \in [-1, 1]$,

$$\|v\|_{H_{\tilde{p},\tilde{\theta}}^{\nu}(\mathcal{O})} \asymp \|v \circ \phi^{-1}\|_{H_{\tilde{p},\tilde{\theta}}^{\nu}(G^{(2)})}, \qquad v \in \mathcal{D}'(\mathcal{O}), \ \mathrm{supp}\, v \subseteq B_r(x_0).$$

Thus, denoting $\tilde{u} := u \circ \phi^{-1}$, $\tilde{f} := f \circ \phi^{-1}$ and $\tilde{g} := g \circ \phi^{-1}$, by Lemma 6.5 we know that on $G^{(2)}$ we have $d\tilde{u} = \tilde{f}dt + \tilde{g}^k dw_t^k$ in the sense of distributions, see Definition 3.3. Furthermore, since $\rho_{G^{(2)}}(y) = \rho_{\mathbb{R}_+^d}(y)$ for all $y \in \phi(\mathcal{O} \cap B_r(x_0))$, the equivalence

$$\|v \circ \phi^{-1}\|_{H_{\tilde{p},\tilde{\theta}}^{\nu}(G^{(2)})} \asymp \|v \circ \phi^{-1}\|_{H_{\tilde{p},\tilde{\theta}}^{\nu}(\mathbb{R}_+^d)}, \qquad v \in \mathcal{D}'(\mathcal{O}), \ \mathrm{supp}\, v \subseteq B_r(x_0),$$

holds for any $\nu \in [-1, 1]$, where we identify $v \circ \phi^{-1}$ with its extension to $\mathbb{R}_+^d$ by zero. Therefore, by making slight abuse of notation and writing $\tilde{u}$, $\tilde{f}$ and $\tilde{g}$ for the extension by zero on $\mathbb{R}_+^d$ of $\tilde{u}$, $\tilde{f}$ and $\tilde{g}$ respectively, we have

$$\tilde{u} \in \mathbb{H}_{p,\theta-p}^{1,q}(\mathbb{R}_+^d, T), \quad \tilde{u}(0) \in U_{p,\theta}^{1,q}(\mathbb{R}_+^d), \quad \tilde{f} \in \mathbb{H}_{p,\theta+p}^{-1,q}(\mathbb{R}_+^d, T), \quad \tilde{g} \in \mathbb{H}_{p,\theta}^{0,q}(\mathbb{R}_+^d, T; \ell_2),$$

and $d\tilde{u} = \tilde{f}dt + \tilde{g}^k dw_t^k$ is fulfilled on $\mathbb{R}_+^d$ in the sense of distributions, see Definition 3.3. Thus, we can apply [83, Theorem 4.1] and use the equivalences above to obtain Estimate (6.5) in the following way:

$$\mathbb{E}[u]_{\mathcal{C}^{\tilde{\beta}/2-1/q}([0,T];H_{p,\theta-(1-\beta)p}^{1-\beta}(\mathcal{O}))}^q$$
$$\leq C\, \mathbb{E}[\tilde{u}]_{\mathcal{C}^{\tilde{\beta}/2-1/q}([0,T];H_{p,\theta+p(\beta-1)}^{1-\beta}(\mathbb{R}_+^d))}^q$$
$$\leq C\, T^{(\beta-\tilde{\beta})q/2}\left(\|\tilde{u}\|_{\mathbb{H}_{p,\theta-p}^{1,q}(\mathbb{R}_+^d, T)}^q + \|\tilde{f}\|_{\mathbb{H}_{p,\theta+p}^{-1,q}(\mathbb{R}_+^d, T)}^q + \|\tilde{g}\|_{\mathbb{H}_{p,\theta}^{0,q}(\mathbb{R}_+^d, T; \ell_2)}^q\right)$$
$$\leq C\, T^{(\beta-\tilde{\beta})q/2}\left(\|u\|_{\mathbb{H}_{p,\theta-p}^{1,q}(\mathcal{O}, T)}^q + \|f\|_{\mathbb{H}_{p,\theta+p}^{-1,q}(\mathcal{O}, T)}^q + \|g\|_{\mathbb{H}_{p,\theta}^{0,q}(\mathcal{O}, T; \ell_2)}^q\right).$$

Now let us give up the assumption on the supports of u, f and g. Let $\xi_0, \xi_1, \ldots, \xi_m$, be a finite partition of unity on $\mathcal{O}$, such that $\xi_0 \in \mathcal{C}_0^{\infty}(\mathcal{O})$, and, for $i = 1, \ldots, m$, $\xi_i \in \mathcal{C}_0^{\infty}(B_r(x_i))$ with $x_i \in \partial\mathcal{O}$. Obviously, $\mathrm{d}(\xi_i u) = \xi_i f dt + \xi_i g_t^k dw_t^k$ for $i = 0, \ldots, m$. Since

$$\mathbb{E}[u]_{\mathcal{C}^{\tilde{\beta}/2-1/q}([0,T];H_{p,\theta-(1-\beta)p}^{1-\beta}(\mathcal{O}))}^q \leq C(m,q) \sum_{i=0}^{m} \mathbb{E}[(\xi_i u)]_{\mathcal{C}^{\tilde{\beta}/2-1/q}([0,T];H_{p,\theta-(1-\beta)p}^{1-\beta}(\mathcal{O}))}^q,$$

we just have to estimate $\mathbb{E}[\xi_i u]^q_{\mathcal{C}^{\tilde{\beta}/2-1/q}([0,T];H^{1-\beta}_{p,\theta-(1-\beta)p}(\mathcal{O}))}$ for each $i \in \{0,\dots,m\}$. For $i \geq 1$ one obtains the required estimate as before, using the fact that $\mathcal{C}_0^\infty(\mathcal{O})$-functions are pointwise multipliers in all spaces $H^\nu_{\tilde{p},\tilde{\theta}}(\mathcal{O})$, $\nu,\tilde{\theta} \in \mathbb{R}$, $\tilde{p} > 1$, see, e.g., [93, Theorem 3.1]. The case $i = 0$ can be treated as follows: Since ξ_0 has compact support in $\mathcal{O}$, for all $\nu,\tilde{\theta} \in \mathbb{R}$ and $\tilde{p} > 1$, we have

$$\|v\xi_0\|_{H^\nu_{\tilde{p},\tilde{\theta}}(\mathcal{O})} \asymp \|v\xi_0\|_{H^\nu_{\tilde{p}}(\mathbb{R}^d)}, \qquad v \in \mathcal{D}'(\mathcal{O}), \tag{6.22}$$

and consequently

$$\mathbb{E}[\xi_0 u]^q_{\mathcal{C}^{\tilde{\beta}/2-1/q}([0,T];H^{1-\beta}_{p,\theta-(1-\beta)p}(\mathcal{O}))} \asymp \mathbb{E}[\xi_0 u]^q_{\mathcal{C}^{\tilde{\beta}/2-1/q}([0,T];H^{1-\beta}_p(\mathbb{R}^d))}.$$

By [83, Theorem 4.11], a further application of (6.22) and the fact that $\mathcal{C}_0^\infty(\mathcal{O})$-functions are pointwise multipliers in all spaces $H^\nu_{\tilde{p},\tilde{\theta}}(\mathcal{O})$, we obtain

$$\mathbb{E}[\xi_0 u]^q_{\mathcal{C}^{\tilde{\beta}/2-1/q}([0,T];H^{1-\beta}_p(\mathbb{R}^d))}$$
$$\leq CT^{(\beta-\tilde{\beta})q/2}\Big(\|\xi_0 u\|^q_{\mathbb{H}^{1,q}_p(\mathbb{R}^d,T)} + \|\xi_0 f\|^q_{\mathbb{H}^{-1,q}_p(\mathbb{R}^d,T)} + \|\xi_0 g\|^q_{\mathbb{H}^{0,q}_p(\mathbb{R}^d,T;\ell_2)}\Big)$$
$$\leq CT^{(\beta-\tilde{\beta})q/2}\Big(\|\xi_0 u\|^q_{\mathbb{H}^{1,q}_{p,\theta-p}(\mathcal{O},T)} + \|\xi_0 f\|^q_{\mathbb{H}^{-1,q}_{p,\theta+p}(\mathcal{O},T)} + \|\xi_0 g\|^q_{\mathbb{H}^{0,q}_{p,\theta}(\mathcal{O},T;\ell_2)}\Big)$$
$$\leq CT^{(\beta-\tilde{\beta})q/2}\Big(\|u\|^q_{\mathbb{H}^{1,q}_{p,\theta-p}(\mathcal{O},T)} + \|f\|^q_{\mathbb{H}^{-1,q}_{p,\theta+p}(\mathcal{O},T)} + \|g\|^q_{\mathbb{H}^{0,q}_{p,\theta}(\mathcal{O},T;\ell_2)}\Big).$$

This finishes the proof of estimate (6.5) for the case $\gamma = 1$.

Next, let us move to the inductive step and assume that the assertion is true for some $\gamma = n \in \mathbb{N}$. Fix $u \in \mathfrak{H}_{p,\theta}^{n+1,q}(\mathcal{O},T)$ and let ψ denote an infinitely differentiable function on $\mathcal{O}$ fulfilling (2.25). Then $v := \psi u_x \in \mathfrak{H}_{p,\theta}^{n,q}(\mathcal{O},T)$ and $dv = \psi f_x dt + \psi g_x^k dw_t^k$ (component-wise). Also, by Lemma 2.45(iii) and (iv),

$$\mathbb{E}[u]^q_{\mathcal{C}^{\tilde{\beta}/2-1/q}([0,T];H^{n+1-\beta}_{p,\theta-(1-\beta)p}(\mathcal{O}))} \leq C\Big(\mathbb{E}[u]^q_{\mathcal{C}^{\tilde{\beta}/2-1/q}([0,T];H^{n-\beta}_{p,\theta-(1-\beta)p}(\mathcal{O}))} +$$
$$\mathbb{E}[v]^q_{\mathcal{C}^{\tilde{\beta}/2-1/q}([0,T];H^{n-\beta}_{p,\theta-(1-\beta)p}(\mathcal{O}))}\Big).$$

Using the induction hypothesis and applying Lemma 2.45(iii) and (iv) once more together with Lemma 2.55(i) and (ii), we see that the induction goes through. $\qquad\square$

6.2 The spaces $\mathfrak{H}_{p,\theta}^{\gamma,q}(\mathcal{O},T)$ and SPDEs

In this section we are concerned with one aspect of the 'suitability' of the spaces $\mathfrak{H}_{p,\theta}^{\gamma,q}(\mathcal{O},T)$ for the regularity analysis of SPDEs. We prove that, if we know that the equation

$$\left.\begin{array}{l} du = \left(a^{ij}u_{x^i x^j} + f\right) dt + \left(\sigma^{ik}u_{x^i} + g^k\right) dw_t^k \quad \text{on } \Omega_T \times \mathcal{O}, \\ u(0) = 0 \quad \text{on } \Omega \times \mathcal{O}. \end{array}\right\} \tag{6.23}$$

has a solution $u \in \mathfrak{H}_{p,\theta}^{\gamma,q}(\mathcal{O},T)$, and f and $g = (g^k)_{k\in\mathbb{N}}$ are smooth enough, then we can lift up the regularity of the solution in the scale $\mathfrak{H}_{p,\theta}^{\nu,q}(\mathcal{O},T)$, $\nu \geq \gamma$, of parabolic weighted Sobolev spaces. For simplicity, in this section we make the following restrictions.

Assumption 6.6. (i) The coefficients a^{ij} and σ^{ik} do not depend on $x \in \mathcal{O}$ and fulfil Assumption 3.1 with vanishing b^i, c, and μ^k, $i, j \in \{1, \ldots, d\}$ and $k \in \mathbb{N}$. That is, a^{ij} and σ^{ik} are real valued predictable stochastic processes and there exist constants $\delta_0, K > 0$ such that for any $(\omega, t) \in \Omega_T$ and $\lambda \in \mathbb{R}^d$,

$$\delta_0 |\lambda|^2 \leq \tilde{a}^{ij}(\omega, t) \lambda^i \lambda^j \leq a^{ij}(\omega, t) \lambda^i \lambda^j \leq K|\lambda|^2,$$

where $\tilde{a}^{ij}(\omega, t) := a^{ij}(\omega, t) - \frac{1}{2} \langle \sigma^{i\cdot}(\omega, t), \sigma^{j\cdot}(\omega, t) \rangle_{\ell_2}$, with $\sigma^{i\cdot}(\omega, t) := \left(\sigma^{ik}(\omega, t) \right)_{k \in \mathbb{N}} \in \ell_2$.

(ii) (a^{ij}) is symmetric, i.e., $a^{ij} = a^{ji}$ for $i, j \in \{1, \ldots, d\}$.

Under these assumptions we can prove the following result.

Theorem 6.7. *Let $\mathcal{O}$ be a bounded Lipschitz domain in $\mathbb{R}^d$ and let a^{ij} and σ^{ik}, $i, j \in \{1, \ldots, d\}$, $k \in \mathbb{N}$, be given coefficients satisfying Assumption 6.6. Fix $\gamma \in \mathbb{R}$, $p \in [2, \infty)$ and $q := mp$ for some $m \in \mathbb{N}$. Furthermore, assume that $u \in \mathfrak{H}_{p,\theta}^{\gamma+1,q}(\mathcal{O}, T)$ is a solution to Eq. (6.23) with $f \in \mathbb{H}_{p,\theta+p}^{\gamma,q}(\mathcal{O}, T)$ and $g \in \mathbb{H}_{p,\theta}^{\gamma+1,q}(\mathcal{O}, T; \ell_2)$. Then, $u \in \mathfrak{H}_{p,\theta}^{\gamma+2,q}(\mathcal{O}, T)$, and*

$$\|u\|_{\mathbb{H}_{p,\theta-p}^{\gamma+2,q}(\mathcal{O},T)}^q \leq C \left(\|u\|_{\mathbb{H}_{p,\theta-p}^{\gamma+1,q}(\mathcal{O},T)}^q + \|f\|_{\mathbb{H}_{p,\theta+p}^{\gamma,q}(\mathcal{O},T)}^q + \|g\|_{\mathbb{H}_{p,\theta}^{\gamma+1,q}(\mathcal{O},T;\ell_2)}^q \right),$$

where the constant $C \in (0, \infty)$ does not depend on u, f and g.

In order to prove this result, we will use the following lemma taken from [82, Lemma 2.3]. Recall that the spaces $\mathcal{H}_p^\gamma(T)$ are the $\mathbb{R}^d$-counterparts of the spaces $\mathfrak{H}_{p,\theta}^{\gamma,p}(G, T)$, compare Remark 3.4.

Lemma 6.8. *Let $p \geq 2$, $m \in \mathbb{N}$, and, for $i = 1, 2, \ldots, m$,*

$$\lambda_i \in (0, \infty), \quad \gamma_i \in \mathbb{R}, \quad u^{(i)} \in \mathcal{H}_p^{\gamma_i+2}(T), \quad u^{(i)}(0, \cdot) = 0.$$

Denote $\Lambda_i := (\lambda_i - \Delta)^{\gamma_i/2}$. Then

$$\mathbb{E}\left[\int_0^T \prod_{i=1}^m \|\Lambda_i \Delta u^{(i)}\|_{L_p}^p \, dt \right]$$

$$\leq C \sum_{i=1}^m \mathbb{E}\left[\int_0^T \left(\|\Lambda_i f^{(i)}\|_{L_p}^p + \|\Lambda_i g_x^{(i)}\|_{L_p(\mathbb{R}^d;\ell_2)}^p \right) \prod_{\substack{j=1 \\ j \neq i}}^m \|\Lambda_j \Delta u^{(j)}\|_{L_p}^p \, dt \right]$$

$$+ C \sum_{1 \leq i < j \leq m} \mathbb{E}\left[\int_0^T \|\Lambda_i g_x^{(i)}\|_{L_p(\mathbb{R}^d;\ell_2)}^p \|\Lambda_j g_x^{(j)}\|_{L_p(\mathbb{R}^d;\ell_2)}^p \prod_{\substack{k=1 \\ k \neq i,j}}^m \|\Lambda_k \Delta u^{(k)}\|_{L_p}^p \, dt \right],$$

where $f^{(i)} := \mathbb{D}u^{(i)} - a^{rs} u_{x^r x^s}^{(i)}$, $g^{(i)k} := \mathbb{S}^k u^{(i)} - \sigma^{rk} u_{x^r}^{(i)}$ and $L_p(\ell_2) := H_p^0(\ell_2)$. The constant C depends only on m, d, p, δ_0, and K.

Now we prove the main result of this section.

Proof of Theorem 6.7. The case $m = 1$, i.e., $p = q$, is covered by [75, Lemma 3.2]. Therefore, let $m \geq 2$. According to Remark 2.46 it is enough to show that

$$\|\Delta u\|_{\mathbb{H}_{p,\theta+p}^{\gamma,q}(\mathcal{O},T)}^q \leq C \left(\|u\|_{\mathbb{H}_{p,\theta-p}^{\gamma+1,q}(\mathcal{O},T)}^q + \|f\|_{\mathbb{H}_{p,\theta+p}^{\gamma,q}(\mathcal{O},T)}^q + \|g\|_{\mathbb{H}_{p,\theta}^{\gamma+1,q}(\mathcal{O},T;\ell_2)}^q \right).$$

Using the definition of weighted Sobolev spaces from Subsection 2.3.3, we observe that

$$\|\Delta u\|_{\mathbb{H}_{p,\theta+p}^{\gamma,q}(\mathcal{O},T)}^q = \mathbb{E}\left[\int_0^T \left(\sum_{n\in\mathbb{Z}} e^{n(\theta+p)}\|(\zeta_{-n}\Delta u(t))(e^n\cdot)\|_{H_p^\gamma}^p\right)^m dt\right]$$

$$\leq C\,\mathbb{E}\left[\int_0^T \left(\sum_{n\in\mathbb{Z}} e^{n(\theta+p)}\left(\|\Delta(\zeta_{-n}u(t))(e^n\cdot)\|_{H_p^\gamma}^p\right.\right.\right.$$

$$\left.\left.\left. + \|(\Delta\zeta_{-n}u(t))(e^n\cdot)\|_{H_p^\gamma}^p + \|(\zeta_{-nx}u_x(t))(e^n\cdot)\|_{H_p^\gamma}^p\right)\right)^m dt\right].$$

(Here $\zeta_{-nx}u_x$ is meant to be the scalar product in $\mathbb{R}^d$.) Now we can use Jensen's inequality and Remark 2.48(i) to obtain

$$\|\Delta u\|_{\mathbb{H}_{p,\theta+p}^{\gamma,q}(\mathcal{O},T)}^q \leq C\,\mathbb{E}\left[\int_0^T \left(\sum_{n\in\mathbb{Z}} e^{n(\theta+p)}\|\Delta(\zeta_{-n}u(t))(e^n\cdot)\|_{H_p^\gamma}^p\right)^m\right.$$

$$\left. + \|u(t)\|_{H_{p,\theta-p}^\gamma(\mathcal{O})}^q + \|u_x(t)\|_{H_{p,\theta}^\gamma(\mathcal{O})}^q\, dt\right].$$

An application of Lemma 2.45(iii) and (iv) leads to

$$\|\Delta u\|_{\mathbb{H}_{p,\theta+p}^{\gamma,q}(\mathcal{O},T)}^q \leq C\,\mathbb{E}\left[\int_0^T \left(\sum_{n\in\mathbb{Z}} e^{n(\theta+p)}\|\Delta(\zeta_{-n}u(t))(e^n\cdot)\|_{H_p^\gamma}^p\right)^m dt\right] + C\,\|u\|_{\mathbb{H}_{p,\theta-p}^{\gamma+1,q}(\mathcal{O},T)}^q.$$

Therefore, it is enough to estimate the first term on the right hand side, i.e.,

$$\mathbb{E}\left[\int_0^T \left(\sum_{n\in\mathbb{Z}} e^{n(\theta+p)}\|\Delta(\zeta_{-n}u(t))(e^n\cdot)\|_{H_p^\gamma}^p\right)^m dt\right]$$

$$= \mathbb{E}\left[\int_0^T \sum_{n_1,\ldots,n_m\in\mathbb{Z}} e^{\left(\sum_{i=1}^m n_i\right)(\theta+p)}\prod_{i=1}^m \|\Delta(\zeta_{-n_i}u(t))(e^{n_i}\cdot)\|_{H_p^\gamma}^p\, dt\right].$$

Tonelli's theorem together with the relation

$$\|u(c\cdot)\|_{H_p^\gamma}^p = c^{p\gamma-d}\|(c^{-2}-\Delta)^{\gamma/2}u\|_{L_p}^p \quad \text{for } c\in(0,\infty), \tag{6.24}$$

applied to $\Delta u^{(n_i)}$ with $u^{(n)} := \zeta_{-n}u$ for $n\in\mathbb{Z}$, show that we only have to handle

$$\sum_{n_1,\ldots,n_m\in\mathbb{Z}} e^{\left(\sum_{i=1}^m n_i\right)(\theta+p+p\gamma-d)}\,\mathbb{E}\left[\int_0^T \prod_{i=1}^m \|(e^{-2n_i}-\Delta)^{\gamma/2}\Delta u^{(n_i)}(t)\|_{L_p}^p\, dt\right].$$

Note that since $u\in\mathfrak{H}_{p,\theta}^{\gamma+1,q}(\mathcal{O},T)$ solves Eq. (6.23) with vanishing initial value,

$$du^{(n)} = (a^{rs}u_{x^r x^s}^{(n)} + f^{(n)})\,dt + (\sigma^{rk}u_{x^r}^{(n)} + g^{(n)k})\,dw_t^k,$$

in the sense of distributions on $\mathbb{R}^d$, see Definition 3.3, where

$$f^{(n)} = -2a^{rs}(\zeta_{-n})_{x^s}u_{x^r} - a^{rs}(\zeta_{-n})_{x^r x^s}u + \zeta_{-n}f, \qquad g^{(n)k} = -\sigma^{rk}(\zeta_{-n})_{x^r}u + \zeta_{-n}g^k,$$

and $u^{(n)}(0) = 0$. Furthermore, applying [80, Theorem 4.10], we have $u^{(n)}\in\mathcal{H}_p^{\gamma+2}(T)$. Thus, we can use Lemma 6.8 to obtain

$$\mathbb{E}\left[\int_0^T \prod_{i=1}^m \|(e^{-2n_i}-\Delta)^{\gamma/2}\Delta u^{(n_i)}(t)\|_{L_p}^p\, dt\right] \leq C\sum_{i=1}^m \left(I_{n_i} + I\!I_{n_i}\right) + C\sum_{1\leq i<j\leq m} I\!I\!I_{n_i n_j}$$

where we denote

$$
I_{n_i} := \mathbb{E}\left[\int_0^T \|\Lambda_{n_i} f^{(n_i)}(t)\|_{L_p}^p \prod_{\substack{j=1 \\ j\neq i}}^m \|\Lambda_{n_j} \Delta u^{(n_j)}(t)\|_{L_p}^p \, dt\right],
$$

$$
I\!I_{n_i} := \mathbb{E}\left[\int_0^T \|\Lambda_{n_i} g_x^{(n_i)}(t)\|_{L_p(\mathbb{R}^d;\ell_2)}^p \prod_{\substack{j=1 \\ j\neq i}}^m \|\Lambda_{n_j} \Delta u^{(n_j)}(t)\|_{L_p}^p \, dt\right],
$$

$$
I\!I\!I_{n_i n_j} := \mathbb{E}\left[\int_0^T \|\Lambda_{n_i} g_x^{(n_i)}(t)\|_{L_p(\mathbb{R}^d;\ell_2)}^p \|\Lambda_{n_j} g_x^{(n_j)}(t)\|_{L_p(\mathbb{R}^d;\ell_2)}^p \prod_{\substack{k=1 \\ k\neq i,j}}^m \|\Lambda_{n_k} \Delta u^{(n_k)}(t)\|_{L_p}^p \, dt\right],
$$

with $\Lambda_n := (e^{-2n} - \Delta)^{\gamma/2}$. Thus, it is enough to find a proper estimate for

$$
\sum_{n_1,\ldots,n_m \in \mathbb{Z}} e^{\left(\sum_{i=1}^m n_i\right)(\theta + p + p\gamma - d)} \left(\sum_{i=1}^m \left(I_{n_i} + I\!I_{n_i}\right) + \sum_{1\le i<j\le m} I\!I\!I_{n_i n_j}\right).
$$

Applying (6.24) first, followed by Tonelli's theorem, then Hölder's and Young's inequality, leads to

$$
\sum_{n_1,\ldots,n_m \in \mathbb{Z}} e^{\left(\sum_{i=1}^m n_i\right)(\theta + p + p\gamma - d)} \sum_{i=1}^m I_{n_i}
$$

$$
= \sum_{n_1,\ldots,n_m \in \mathbb{Z}} e^{\left(\sum_{i=1}^m n_i\right)(\theta + p)} \sum_{i=1}^m \mathbb{E}\left[\int_0^T \|f^{(n_i)}(t, e^{n_i}\cdot)\|_{H_p^\gamma}^p \prod_{\substack{j=1 \\ j\neq i}}^m \|\Delta u^{(n_j)}(t, e^{n_j}\cdot)\|_{H_p^\gamma}^p \, dt\right]
$$

$$
\le C \mathbb{E}\left[\int_0^T \left(\sum_{n\in\mathbb{Z}} e^{n(\theta+p)} \|f^{(n)}(t, e^n\cdot)\|_{H_p^\gamma}^p\right)\left(\sum_{n\in\mathbb{Z}} e^{n(\theta+p)} \|\Delta u^{(n)}(t, e^n\cdot)\|_{H_p^\gamma}^p\right)^{m-1} dt\right]
$$

$$
\le C(\varepsilon) \mathbb{E}\left[\int_0^T \left(\sum_{n\in\mathbb{Z}} e^{n(\theta+p)} \|f^{(n)}(t, e^n\cdot)\|_{H_p^\gamma}^p\right)^{\frac{q}{p}} dt\right]
$$

$$
+ \varepsilon \mathbb{E}\left[\int_0^T \left(\sum_{n\in\mathbb{Z}} e^{n(\theta+p)} \|\Delta u^{(n)}(t, e^n\cdot)\|_{H_p^\gamma}^p\right)^{\frac{q}{p}} dt\right].
$$

Using the definition of $f^{(n)}$ and arguing as at the beginning of the proof, we obtain

$$
\sum_{n\in\mathbb{Z}} e^{n(\theta+p)} \|f^{(n)}(t, e^n\cdot)\|_{H_p^\gamma}^p \le C\left(\|u_x(t)\|_{H_{p,\theta}^\gamma(\mathcal{O})}^p + \|u(t)\|_{H_{p,\theta-p}^\gamma(\mathcal{O})}^p + \|f(t)\|_{H_{p,\theta+p}^\gamma(\mathcal{O})}^p\right)
$$

$$
\le C\left(\|u(t)\|_{H_{p,\theta-p}^{\gamma+1}(\mathcal{O})}^p + \|f(t)\|_{H_{p,\theta+p}^\gamma(\mathcal{O})}^p\right).
$$

Moreover,

$$\sum_{n\in\mathbb{Z}} e^{n(\theta+p)}\|\Delta u^{(n)}(t,e^n\cdot)\|_{H_p^\gamma}^p$$

$$\leq \sum_{n\in\mathbb{Z}} e^{n(\theta+p)}\|(\Delta\zeta_{-n}u(t))(e^n\cdot)\|_{H_p^\gamma}^p$$

$$+ \sum_{n\in\mathbb{Z}} e^{n(\theta+p)}\|(\zeta_{-nx}u_x(t))(e^n\cdot)\|_{H_p^\gamma}^p + \sum_{n\in\mathbb{Z}} e^{n(\theta+p)}\|(\zeta_{-n}\Delta u(t))(e^n\cdot)\|_{H_p^\gamma}^p$$

$$\leq C\left(\|u(t)\|_{H_{p,\theta-p}^\gamma(\mathcal{O})}^p + \|u_x(t)\|_{H_{p,\theta}^\gamma(\mathcal{O})}^p + \|\Delta u\|_{H_{p,\theta+p}^\gamma(\mathcal{O})}^p\right)$$

$$\leq C\left(\|u(t)\|_{H_{p,\theta-p}^{\gamma+1}(\mathcal{O})}^p + \|\Delta u\|_{H_{p,\theta+p}^\gamma(\mathcal{O})}^p\right).$$

Combining the last three estimates, we obtain for any $\varepsilon > 0$ a constant $C(\varepsilon)\in(0,\infty)$, such that

$$\sum_{n_1,\ldots,n_m\in\mathbb{Z}} e^{\left(\sum_{i=1}^m n_i\right)(\theta+p+p\gamma-d)}\sum_{i=1}^m I_{n_i}$$

$$\leq \varepsilon\|\Delta u\|_{\mathbb{H}_{p,\theta+p}^{\gamma,q}(\mathcal{O},T)}^q + C(\varepsilon)\left(\|f\|_{\mathbb{H}_{p,\theta+p}^{\gamma,q}(\mathcal{O},T)}^q + \|u\|_{\mathbb{H}_{p,\theta-p}^{\gamma+1,q}(\mathcal{O},T)}^q\right).$$

Similar arguments yield also

$$\sum_{n_1,\ldots,n_m\in\mathbb{Z}} e^{\left(\sum_{i=1}^m n_i\right)(\theta+p+p\gamma-d)}\left(\sum_{i=1}^m II_{n_i} + \sum_{1\leq i<j\leq m} III_{n_in_j}\right)$$

$$\leq \varepsilon\|\Delta u\|_{\mathbb{H}_{p,\theta+p}^{\gamma,q}(\mathcal{O},T)}^q + C(\varepsilon)\left(\|g\|_{\mathbb{H}_{p,\theta}^{\gamma+1,q}(\mathcal{O},T;\ell_2)}^q + \|u\|_{\mathbb{H}_{p,\theta-p}^{\gamma+1,q}(\mathcal{O},T)}^q\right),$$

which finishes the proof. $\qquad\square$

Iterating Theorem 6.7 and using the properties from Lemma 2.45 of the weighted Sobolev spaces leads to the following result.

Corollary 6.9. *Let $\mathcal{O}$ be a bounded Lipschitz domain in $\mathbb{R}^d$ and let a^{ij} and σ^{ik}, $i,j\in\{1,\ldots,d\}$, $k\in\mathbb{N}$, be given coefficients satisfying Assumption 6.6. Fix $\gamma\geq 1$, $p\in[2,\infty)$ and $q:=mp$ for some $m\in\mathbb{N}$. Furthermore, assume that $u\in\mathfrak{H}_{p,\theta}^{0,q}(\mathcal{O},T)$ is a solution to Eq. (6.23) with $f\in\mathbb{H}_{p,\theta+p}^{\gamma-2,q}(\mathcal{O},T)$ and $g\in\mathbb{H}_{p,\theta}^{\gamma-1,q}(\mathcal{O},T;\ell_2)$. Then $u\in\mathfrak{H}_{p,\theta}^{\gamma,q}(\mathcal{O},T)$, and*

$$\|u\|_{\mathfrak{H}_{p,\theta}^{\gamma,q}(\mathcal{O},T)}^q \leq C\left(\|u\|_{\mathbb{H}_{p,\theta-p}^{0,q}(\mathcal{O},T)}^q + \|f\|_{\mathbb{H}_{p,\theta+p}^{\gamma-2,q}(\mathcal{O},T)}^q + \|g\|_{\mathbb{H}_{p,\theta}^{\gamma-1,q}(\mathcal{O},T;\ell_2)}^q\right)^q,$$

where the constant $C\in(0,\infty)$ does not depend on u, f and g.

Remark 6.10. The assertion of Theorem 6.7 (and Corollary 6.9) can be proved in the same way for arbitrary domains $G\subset\mathbb{R}^d$ with non-empty boundary instead of the bounded Lipschitz domain $\mathcal{O}\subset\mathbb{R}^d$, see [26, Theorem 3.8]. Arguing along the lines of [73,75], it can also be extended to the case where the coefficients depend on the space variable $x\in\mathcal{O}$. Also, the symmetry of (a^{ij}) can be dropped. Since we are mainly interested in the stochastic heat equation on bounded Lipschitz domains, we do not consider these cases in this thesis.

6.3 The stochastic heat equation in $\mathfrak{H}_{p,\theta}^{\gamma,q}(\mathcal{O},T)$

In this section we develop a first $L_q(L_p)$-theory for the stochastic heat equation on bounded Lipschitz domains. We prove that under suitable conditions on the free terms, the stochastic heat equation possesses a unique solution in the class $\mathfrak{H}^{\gamma,q}_{p,d}(\mathcal{O},T)$ with $q > p \geq 2$ (Subsection 6.3.1). This has important consequences for the space time regularity of the solution process (topic (T2) in the introduction). We collect and discuss them in Subsection 6.3.2. In particular, we reach our main goal in this chapter and prove a result on the Hölder regularity of the paths of the solution to the heat equation, considered as a stochastic process taking values in Besov spaces from the non-linear approximation scale ($*$) (Theorem 6.17).

6.3.1 A result on the $L_q(L_p)$-regularity

We have already seen in Theorem 3.13 that the stochastic heat equation (6.1) has a solution u in the class $\mathfrak{H}^{\gamma}_{p,d}(\mathcal{O},T) = \mathfrak{H}^{\gamma,p}_{p,d}(\mathcal{O},T)$, provided the free terms f and g fulfil adequate conditions. In this subsection we want to extend this result and prove the existence of solutions to Eq. (6.1) in the classes $\mathfrak{H}^{\gamma,q}_{p,d}(\mathcal{O},T)$ with $q > p \geq 2$ on general bounded Lipschitz domains $\mathcal{O} \subset \mathbb{R}^d$. Our main goal is to prove the following statement.

Theorem 6.11. *Let $\mathcal{O}$ be a bounded Lipschitz domain in $\mathbb{R}^d$ and let $\gamma \geq 0$. There exists an exponent p_0 with $p_0 > 3$ when $d \geq 3$ and $p_0 > 4$ when $d = 2$, such that for $p \in [2,p_0)$ and $p \leq q < \infty$, Eq. (6.1) has a unique solution $u \in \mathfrak{H}^{\gamma+2,q}_{p,d}(\mathcal{O},T)$, provided*

$$f \in \mathbb{H}^{0,q}_{p,d}(\mathcal{O},T) \cap \mathbb{H}^{\gamma,q}_{p,d+p}(\mathcal{O},T) \qquad and \qquad g \in \mathbb{H}^{1,q}_{p,d-p}(\mathcal{O},T;\ell_2) \cap \mathbb{H}^{\gamma+1,q}_{p,d}(\mathcal{O},T;\ell_2). \quad (6.25)$$

Moreover, there exists a constant $C \in (0,\infty)$, which does not depend on u, f and g, such that

$$\|u\|^q_{\mathfrak{H}^{\gamma+2,q}_{p,d}(\mathcal{O},T)} \leq C\big(\|f\|^q_{\mathbb{H}^{0,q}_{p,d}(\mathcal{O},T)} + \|f\|^q_{\mathbb{H}^{\gamma,q}_{p,d+p}(\mathcal{O},T)} + \|g\|^q_{\mathbb{H}^{1,q}_{p,d-p}(\mathcal{O},T;\ell_2)} + \|g\|^q_{\mathbb{H}^{\gamma+1,q}_{p,d}(\mathcal{O},T;\ell_2)}\big), \quad (6.26)$$

and if $q = mp$ for some $m \in \mathbb{N}$,

$$\|u\|^q_{\mathfrak{H}^{\gamma+2,q}_{p,d}(\mathcal{O},T)} \leq C\big(\|f\|^q_{\mathbb{H}^{0,q}_{p,d}(\mathcal{O},T)} + \|f\|^q_{\mathbb{H}^{\gamma,q}_{p,d+p}(\mathcal{O},T)} + \|g\|^q_{\mathbb{H}^{\gamma+1,q}_{p,d}(\mathcal{O},T;\ell_2)}\big). \quad (6.27)$$

Furthermore, for arbitrary $d \geq 2$, if $\mathcal{O}$ additionally fulfils a uniform outer ball condition, the assertion holds with $p_0 = \infty$.

For bounded $\mathcal{C}^1_u$-domains, this theorem is covered by [74, Theorem 2.17]. Unfortunately, the proof techniques used there do not work if the boundary is assumed to be only Lipschitz continuous. Therefore, we use a different strategy. In a first step we use the stochastic maximal regularity results from [121, 122] to prove that there exists a solution of the stochastic heat equation in $\mathfrak{H}^{1,q}_{p,d}(\mathcal{O},T)$ with $q > p$, i.e., we prove the following statement.

Proposition 6.12. *Let $\mathcal{O}$ be a bounded Lipschitz domain in $\mathbb{R}^d$. There exists an exponent p_0 with $p_0 > 3$ when $d \geq 3$ and $p_0 > 4$ when $d = 2$, such that for $p \in [2,p_0)$ and $p \leq q < \infty$, Eq. (6.1) has a unique solution $u \in \mathfrak{H}^{1,q}_{p,d}(\mathcal{O},T)$, provided*

$$f \in \mathbb{H}^{0,q}_{p,d}(\mathcal{O},T) \qquad and \qquad g \in \mathbb{H}^{1,q}_{p,d-p}(\mathcal{O},T;\ell_2).$$

Moreover, there exists a constant $C \in (0,\infty)$, which does not depend on u, f and g, such that

$$\|u\|^q_{\mathbb{H}^{1,q}_{p,d-p}(\mathcal{O},T)} \leq C\big(\|f\|^q_{\mathbb{H}^{0,q}_{p,d}(\mathcal{O},T)} + \|g\|^q_{\mathbb{H}^{0,q}_{p,d}(\mathcal{O},T;\ell_2)}\big). \quad (6.28)$$

Furthermore, for arbitrary $d \geq 2$, if $\mathcal{O}$ additionally fulfils a uniform outer ball condition, the assertion holds with $p_0 = \infty$.

As already mentioned, we want to apply the maximal regularity theory from [121, 122] in order to prove this result. Therefore, we have to rewrite our equation as a Banach space valued ordinary SDE of the form

$$\left.\begin{aligned} \mathrm{d}u(t) + Au(t)\,\mathrm{d}t = f(t)\,\mathrm{d}t + b(t)\,\mathrm{d}W_H(t), \qquad t \in [0, T] \\ u(0) = 0, \end{aligned}\right\}$$

where A is a suitable unbounded operator on some Banach space and W_H is an H-cylindrical Brownian motion on a Hilbert space H. Thus, before we start with the proof of Proposition 6.12, we introduce a proper operator and check its relevant properties.

Let $\mathcal{O}$ be a bounded Lipschitz domain in $\mathbb{R}^d$. As in [126, Definition 3.1], for arbitrary $p \in (1, \infty)$, we define the *weak Dirichlet-Laplacian* $\Delta_{p,w}^D$ on $L_p(\mathcal{O})$ as follows:

$$D(\Delta_{p,w}^D) := \left\{ u \in \mathring{W}_p^1(\mathcal{O}) \ : \ \Delta u \in L_p(\mathcal{O}) \right\},$$

$$\Delta_{p,w}^D u := \Delta u = \sum_{i,j=1}^d \delta_{i,j} u_{x^i x^j}, \qquad u \in D(\Delta_{p,w}^D).$$

Fix $p \in [2, p_0)$ with either

[C1] $p_0 = 3 + \delta$ when $d \geq 3$, or

[C2] $p_0 = 4 + \delta$ when $d = 2$, or

[C3] $p_0 = \infty$ when $d \geq 2$ and $\mathcal{O}$ additionally fulfils a uniform outer ball condition,

where $\delta > 0$ is taken from [126, Proposition 4.1]. Then, the unbounded operator $\Delta_{p,w}^D$ generates a positive analytic contraction semigroup $\{S_p(t)\}_{t \geq 0}$ of negative type on $L_p(\mathcal{O})$, see [126, Theorem 3.8, Corollary 4.2, Lemma 4.4, and Corollary 4.5]. Therefore, the positive operator $A := -\Delta_{p,w}^D$ admits an H^∞-calculus of angle less than $\pi/2$ and possesses bounded imaginary powers (see Theorem 2.69 and Theorem 2.70). Thus, by [116, Theorem 1.15.3],

$$\left[L_p(\mathcal{O}), D(-\Delta_{p,w}^D) \right]_{\frac{1}{2}} = D((-\Delta_{p,w}^D)^{\frac{1}{2}}), \tag{6.29}$$

where the square root of the negative of the weak Dirichlet-Laplacian $(-\Delta_{p,w}^D)^{\frac{1}{2}}$ is defined as the inverse of the operator

$$(-\Delta_{p,w}^D)^{-\frac{1}{2}} := \pi^{-\frac{1}{2}} \int_0^\infty t^{-\frac{1}{2}} S_p(t)\,\mathrm{d}t : L_p(\mathcal{O}) \to L_p(\mathcal{O}) \tag{6.30}$$

with domain

$$D((-\Delta_{p,w}^D)^{\frac{1}{2}}) := \mathrm{Range}((-\Delta_{p,w}^D)^{-\frac{1}{2}}),$$

see [102, Chapter 2.6]. Endowed with the norm

$$\|u\|_{D((-\Delta_{p,w}^D)^{\frac{1}{2}})} := \|(-\Delta_{p,w}^D)^{\frac{1}{2}} u\|_{L_p(\mathcal{O})}, \quad u \in D((-\Delta_{p,w}^D)^{\frac{1}{2}}),$$

$D((-\Delta_{p,w}^D)^{\frac{1}{2}})$ becomes a Banach space. Exploiting the fundamental results from [126] and [67], we can prove the following identity, which is crucial if we want to apply the results from [121] in our setting.

Lemma 6.13. *Let $\mathcal{O}$ be a bounded Lipschitz domain in $\mathbb{R}^d$. There is an exponent p_0 with $p_0 > 4$ when $d = 2$ and $p_0 > 3$ when $d \geq 3$ such that if $p \in [2, p_0)$, then*

$$D((-\Delta^D_{p,w})^{\frac{1}{2}}) = \mathring{W}^1_p(\mathcal{O}) \tag{6.31}$$

with equivalent norms. Furthermore, for arbitrary $d \geq 2$, if $\mathcal{O}$ additionally fulfils a uniform outer ball condition, (6.31) holds for arbitrary $p \in [2, \infty)$ with equivalent norms.

Proof. We fix $p \in [2, p_0)$ with p_0 as in [C1], [C2], or [C3] with $\delta > 0$ from [126, Proposition 4.1]. As a consequence of [67, Theorem 7.5] we have

$$(-\Delta^D_{2,w})^{-\frac{1}{2}} L_p(\mathcal{O}) = \mathring{W}^1_p(\mathcal{O}) \tag{6.32}$$

and

$$\left\| (-\Delta^D_{2,w})^{\frac{1}{2}} u \right\|_{L_p(\mathcal{O})} \asymp \|Du\|_{L_p(\mathcal{O})}, \quad u \in \mathring{W}^1_p(\mathcal{O}). \tag{6.33}$$

Moreover, by [126, Proposition 4.1] the semigroups $\{S_2(t)\}_{t \geq 0}$ and $\{S_p(t)\}_{t \geq 0}$ are consistent, i.e.,

$$S_2(t)f = S_p(t)f, \quad f \in L_p(\mathcal{O}), \, t \geq 0,$$

and therefore

$$(-\Delta^D_{p,w})^{-\frac{1}{2}} f = (-\Delta^D_{2,w})^{-\frac{1}{2}} f, \quad f \in L_p(\mathcal{O})$$

according to (6.30). Thus, by (6.32), $\mathring{W}^1_p(\mathcal{O}) = \text{Range}((-\Delta^D_{p,w})^{-\frac{1}{2}}) = D((-\Delta^D_{p,w})^{\frac{1}{2}})$, and the norm equivalence follows immediately from (6.33). $\qquad \square$

Remark 6.14. The comparison of the L_p-norms of $L^{1/2}u$ and Du for second order elliptic operators $(-L)$ is known as Kato's square root problem in L_p. On the whole space $\mathbb{R}^d$ and for $p = 2$, equivalence of the norms for uniformly complex elliptic operators in divergence form with bounded measurable coefficients has been established in the seminal work [10]. Also, on bounded Lipschitz domains it has been proven in [12], among other themes, that for symmetric real-valued elliptic operators the equivalence $\|(-L)^{1/2} \cdot\|_{L_p(\mathcal{O})} \asymp \|D\cdot\|_{L_p(\mathcal{O})}$ holds for certain $p \geq 2$. We expect that, if the results concerning the semigroup generated by $\Delta^D_{p,w}$ from [126], which we use in the proof of Lemma 6.13 and in the proof of Proposition 6.12 below, extend to second order elliptic operators, then Proposition 6.12 and Theorem 6.11 remain valid for equations of the type

$$du = (Lu + f) \, dt + g^k \, dw^k_t, \quad u(0) = 0.$$

In order to keep the exposition at a reasonable level we do not go into details here.

After clarifying these properties of $A = -\Delta^D_{p,w}$, we are ready to prove the existence of a solution $u \in \mathfrak{H}^{1,q}_{p,d}(\mathcal{O}, T)$ to the stochastic heat equation.

Proof of Proposition 6.12. As in the proof of Lemma 6.13 we fix $p \in [2, p_0)$ with p_0 satisfying either [C1], [C2], or [C3] with $\delta > 0$ from [126, Proposition 4.1]. Furthermore, we fix $q \geq p$ and assume that $f \in \mathbb{H}^{0,q}_{p,d}(\mathcal{O}, T)$ and $g \in \mathbb{H}^{1,q}_{p,d-p}(\mathcal{O}, T; \ell_2)$. We will write $W_{\ell_2} = (W_{\ell_2}(t))_{t \in [0,T]}$ for the ℓ_2-cylindrical Brownian motion defined by

$$\ell_2 \ni h \mapsto W_{\ell_2}(t)h := \sum_{k=1}^{\infty} w^k_t \langle e_k, h \rangle_{\ell_2} \in L_2(\Omega), \quad t \in [0, T],$$

where $\{e_k : k \in \mathbb{N}\}$ denotes the standard orthonormal basis of ℓ_2, see also Example 2.22. Let Φ be the isomorphism between $H_{p,d-p}^1(\mathcal{O};\ell_2)$ and $\Gamma(\ell_2, H_{p,d-p}^1(\mathcal{O}))$ from Theorem 2.54. Then, since $H_{p,d-p}^1(\mathcal{O}) = \mathring{W}_p^1(\mathcal{O})$, see Lemma 2.51,

$$\Phi_g := \Phi(g) \in L_q(\Omega_T, \mathcal{P}_T, \mathbb{P}_T; \Gamma(\ell_2, \mathring{W}_p^1(\mathcal{O}))). \tag{6.34}$$

Thus, by Lemma 6.13,

$$\Phi_g \in L_q(\Omega_T, \mathcal{P}_T, \mathbb{P}_T; \Gamma(\ell_2, X_{\frac{1}{2}}))$$

with

$$X_{\frac{1}{2}} := [L_p(\mathcal{O}), D(-\Delta_{p,w}^D)]_{\frac{1}{2}} = D((-\Delta_{p,w}^D)^{\frac{1}{2}}),$$

see also (6.29). Moreover, as already mentioned, $-\Delta_{p,w}^D$ admits an H^∞-calculus of angle less than $\pi/2$, and $X_1 := D(\Delta_{p,w}^D) \hookrightarrow X_0 := L_p(\mathcal{O})$ densely, since $C_0^\infty(\mathcal{O})$ is contained in $D(\Delta_{p,w}^D)$. Using all these facts, we can apply [121, Theorem 4.5(ii)] and obtain the existence of a stochastic process

$$u \in L_q(\Omega_T, \mathcal{P}_T, \mathbb{P}_T; D(-\Delta_{p,w}^D)) \tag{6.35}$$

solving the infinite dimensional ordinary SDE

$$\left. \begin{aligned} \mathrm{d}u(t) - \Delta_{p,w}^D u(t)\,\mathrm{d}t &= f(t)\,\mathrm{d}t + \Phi_g(t)\,\mathrm{d}W_{\ell_2}(t), \qquad t \in [0,T], \\ u(0) &= 0, \end{aligned} \right\}$$

in the sense of [121, Definition 4.2]. In particular, there exists a modification $\tilde{u}$ of u, such that, with probability one, the equality

$$\tilde{u}(t) = \int_0^t \Delta\tilde{u}(s)\,\mathrm{d}s + \int_0^t f(s)\,\mathrm{d}s + \int_0^t \Phi_g(s)\,\mathrm{d}W_{\ell_2}(s) \qquad (\text{in } L_p(\mathcal{O})) \tag{6.36}$$

holds for all $t \in [0,T]$. Note that, since (6.34) holds and $\mathring{W}_p^1(\mathcal{O}) = H_{p,d-p}^1(\mathcal{O})$ is a UMD Banach space with type 2 (see Lemma 2.50), the stochastic integral on the right hand side is well-defined in the sense of [120] as a $\mathring{W}_p^1(\mathcal{O})$-valued stochastic processes, see Theorem 2.32. Fix $\varphi \in C_0^\infty(\mathcal{O})$. Then, $\mathbb{P}$-a.s.,

$$(\tilde{u}(t,\cdot),\varphi) = \int_0^t (\Delta\tilde{u}(s,\cdot),\varphi)\,\mathrm{d}s + \int_0^t (f(s,\cdot),\varphi)\,\mathrm{d}s + \left(\int_0^t \Phi_g(s)\,\mathrm{d}W_{\ell_2}(s),\varphi\right), \qquad t \in [0,T],$$

since $\mathbb{P}$-a.s. Equality (6.36) holds for all $t \in [0,T]$. Furthermore, since

$$\sum_{k=1}^\infty \int_0^\cdot (g^k(s,\cdot),\varphi)\,\mathrm{d}w_s^k = \left(\int_0^\cdot \Phi_g(s)\,\mathrm{d}W_{\ell_2}(s),\varphi\right) \qquad \mathbb{P}\text{-a.s.}$$

in $\mathcal{C}([0,T];\mathbb{R})$, cf. Proposition 3.6, the identity

$$(\tilde{u}(t,\cdot),\varphi) = \int_0^t (\Delta\tilde{u}(s,\cdot),\varphi)\,\mathrm{d}s + \int_0^t (f(s,\cdot),\varphi)\,\mathrm{d}s + \sum_{k=1}^\infty \int_0^t (g^k(s,\cdot),\varphi)\,\mathrm{d}w_s^k, \qquad t \in [0,T],$$

holds with probability one. Therefore, and by (6.35), since $D(-\Delta_{p,w}^D) \hookrightarrow H_{p,d-p}^1(\mathcal{O}) = \mathring{W}_p^1(\mathcal{O})$, u belongs to $\mathfrak{H}_{p,d}^{1,q}(\mathcal{O},T)$ and solves Eq. (6.1) in the sense of Definition 3.10. Since $\mathfrak{H}_{p,d}^{1,q}(\mathcal{O},T) \hookrightarrow \mathfrak{H}_{2,d}^{1,2}(\mathcal{O},T)$, the uniqueness follows from Theorem 3.13. Thus, in order to finish the proof, we show

Estimate (6.28). To this end we will use the fact that the stochastic process $V : \Omega_T \to L_p(\mathcal{O})$ defined as

$$V(t) := \int_0^t S_p(t-s)f(s) \, ds + \int_0^t S_p(t-s)\Phi_g(s) \, dW_{\ell_2}(s), \qquad t \in [0,T],$$

is a modification of u, see [121, Proposition 4.4]. Since $-\Delta_{p,w}^D$ has the (deterministic) maximal regularity property (see [126, Proposition 6.1]) and $D(-\Delta_{p,w}^D) \hookrightarrow D((-\Delta_{p,w}^D)^{\frac{1}{2}}) = H_{p,d-p}^1(\mathcal{O})$, we obtain

$$\mathbb{E}\left[\left\|t \mapsto \int_0^t S_p(t-s)f(s) \, ds\right\|_{L_q([0,T];H_{p,d-p}^1(\mathcal{O}))}^q\right] \leq C \|f\|_{\mathbb{H}_{p,d}^{0,q}(\mathcal{O},T)}^q. \tag{6.37}$$

Simultaneously, notice that $A := -\Delta_{p,w}^D$ and g (respectively Φ_g) fulfil the assumptions of [122, Theorem 1.1]; we have already checked them in our explanations above. Thus, applying this result, we obtain

$$\mathbb{E}\left[\left\|t \mapsto \int_0^t S_p(t-s)\Phi_g(s) \, dW_{\ell_2}(s)\right\|_{L_q([0,T];H_{p,d-p}^1(\mathcal{O}))}^q\right] \leq C \|g\|_{\mathbb{H}_{p,d}^{0,q}(\mathcal{O},T;\ell_2)}^q. \tag{6.38}$$

The constants in (6.37) and (6.38) do not depend on f and g. Therefore, using the last two estimates we obtain the existence of a constant C, independent of f or g, such that

$$\|V\|_{\mathbb{H}_{p,d-p}^{1,q}(\mathcal{O},T)}^q \leq C \left(\|f\|_{\mathbb{H}_{p,d}^{0,q}(\mathcal{O},T)}^q + \|g\|_{\mathbb{H}_{p,d}^{0,q}(\mathcal{O},T;\ell_2)}^q\right).$$

Since V is just a modification of the solution u, Estimate (6.28) follows. $\qquad\square$

Now using the lifting argument from Section 6.2 and interpolation theory we can prove the main result of this subsection.

Proof of Theorem 6.11. Let $\gamma \geq 0$. Again, as in the proof of Lemma 6.13, let $p \in [2, p_0)$ with p_0 satisfying [C1], or [C2], or [C3] with $\delta > 0$ as in Theorem [126, Proposition 4.1]. We first consider the case $q = mp$ for some fixed $m \in \mathbb{N}$. Assume that f and g fulfil (6.25). Then, by Proposition 6.12 there exists a unique solution $u \in \mathfrak{H}_{p,d}^{1,q}(\mathcal{O},T)$. An application of Corollary 6.9 yields the estimate

$$\|u\|_{\mathfrak{H}_{p,d}^{\gamma+2,q}(\mathcal{O},T)}^q \leq C \left(\|u\|_{\mathbb{H}_{p,d-p}^{0,q}(\mathcal{O},T)}^q + \|f\|_{\mathbb{H}_{p,d+p}^{\gamma,q}(\mathcal{O},T)}^q + \|g\|_{\mathbb{H}_{p,d}^{\gamma+1,q}(\mathcal{O},T;\ell_2)}^q\right).$$

Thus, $u \in \mathfrak{H}_{p,d}^{\gamma+2,q}(\mathcal{O},T)$, and using Estimate (6.28), leads to

$$\|u\|_{\mathfrak{H}_{p,d}^{\gamma+2,q}(\mathcal{O},T)}^q \leq C \left(\|f\|_{\mathbb{H}_{p,d}^{0,q}(\mathcal{O},T)}^q + \|g\|_{\mathbb{H}_{p,d}^{0,q}(\mathcal{O},T;\ell_2)}^q + \|f\|_{\mathbb{H}_{p,d+p}^{\gamma,q}(\mathcal{O},T)}^q + \|g\|_{\mathbb{H}_{p,d}^{\gamma+1,q}(\mathcal{O},T;\ell_2)}^q\right). \tag{6.39}$$

Hence, we have proven estimate (6.27), since $H_{p,d}^{\gamma+1}(\mathcal{O};\ell_2) \hookrightarrow H_{p,d}^0(\mathcal{O};\ell_2)$. In order to get rid of the restriction $q = mp$ with $m \in \mathbb{N}$ and prove the assertion for general $q \geq p$ we argue by following the lines of [82, Proof of Theorem 2.1, page 7]. Let f and g fulfil (6.25) with a fixed $q \geq p$. Denote $E^\gamma := \left(H_{p,d+p}^\gamma(\mathcal{O}) \cap H_{p,d}^0(\mathcal{O})\right) \times \left(H_{p,d}^{\gamma+1}(\mathcal{O};\ell_2) \cap H_{p,d-p}^1(\mathcal{O};\ell_2)\right)$. By (6.39) and since $H_{p,d-p}^1(\mathcal{O};\ell_2) \hookrightarrow H_{p,d}^0(\mathcal{O};\ell_2)$, for any $m \in \mathbb{N}$, the operator

$$\mathcal{R}_m : L_{mp}(\Omega_T, \mathcal{P}_T, \mathbb{P}_T; E^\gamma) \to L_{mp}(\Omega_T, \mathcal{P}_T, \mathbb{P}_T; H_{p,d-p}^{\gamma+2}(\mathcal{O}))$$

$$(f,g) \mapsto \mathcal{R}_m(f,g),$$

where $\mathcal{R}_m(f,g)$ is the unique solution in the class $\mathfrak{H}_{p,d}^{\gamma+2,mp}(\mathcal{O},T)$ of the corresponding stochastic heat equation (6.1), is well-defined. Moreover, it is a linear and bounded operator, and, because of the uniqueness of the solution, $\mathcal{R} = \mathcal{R}_m$ is independent of $m \in \mathbb{N}$. Therefore, using interpolation results like, e.g., [13, Theorem 5.1.2], shows that $\mathcal{R}$ is a well-defined linear and bounded operator from $L_q(\Omega_T, \mathcal{P}_T, \mathbb{P}_T; E^\gamma)$ to $L_q(\Omega_T, \mathcal{P}_T, \mathbb{P}_T; H_{p,d-p}^{\gamma+2}(\mathcal{O}))$ mapping any couple $(f,g) \in L_q(\Omega_T, \mathcal{P}_T, \mathbb{P}_T; E^\gamma)$ to the unique solution $\mathcal{R}(f,g) = u \in \mathfrak{H}_{p,d}^{\gamma+2,q}(\mathcal{O},T)$ of Eq. (6.1).   □

When proving Hölder regularity of the solution, considered as a stochastic process taking values in Besov spaces from the scale $(*)$, we will mainly use the following consequence of Theorem 6.11.

Corollary 6.15. *Let $\mathcal{O}$ be a bounded Lipschitz domain in $\mathbb{R}^d$. There exists an exponent p_0 with $p_0 > 3$ when $d \geq 3$ and $p_0 > 4$ when $d = 2$, such that for $p \in [2,p_0)$ and $p \leq q < \infty$, Eq. (6.1) has a unique solution $u \in \mathfrak{H}_{p,d}^{2,q}(\mathcal{O},T)$, provided*

$$f \in \mathbb{H}_{p,d}^{0,q}(\mathcal{O},T) \qquad and \qquad g \in \mathbb{H}_{p,d-p}^{1,q}(\mathcal{O},T;\ell_2). \tag{6.40}$$

Moreover, there exists a constant $C \in (0,\infty)$, which does not depend on u, f and g, such that

$$\|u\|_{\mathfrak{H}_{p,d}^{2,q}(\mathcal{O},T)}^q \leq C\big(\|f\|_{\mathbb{H}_{p,d}^{0,q}(\mathcal{O},T)}^q + \|g\|_{\mathbb{H}_{p,d-p}^{1,q}(\mathcal{O},T;\ell_2)}^q\big),$$

and if $q = mp$ for some $m \in \mathbb{N}$,

$$\|u\|_{\mathfrak{H}_{p,d}^{2,q}(\mathcal{O},T)}^q \leq C\big(\|f\|_{\mathbb{H}_{p,d}^{0,q}(\mathcal{O},T)}^q + \|g\|_{\mathbb{H}_{p,d-p}^{1,q}(\mathcal{O},T;\ell_2)}^q\big).$$

Furthermore, for $d \geq 2$, if $\mathcal{O}$ additionally fulfils a uniform outer ball condition, the assertion holds with $p_0 = \infty$.

Proof. Since $H_{p,d}^0(\mathcal{O}) \hookrightarrow H_{p,d+p}^0(\mathcal{O})$ and $H_{p,d-p}^1(\mathcal{O};\ell_2) \hookrightarrow H_{p,d}^1(\mathcal{O};\ell_2)$, condition (6.40) implies (6.25) with $\gamma = 0$, and the assertion follows immediately from Theorem 6.11.   □

6.3.2 Space time regularity

In this subsection we collect the fruits of our work and present new results concerning the Hölder regularity of the paths of the solution to the stochastic heat equation (6.1) on general bounded Lipschitz domains. We start with a Hölder-Sobolev regularity result, i.e., with a result concerning the Hölder regularity of the paths of the solution to the stochastic heat equation, considered as a stochastic process taking values in weighted Sobolev spaces.

Theorem 6.16. *Let $\mathcal{O}$ be a bounded Lipschitz domain in $\mathbb{R}^d$ and fix $\gamma \in \mathbb{N}_0$. Assume that $u \in \mathfrak{H}_{p,d}^{\gamma+2,q}(\mathcal{O},T)$ is the unique solution of Eq. (6.1) with $f \in \mathbb{H}_{p,d}^{0,q}(\mathcal{O},T) \cap \mathbb{H}_{p,d+p}^{\gamma,q}(\mathcal{O},T)$ and $g \in \mathbb{H}_{p,d-p}^{1,q}(\mathcal{O},T;\ell_2) \cap \mathbb{H}_{p,d}^{\gamma+1,q}(\mathcal{O},T;\ell_2)$, where $p \leq q < \infty$ and $p \in [2,p_0)$ with*

(i)   *$p_0 > 3$ when $d \geq 3$ and $p_0 > 4$ when $d = 2$,*

or, alternatively,

(ii)   *$p_0 = \infty$ for $d \geq 2$, if $\mathcal{O}$ additionally fulfils a uniform outer ball condition,*

as in Theorem 6.11. Furthermore, fix

$$\frac{2}{q} < \tilde{\beta} < \beta \leq 1.$$

Then there exists a constant $C \in (0, \infty)$, which does not depend on u, f and g, such that

$$\mathbb{E}\|u\|_{C^{\tilde{\beta}/2 - 1/q}([0,T]; H_{p,d-(1-\beta)p}^{\gamma+2-\beta}(\mathcal{O}))}^q$$
$$\leq C\big(\|f\|_{\mathbb{H}_{p,d}^{0,q}(\mathcal{O},T)}^q + \|f\|_{\mathbb{H}_{p,d+p}^{\gamma,q}(\mathcal{O},T)}^q + \|g\|_{\mathbb{H}_{p,d-p}^{1,q}(\mathcal{O},T;\ell_2)}^q + \|g\|_{\mathbb{H}_{p,d}^{\gamma+1,q}(\mathcal{O},T;\ell_2)}^q\big),$$

and if $q = mp$ for some $m \in \mathbb{N}$,

$$\mathbb{E}\|u\|_{C^{\tilde{\beta}/2 - 1/q}([0,T]; H_{p,d-(1-\beta)p}^{\gamma+2-\beta}(\mathcal{O}))}^q$$
$$\leq C\big(\|f\|_{\mathbb{H}_{p,d}^{0,q}(\mathcal{O},T)}^q + \|f\|_{\mathbb{H}_{p,d+p}^{\gamma,q}(\mathcal{O},T)}^q + \|g\|_{\mathbb{H}_{p,d-p}^{\gamma+1,q}(\mathcal{O},T;\ell_2)}^q\big).$$

Proof. The assertion is an immediate consequence of Theorem 6.11 and Theorem 6.1. $\qquad \square$

Now we look at the solution of the stochastic heat equation as a stochastic process taking values in the Besov spaces from the scale $(*)$. Given the setting of Theorem 6.11, an application of embedding (4.13) shows that the solution $u \in \mathfrak{H}_{p,d}^{\gamma+2,q}(\mathcal{O},T)$, $\gamma \geq 0$, of the stochastic heat equation fulfils

$$u \in L_q(\Omega_T, \mathcal{P}_T, \mathbb{P}_T; B_{\tau,\tau}^\alpha(\mathcal{O})), \qquad \frac{1}{\tau} = \frac{\alpha}{d} + \frac{1}{p}, \qquad \text{for all} \quad 0 < \alpha < \frac{d}{d-1}.$$

We are interested in the Hölder regularity of the paths of this $B_{\tau,\tau}^\alpha(\mathcal{O})$-valued process.

Theorem 6.17. *Let $\mathcal{O}$ be a bounded Lipschitz domain in $\mathbb{R}^d$ and let the setting of Corollary 6.15 be given. That is, let $u \in \mathfrak{H}_{p,d}^{2,q}(\mathcal{O},T)$ be the unique solution of Eq. (6.1) with $f \in \mathbb{H}_{p,d}^{0,q}(\mathcal{O},T)$ and $g \in \mathbb{H}_{p,d-p}^{1,q}(\mathcal{O},T;\ell_2)$, where $p \leq q < \infty$ and $p \in [2, p_0)$ with*

    (i) $p_0 > 3$ *when $d \geq 3$ and $p_0 > 4$ when $d = 2$,*

or, alternatively,

    (ii) $p_0 = \infty$ *for $d \geq 2$, if $\mathcal{O}$ additionally fulfils a uniform outer ball condition,*

as in Corollary 6.15. Furthermore, fix

$$\frac{2}{q} < \tilde{\beta} < 1.$$

Then, for all α and τ with

$$\frac{1}{\tau} = \frac{\alpha}{d} + \frac{1}{p}, \qquad \text{and} \quad 0 < \alpha < (1 - \tilde{\beta})\frac{d}{d-1}, \tag{6.41}$$

there exists a constant $C \in (0, \infty)$ which does not depend on u, f and g such that

$$\mathbb{E}\|u\|_{C^{\tilde{\beta}/2 - 1/q}([0,T]; B_{\tau,\tau}^\alpha(\mathcal{O}))}^q \leq C\big(\|f\|_{\mathbb{H}_{p,d}^{0,q}(\mathcal{O},T)}^q + \|g\|_{\mathbb{H}_{p,d-p}^{1,q}(\mathcal{O},T;\ell_2)}^q\big), \tag{6.42}$$

and if $q = mp$ for some $m \in \mathbb{N}$,

$$\mathbb{E}\|u\|_{C^{\tilde{\beta}/2 - 1/q}([0,T]; B_{\tau,\tau}^\alpha(\mathcal{O}))}^q \leq C\big(\|f\|_{\mathbb{H}_{p,d}^{0,q}(\mathcal{O},T)}^q + \|g\|_{\mathbb{H}_{p,d}^{1,q}(\mathcal{O},T;\ell_2)}^q\big). \tag{6.43}$$

Proof. The assertion follows immediately from Theorem 6.2 and Corollary 6.15. $\qquad \square$

Remark 6.18. Since $\tilde{\beta} < 1$ is assumed in Theorem 6.17, the Hölder regularity of the paths of the solution process determined in (6.42) and (6.43) is always strictly less than $\frac{1}{2}$. Moreover, we have a typical trade-off between time and space regularity: the higher the Hölder regularity in time, the more restrictive condition (6.41), and therefore, the less the Besov regularity α in space. If we rise the Hölder regularity in time direction by $\varepsilon > 0$, we lose $2\varepsilon\frac{d}{d-1}$ from the Besov regularity α in space.

Example 6.19. Let $\mathcal{O}$ be a bounded Lipschitz domain in $\mathbb{R}^d$. Let $p \in [2, p_0)$ with p_0 satisfying condition (i) from Theorem 6.17 above or, alternatively, let $p \in [2, \infty)$ if $\mathcal{O}$ additionally fulfils a uniform outer ball condition. Furthermore, assume that

$$f \in L_\infty(\Omega_T, \mathcal{P}_T, \mathbb{P}_T; H_{p,d}^0(\mathcal{O})) \quad \text{and} \quad g \in L_\infty(\Omega_T, \mathcal{P}_T, \mathbb{P}_T; H_{p,d-p}^1(\mathcal{O}; \ell_2)).$$

Then, for any $q \geq p$, $f \in \mathbb{H}_{p,d}^{0,q}(\mathcal{O},T)$ and $g \in \mathbb{H}_{p,d-p}^1(\mathcal{O},T;\ell_2)$, and, by Corollary 6.15, there exists a unique solution $u \in \mathfrak{H}_{p,d}^{2,q}(\mathcal{O},T)$ to the stochastic heat equation (6.1). Chose an arbitrary $\alpha > 0$ such that

$$0 < \alpha < \frac{d}{d-1}.$$

Then there exists a $\tilde{\beta} = \tilde{\beta}(\alpha) > 0$ and a corresponding $m = m(\alpha) \in \mathbb{N}$ such that simultaneously

$$\frac{2}{mp} < \tilde{\beta} < 1 \quad \text{and} \quad \alpha < (1-\tilde{\beta})\frac{d}{d-1}.$$

Therefore, an application of Theorem 6.17 yields

$$\mathbb{E}\|u\|_{\mathcal{C}^\varepsilon([0,T];B_{\tau,\tau}^\alpha(\mathcal{O}))}^q < \infty, \quad \frac{1}{\tau} = \frac{\alpha}{d} + \frac{1}{p},$$

with $\varepsilon = \varepsilon(\alpha) := \frac{\tilde{\beta}}{2} - \frac{1}{mp} > 0$. Thus, for all α and τ with

$$\frac{1}{\tau} = \frac{\alpha}{d} + \frac{1}{p}, \quad \text{and} \quad 0 < \alpha < \frac{d}{d-1},$$

we have

$$u \in \mathcal{C}([0,T];B_{\tau,\tau}^\alpha(\mathcal{O})) \quad \mathbb{P}\text{-a.s.}$$

Zusammenfassung

In der vorliegenden Arbeit wird die Regularität von Lösungen (semi)linearer parabolischer stochastischer partieller Differentialgleichungen (in der Arbeit stets mit SPDEs abgekürzt) auf beschränkten Lipschitz-Gebieten untersucht. Es werden Itô-Differentialgleichungen zweiter Ordnung mit Dirichlet-Nullrandbedingung betrachtet. Sie haben die allgemeine Form:

$$
\begin{aligned}
\mathrm{d}u &= \left(\sum_{i,j=1}^{d} a^{ij} u_{x^i x^j} + \sum_{i=1}^{d} b^i u_{x^i} + cu + f + L(u) \right) \mathrm{d}t \\
&\quad + \sum_{k=1}^{\infty} \left(\sum_{i=1}^{d} \sigma^{ik} u_{x^i} + \mu^k u + g^k + (\Lambda(u))^k \right) \mathrm{d}w_t^k \qquad \text{on } \Omega \times [0,T] \times \mathcal{O}, \\
u &= 0 \qquad \text{on } \Omega \times (0,T] \times \partial\mathcal{O}, \\
u(0) &= u_0 \qquad \text{on } \Omega \times \mathcal{O}.
\end{aligned}
\tag{1}
$$

Hierbei bezeichnet $\mathcal{O}$ ein beschränktes Lipschitz-Gebiet in $\mathbb{R}^d$ ($d \geq 2$), während $T \in (0,\infty)$ für den Endzeitpunkt steht. Mit $(w_t^k)_{t \in [0,T]}$, $k \in \mathbb{N}$, wird eine Folge unabhängiger reellwertiger Standard-Brown'scher Bewegungen bezüglich einer normalen Filtration $(\mathcal{F}_t)_{t \in [0,T]}$ auf einem vollständigen Wahrscheinlichkeitsraum $(\Omega, \mathcal{F}, \mathbb{P})$ bezeichnet und $\mathrm{d}u$ ist im Sinne von Itôs stochastischem Differential nach der Zeit $t \in [0,T]$ zu verstehen. Die Koeffizienten a^{ij}, b^i, c, σ^{ik} und μ^k, mit $i,j \in \{1,\dots,d\}$ und $k \in \mathbb{N}$, sind reellwertige Funktionen auf $\Omega \times [0,T] \times \mathcal{O}$, welche bestimmten Bedingungen genügen. Diese sind in Kapitel 3 genau formuliert, siehe insbesondere Assumption 3.1. Bei den Nichtlinearitäten L und Λ wird davon ausgegangen, dass sie in geeigneten Räumen Lipschitz-stetig sind. Wie diese genau aussehen, wird in Kapitel 5 präzisiert, siehe insbesondere Assumption 5.9. In dieser Arbeit wird eine funktionalanalytische Sichtweise eingenommen. So wird die Lösung u einer SPDE nicht als eine von $(\omega, t, x) \in \Omega \times [0,T] \times \mathcal{O}$ abhängige reellwertige Funktion betrachtet. Diese wird vielmehr als eine auf $\Omega \times [0,T]$ definierte Funktion mit Werten in dem mit $\mathcal{D}'(\mathcal{O})$ bezeichneten Raum reellwertiger Distributionen aufgefasst.

Eine der bekanntesten Gleichungen der Form (1) ist die stochastische Wärmeleitungsgleichung mit additivem oder multiplikativem Rauschen. Allgemeinere Gleichungen vom Typ (1) mit endlich vielen $(w_t^k)_{t \in [0,T]}$, $k \in \{1,\dots,N\}$, treten beispielsweise in der nichtlinearen (stochastischen) Filtertheorie auf, vgl. [80, Section 8.1]. Werden unendlich viele Brown'sche Bewegungen $(w_t^k)_{t \in [0,T]}$, $k \in \mathbb{N}$, herangezogen, so können auch Systeme, welche einem weißen Rauschen in Raum und Zeit unterliegen, betrachtet werden, vgl. [80, Section 8.1]. Diese Gleichungen werden in der Literatur als mathematische Modelle für Reaktionsdiffusionsgleichungen, welche einem nicht vernachlässigbaren Rauschen unterliegen, vorgeschlagen, vgl. [32, Section 0.7] und die darin angegebene Literatur, insbesondere [9].

Die Frage nach der Existenz und Eindeutigkeit einer Lösung der Gleichung (1) ist weitgehend geklärt, siehe hierzu exemplarisch [75]. Allerdings kann diese Lösung in den überwiegenden Fällen nicht explizit angegeben und ebenso wenig numerisch exakt berechnet werden. Umso wichtiger sind daher numerische Verfahren, die eine konstruktive Approximation der Lösung ermöglichen. Grundsätzlich wird hierbei zwischen *uniformen* und *adaptiven* numerischen Verfahren unterschieden. Letztere versprechen eine effizientere Approximation, deren Implementierung ist allerdings mit einem erheblich höheren Aufwand verbunden. Daher muss vorab geklärt werden, ob die erzielbare Konvergenzrate tatsächlich höher ist als bei uniformen Alternativen. Diese Frage lässt sich in zahlreichen Fällen nach einer rigorosen Analyse der Regularität des zu approximierenden Objekts (hier: die Lösung der SPDE) klären. Dies gilt insbesondere für numerischen Methoden, welche auf Wavelets basieren.

In Abschnitt 1.1 dieser Arbeit werden die Zusammenhänge zwischen Regularität und Konvergenzordnung numerischer Methoden für den Fall auf Wavelets basierender Verfahren zur Approximation einer p-fach Lebesgue-integrierbaren Funktion $u \in L_p(\mathcal{O})$ ausführlich erläutert. Der Fehler wird in der $L_p(\mathcal{O})$-Norm gemessen. In diesem Setting wird, einerseits, die Konvergenzordnung uniformer Verfahren durch die Regularität der Zielfunktion u in der Skala $W_p^s(\mathcal{O})$, $s > 0$, von $L_p(\mathcal{O})$-Sobolev-Räumen bestimmt: $u \in W_p^s(\mathcal{O})$ impliziert, dass uniforme Verfahren die Konvergenzrate s/d erreichen können. Insbesondere gilt folgende Umkehrung: Falls $u \notin W_p^s(\mathcal{O})$ für ein $s > 0$, so wird ein uniformes Verfahren nie eine bessere Konvergenzrate als s/d erreichen. Andererseits, wird die Rate der sogenannten 'best m-term'-Approximation durch die Regularität der Zielfunktion in der speziellen Skala

$$B_{\tau,\tau}^\alpha(\mathcal{O}), \quad \frac{1}{\tau} = \frac{\alpha}{d} + \frac{1}{p}, \quad \alpha > 0, \tag{$*$}$$

von Besov-Räumen bestimmt. Bei dieser Methode wird u für jedes $m \in \mathbb{N}$ durch diejenige Linearkombination von m Termen (hier: Wavelets) approximiert, die den Fehler minimiert. Daher gilt die Rate der 'best m-term'-Approximation als Benchmark für die Konvergenzrate konstruktiver Approximationsverfahren.

Die obigen Resultate haben folgende Konsequenzen für die Entscheidung, welche Klasse von Verfahren bei der Lösung von Gleichungen der Form (1) zum Einsatz kommen sollte: Stimmen die räumliche Sobolev-Regularität und die räumliche Besov-Regularität in der Skala ($*$) der Lösung $u = u(\omega, t, \cdot)$ überein, so sind uniforme Verfahren aufgrund ihrer (relativen) Einfachheit vorzuziehen. Ist dies nicht der Fall, sprich, ist die räumliche Besov-Regularität in der Skala ($*$) höher als die räumliche Sobolev-Regularität, dann besteht die berechtigte Hoffnung, dass durch räumlich adaptiv arbeitende Verfahren eine höhere Konvergenzordnung erreicht werden kann. Vor diesem Hintergrund wird in der vorliegenden Arbeit folgenden beiden Fragestellungen nachgegangen:

(T1) **Räumliche Regularität.** Wie hoch ist die räumliche Regularität der Lösung $u = u(\omega, t, \cdot)$ der Gleichung (1) in der Skala ($*$) von Besov-Räumen? Es wird ein möglichst hohes $\alpha^* > 0$ gesucht, so dass für alle $0 < \alpha < \alpha^*$ und $1/\tau = \alpha/d + 1/p$, die Lösung u als p-fach Bochner-integrierbarer $B_{\tau,\tau}^\alpha(\mathcal{O})$-wertiger stochastischer Prozess aufgefasst werden kann.

(T2) **Raum-zeitliche Regularität.** Angenommen die Lösung u lässt sich als $B_{\tau,\tau}^\alpha(\mathcal{O})$-wertiger stochastischer Prozess mit α und τ wie in ($*$) auffassen. Was kann über die Hölder-Regularität der Pfade dieses Prozesses gesagt werden?

Die Behandlung des Punktes (T1) dient der Beantwortung der Frage, ob der Einsatz räumlich adaptiver numerischer Verfahren zur Lösung von SPDEs gerechtfertigt ist. Sollte sich herausstellen, dass

$$u \in L_p(\Omega \times [0, T]; B_{\tau,\tau}^\alpha(\mathcal{O})), \quad \frac{1}{\tau} = \frac{\alpha}{d} + \frac{1}{p}, \quad \text{mit} \quad \alpha > \tilde{s}_{\max}^{\mathrm{Sob}}(u), \tag{2}$$

wobei

$$\tilde{s}_{\max}^{\mathrm{Sob}}(u) := \sup\left\{s \geq 0 : u \in L_p(\Omega \times [0,T]; W_p^s(\mathcal{O}))\right\},$$

so lässt sich aufgrund der obigen Erläuterungen eine klare Empfehlung für die Entwicklung adaptiver Verfahren formulieren. Hierbei bezeichnet $L_p(\Omega \times [0,T]; E)$ den Raum aller vorhersagbaren, p-fach Bochner-integrierbaren stochastischen Prozesse mit Werten in einem (Quasi-) Banach-Raum $(E, \|\cdot\|_E)$. Die Beantwortung der zweiten Frage (T2) soll bei der Konvergenzanalyse entsprechender numerischer Raum-Zeit-Schemata eingesetzt werden. Eine solche Analyse wurde erst vor kurzem in [24] begonnen und befindet sich derzeit noch in ihren Anfängen.

Nachdem die Ziele formuliert sind und die Motivation erläutert wurde, sollen im Folgenden die erzielten Resultate zusammengefasst werden.

Das Gerüst: Eine geeignete L_p-Theorie für SPDEs

Eine direkte Anwendung abstrakter Ansätze für SPDEs, wie zum Beispiel des Halbgruppenansatzes für SPDEs von Da Prato und Zabczyk [32] sowie dessen Weiterentwicklung in [121, 122] oder aber des von Pardoux begründeten Variationsansatzes für SPDEs [101], liefern keine zufriedenstellenden Antworten auf die unter (T1) und (T2) formulierten Fragen.[1] Daher wird in dieser Arbeit ein indirekter Weg eingeschlagen. Die in [75] entwickelte L_p-Theorie wird als Grundgerüst benutzt (und erweitert). Sie garantiert die Existenz und Eindeutigkeit einer Lösung der Gleichung (1) auf allgemeinen beschränkten Lipschitz-Gebieten $\mathcal{O} \subset \mathbb{R}^d$ – allerdings noch nicht in den passenden Räumen. Daher muss diese Lösung anschließend hinsichtlich der Fragestellungen (T1) und (T2) analysiert werden.

Einbettungen gewichteter Sobolev-Räume in Besov-Räumen

Die in [75] betrachteten Lösungen linearer SPDEs sind Elemente bestimmter Banach-Räume $\mathfrak{H}_{p,\theta}^\gamma(\mathcal{O}, T)$ mit $p \in [2,\infty)$ sowie $\gamma, \theta \in \mathbb{R}$, welche aus stochastischen Prozessen mit Werten in gewichteten Sobolev-Räumen $H_{p,\theta-p}^\gamma(\mathcal{O})$ bestehen. Für $\gamma \in \mathbb{N}$ lässt sich $H_{p,\theta}^\gamma(\mathcal{O})$ als der Raum aller reellwertigen messbaren Funktionen auf $\mathcal{O}$, welche endliche Norm

$$u \mapsto \left(\sum_{|\alpha| \leq \gamma} \int_{\mathcal{O}} \left|\rho_{\mathcal{O}}(x)^{|\alpha|} D^\alpha u(x)\right|^p \rho_{\mathcal{O}}(x)^{\theta-d}\, \mathrm{d}x\right)^{1/p}$$

besitzen, definieren. Hierbei bezeichnet $\rho_{\mathcal{O}}(x)$ die Distanz zwischen einem Punkt $x \in \mathcal{O}$ und dem Rand $\partial\mathcal{O}$ des Gebietes. Für nicht ganzzahlige $\gamma \in (0,\infty) \setminus \mathbb{N}$ können diese Räume mittels komplexer Interpolation gewonnen werden, während für negative $\gamma < 0$ eine Charakterisierung über Dualität möglich ist. Aus der Definition der Banach-Räume $\mathfrak{H}_{p,\theta}^\gamma(\mathcal{O}, T)$ lässt sich unmittelbar schließen, dass diese in dem Raum $L_p(\Omega \times [0,T]; H_{p,\theta-p}^\gamma(\mathcal{O}))$ der p-fach Bochner-integrierbaren, vorhersagbaren $H_{p,\theta-p}^\gamma(\mathcal{O})$-wertigen stochastischen Prozesse stetig linear eingebettet sind. In Formeln:

$$\mathfrak{H}_{p,\theta}^\gamma(\mathcal{O}, T) \hookrightarrow L_p(\Omega \times [0,T]; H_{p,\theta-p}^\gamma(\mathcal{O})). \tag{3}$$

('$\hookrightarrow$' bedeutet 'stetig linear eingebettet'.) Folglich, führt der Nachweis einer Einbettung gewichteter Sobolev-Räume in die Besov-Räume der Skala (∗) unmittelbar zu einer Aussage über die räumliche Besov-Regularität von SPDEs im Sinne von (T1). Die Vermutung, dass eine solche Einbettung tatsächlich nachgewiesen werden kann, ist durch die in [38] bewiesenen Resultate gestützt. Darin wird, unter Ausnutzung gewichteter Sobolev-Normabschätzungen nachgewiesen,

---

[1] Die Gründe dafür werden ausführlich in Abschnitt 1.2 dieser Arbeit diskutiert.

dass die Lösungen bestimmter deterministischer elliptischer Differentialgleichungen eine hohe Besov-Regularität in der Skala (∗) aufweisen. Dies wurde unter anderem dadurch erreicht, dass die Wavelet-Koeffizienten der Lösung mittels gewichteter Sobolev (Halb-)Normen abgeschätzt werden konnten. Die Äquivalenz von Besov-Normen und entsprechenden gewichteten Folgennormen von Wavelet-Koeffizienten lieferten schließlich die gewünschte Abschätzung der Besov-Norm.

Durch den Einsatz ähnlicher Techniken wird in Kapitel 4 dieser Arbeit nachgewiesen, dass für beliebige beschränkte Lipschitz-Gebiete $\mathcal{O} \subset \mathbb{R}^d$ und Parameter $p \in [2, \infty)$ sowie $\gamma, \nu \in (0, \infty)$ Folgendes gilt (vgl. Theorem 4.7):

$$H_{p,d-\nu p}^{\gamma}(\mathcal{O}) \hookrightarrow B_{\tau,\tau}^{\alpha}(\mathcal{O}), \qquad \frac{1}{\tau} = \frac{\alpha}{d} + \frac{1}{p}, \qquad \text{für alle} \qquad 0 < \alpha < \min\left\{\gamma, \nu \frac{d}{d-1}\right\}. \quad (4)$$

Die Beweisführung für den speziellen Fall $\gamma \in \mathbb{N}$ verläuft ähnlich wie in dem Beweis von [38, Theorem 3.2]. Zudem wird auf auf die Tatsache zurückgegriffen, dass unter den gleichen Bedingungen,

$$H_{p,d-\nu p}^{\gamma}(\mathcal{O}) \hookrightarrow \mathring{W}_p^{\gamma \wedge \nu}(\mathcal{O}),$$

wobei $\mathring{W}_p^{s}(\mathcal{O})$ für den Abschluss in $W_p^{s}(\mathcal{O})$ des mit $\mathcal{C}_0^{\infty}(\mathcal{O})$ bezeichneten Raumes der unendlich oft differenzierbaren Funktionen mit kompaktem Träger in $\mathcal{O}$ steht. Diese Aussage wird in Proposition 4.1 bewiesen. Durch den Einsatz der komplexen Interpolationsmethode lässt sich die Einbettung (4) auch auf allgemeine $\gamma > 0$ übertragen (Theorem 4.7).

Folgende Konsequenzen des Theorems 4.7 liegen auf der Hand: Bis zu einem gewissen Grad lässt sich die Untersuchung der räumlichen Regularität der Lösungen von SPDEs in der Skala (∗) auf die Analyse der räumlichen gewichteten Sobolev-Regularität derselben zurückführen. Mit anderen Worten verbirgt sich hinter jedem Resultat zur gewichteten Sobolev-Regularität der Lösungen von SPDEs eine Aussage über deren räumliche Besov-Regularität in der Skala (∗).

(T1) Räumliche Regularität in der Skala (∗) von Besov-Räumen

Wie bereits erwähnt, sind die in dieser Arbeit betrachteten Lösungen von SPDEs der Form (1) Elemente der Banach-Räume $\mathfrak{H}_{p,\theta}^{\gamma}(\mathcal{O}, T)$ mit $p \in [2, \infty)$, $\gamma, \theta \in \mathbb{R}$. Aufgrund der Gleichheit

$$\theta - p = d - \left(1 + \frac{d - \theta}{p}\right)p,$$

folgt aus der Kombination der Einbettungen (3) und (4) dass

$$\mathfrak{H}_{p,\theta}^{\gamma}(\mathcal{O}, T) \hookrightarrow L_p(\Omega \times [0, T]; B_{\tau,\tau}^{\alpha}(\mathcal{O})), \frac{1}{\tau} = \frac{\alpha}{d} + \frac{1}{p}, \text{ für alle } 0 < \alpha < \gamma \wedge \left(1 + \frac{d - \theta}{p}\right)\frac{d}{d-1}. \quad (5)$$

In Kapitel 5 wird diese Einbettung benutzt, um räumliche Regularität in der Skala (∗) für Lösungen linearer und semilinearer SPDEs auf allgemeinen beschränkten Lipschitz-Gebieten $\mathcal{O} \subset \mathbb{R}^d$ nachzuweisen.

Lineare Gleichungen

Die in [75] entwickelte L_p-Theorie garantiert die Existenz und Eindeutigkeit einer Lösung $u \in \mathfrak{H}_{p,\theta}^{\gamma}(\mathcal{O}, T)$ für eine große Klasse linearer Gleichungen der From (1) mit $L = 0$ und $\Lambda = 0$. Die Anwendung der Einbettung (5) zeigt, dass

$$u \in L_p(\Omega \times [0, T]; B_{\tau,\tau}^{\alpha}(\mathcal{O})), \qquad \frac{1}{\tau} = \frac{\alpha}{d} + \frac{1}{p}, \qquad \text{für alle} \qquad 0 < \alpha < \gamma \wedge \left(1 + \frac{d - \theta}{p}\right)\frac{d}{d-1}, \quad (6)$$

siehe hierzu Theorem 5.2. Damit wurde eine Antwort auf die unter (T1) formulierte Frage für
den Fall linearer Gleichungen gefunden: Die Lösung $u \in \mathfrak{H}_{p,\theta}^{\gamma}(\mathcal{O},T)$ lässt sich als p-fach Bochner-
integrierbarer $B_{\tau,\tau}^{\alpha}(\mathcal{O})$-wertiger stochastischer Prozess mit $1/\tau = \alpha/d + 1/p$ auffassen, und zwar
für alle $0 < \alpha < \alpha^{*}$, wobei

$$\alpha^{*} := \min\left\{\gamma, \left(1 + \frac{d-\theta}{p}\right)\frac{d}{d-1}\right\} > 0$$

gewählt werden kann. Die genauen Bedingungen an den Gewichtsparameter $\theta \in \mathbb{R}$, unter denen
(6) erfüllt ist, finden sich in dem Hauptresultat zur räumlichen Regularität der Lösung linearer
SPDEs, Theorem 5.2. Beispielsweise gilt die Aussage (6) für $p = 2$, $\gamma = 2$ und $\theta = d = 2$, so
dass folglich

$$u \in L_2(\Omega \times [0,T]; B_{\tau,\tau}^{\alpha}(\mathcal{O})), \quad \frac{1}{\tau} = \frac{\alpha}{2} + \frac{1}{2}, \quad \text{für alle} \quad 0 < \alpha < 2$$

gilt. In Verbindung mit der in [92] etablierten Schranke für die räumliche Sobolev-Regularität der
Lösungen von SPDEs auf nicht-konvexen polygonalen Gebieten zeigen die erzielten Resultate,
dass, in der Tat die Lösung von SPDEs das durch (2) beschriebene Verhalten aufweisen kann.
Damit haben wir einen klaren Hinweis dafür, dass räumlich adaptiv arbeitende Verfahren für die
Lösung von SPDEs entwickelt werden sollten. Zahlreiche Beispiele, die diese These untermauern
sollen, sowie weiterführende Bemerkungen finden sich in Abschnitt 5.1.

Semilineare Gleichungen

Zahlreiche Phänomene aus der Physik oder aus der Chemie verlangen nach einer Modellierung
durch *nichtlineare* Gleichungen. Es ergibt sich daher die Frage, ob sich die weiter oben erzielten Resultate zur Besov-Regularität der Lösungen linearer SPDEs auf nichtlineare Gleichungen
übertragen lassen. Als einen ersten Schritt in diese Richtung wird in Abschnitt 5.2 eine Klasse
semilinearer SPDEs der Form (1) mit Lipschitz-stetigen Nichtlinearitäten L und Λ daraufhin
untersucht.
Wie zuvor soll die Einbettung (5) für den Nachweis räumlicher Besov-Regularität in der Skala (∗)
herangezogen werden. Da allerdings für semilineare Gleichungen keine entsprechende L_p-Theorie
existiert, muss zunächst die Existenz einer Lösung $u \in \mathfrak{H}_{p,\theta}^{\gamma}(\mathcal{O},T)$ unter geeigneten Bedingungen
nachgewiesen werden. Dies geschieht in Theorem 5.13. Die Nichtlinearitäten L und Λ genügen
bestimmten Lipschitz-Bedingungen (siehe Assumption 5.9), so dass (1) als 'gestörte' lineare
Gleichung interpretiert werden kann. Die Anwendung geeigneter Fixpunkt-Argumente, siehe
Lemma 5.16, liefert dann den Beweis für die Existenz einer Lösung $u \in \mathfrak{H}_{p,\theta}^{\gamma}(\mathcal{O},T)$. Diese erfüllt
nach (5) zwangsläufig auch (6), so dass für die betrachtete Klasse semilinearer Gleichungen eine
Regularitätsaussage in der Skala (∗) bewiesen werden kann, siehe hierzu Theorem 5.15.

(T2) Raum-zeitliche Regularität

Nachdem nachgewiesen werden konnte, dass die Lösung $u \in \mathfrak{H}_{p,\theta}^{\gamma}(\mathcal{O},T)$ linearer und nichtlinearer
SPDEs der Form (1) für $0 < \alpha < \alpha^{*}$ und $1/\tau = \alpha/d + 1/p$ als $B_{\tau,\tau}^{\alpha}(\mathcal{O})$-wertiger stochastischer
Prozess aufgefasst werden kann, wird die zweite große Fragestellung (T2) dieser Arbeit unter-
sucht: Die Hölder-Regularität der Pfade des Lösungsprozesses. Dies geschieht in Kapitel 6.
Die Analyse der Hölder-Regularität der Pfade der in den Banach-Räumen $\mathfrak{H}_{p,\theta}^{\gamma}(\mathcal{O},T)$ enthaltenen
stochastischen Prozesse ist bereits Teil der in [75] entwickelten L_p-Theorie. Ein Element $u \in$
$\mathfrak{H}_{p,\theta}^{\gamma}(\mathcal{O},T)$ wird darin als stochastischer Prozess mit Werten in gewichteten Sobolev-Räumen
aufgefasst. Insbesondere wird nachgewiesen, dass für $2/p < \tilde{\beta} < \beta \le 1$,

$$\|u\|_{\mathcal{C}^{\tilde{\beta}/2 - 1/p}([0,T]; H_{p,\theta-(1-\beta)p}^{\gamma-\beta}(\mathcal{O}))} < \infty \qquad \mathbb{P}\text{-fast sicher.} \tag{7}$$

Hierbei wird, wie üblich, für einen beliebigen (Quasi-)Banach-Raum $(E, \|\cdot\|_E)$, der Raum der κ-Hölder-stetigen E-wertigen Funktionen auf $[0, T]$ mit $(\mathcal{C}^\kappa([0, T]; E), \|\cdot\|_{\mathcal{C}^\kappa([0,T];E)})$ bezeichnet. Auf den ersten Blick sieht es so aus, als ließe sich daraus unmittelbar eine Aussage über Hölder-Regularität der Pfade der Lösungen $u \in \mathfrak{H}_{p,\theta}^\gamma(\mathcal{O}, T)$, aufgefasst als stochastische Prozesse mit Werten in der Skala $(*)$ von Besov-Räumen, herleiten. Eine Anwendung der Einbettung (4) würde genügen. Allerdings sind die sich daraus ergebenden Resultate nicht zufriedenstellend. Dies liegt vorwiegend an der Kombination der folgenden beiden Umstände: Der Hölder-Exponent $\kappa = \tilde{\beta}/2 - 1/p$ hängt von dem Parameter p, der gleichzeitig die Integrabilität in Raumrichtung misst, ab. Gleichzeitig müssen bestimmte Annahmen über den Gewichtsparameter $\theta \in \mathbb{R}$ getroffen werden, um überhaupt die Existenz einer Lösung $u \in \mathfrak{H}_{p,\theta}^\gamma(\mathcal{O})$ zu erhalten.

Um trotz dieser Hürden geeignete Resultate zu erzielen, bedienen wir uns folgender Strategie. Zunächst wird die Hölder-Regularität der Pfade der Klasse $\mathfrak{H}_{p,\theta}^{\gamma,q}(\mathcal{O}, T)$ stochastischer Prozesse untersucht. Die Elemente dieser Banach-Räume sind q-fach Bochner-integrierbare $H_{p,\theta-p}^\gamma(\mathcal{O})$-wertige stochastische Prozesse, welche bestimmten Bedingungen genügen. Im Grunde genommen sind es Erweiterungen der Klassen $\mathfrak{H}_{p,\theta}^\gamma(\mathcal{O}, T)$, wobei jetzt der Parameter q, der die Integrabilität nach der Zeit (und nach $\omega \in \Omega$) misst, sich ausdrücklich von dem Parameter p, der die Integrabilität in Raumrichtung angibt, unterscheiden darf. Es lässt sich nachweisen, dass für $u \in \mathfrak{H}_{p,\theta}^{\gamma,q}(\mathcal{O}, T)$ mit $2 \le p \le q < \infty$, $\gamma \in \mathbb{N}$ und $2/q < \tilde{\beta} < \beta \le 1$, gilt:

$$\|u\|_{\mathcal{C}^{\tilde{\beta}/2-1/q}([0,T];H_{p,\theta-(1-\beta)p}^{\gamma-\beta}(\mathcal{O}))} < \infty \qquad \mathbb{P}\text{-fast sicher,}$$

siehe Theorem 6.1. Insbesondere, hängt jetzt der Hölder-Exponent nicht mehr vom räumlichen Integrabilitätsparameter ab. Daher ergibt die Anwendung der Einbettung (7) brauchbare Resultate zur Hölder-Besov-Regularität stochastischer Prozesse aus $\mathfrak{H}_{p,\theta}^{\gamma,q}(\mathcal{O}, T)$ – auch für den Fall, dass der Gewichtsparameter $\theta \in \mathbb{R}$ den oben erwähnten Einschränkungen genügen muss.

Diese Resultate lassen sich nur dann für die Beantwortung der unter (T2) formulierten Frage heranziehen, wenn nachgewiesen werden kann, dass die Lösungen zu den SPDEs der Form (1) in der Klasse $\mathfrak{H}_{p,\theta}^{\gamma,q}(\mathcal{O}, T)$ mit $q \ne p$ enthalten sind. Das bedeutet, dass die in [75] entwickelte L_p-Theorie soweit wie möglich zu einer $L_q(L_p)$-Theorie ausgebaut werden muss. In dieser Arbeit wird eine erste $L_q(L_p)$-Theorie für die stochastische Wärmeleitungsgleichung mit additivem Rauschen auf allgemeinen beschränkten Lipschitz-Gebieten entwickelt, siehe hierzu insbesondere Theorem 6.11. Die Beweise basieren auf einer Kombination von Resultaten aus dem Halbgruppenansatz mit Techniken aus dem von N.V. Krylov begründeten analytischen Ansatz für SPDEs. Aus dem Halbgruppenansatz kann die Existenz einer Lösung, die allerdings geringe räumliche gewichtete Sobolev-Regularität aufweist, gezeigt werden, siehe Theorem 6.12. Indem Techniken aus dem analytischen Ansatz benutzt werden, kann anschließend nachgewiesen werden, dass diese Regularität anwächst, sobald die Koeffizienten der Gleichung eine höhere Regularität haben, siehe Theorem 6.7. Um diese beiden Ansätze zusammenbringen zu können, muss zunächst nachgewiesen werden, dass die jeweiligen Lösungsbegriffe übereinstimmen. Dazu werden im Laufe der Arbeit einzelne Hilfsresultate bewiesen, siehe etwa Theorem 2.54 sowie Proposition 3.6.

Die in Kapitel 6 durchgeführte Analyse führt schließlich zu einer zufriedenstellenden Aussage über die Hölder Regularität der Pfade der Lösung der Wärmeleitungsgleichung $u \in \mathfrak{H}_{p,d}^{\gamma,q}(\mathcal{O}, T)$, aufgefasst als stochastischer Prozess mit Werten in den Besov-Räumen aus der Skala $(*)$, siehe Theorem 6.17. Insbesondere lässt sich unter geeigneten Bedingungen an die Komponenten der Gleichung nachweisen, dass für alle Parameter, die der Bedingung

$$\frac{2}{q} < \tilde{\beta} < 1, \qquad \frac{1}{\tau} = \frac{\alpha}{d} + \frac{1}{p}, \quad \text{und} \quad 0 < \alpha < \left(1 - \tilde{\beta}\right)\frac{d}{d-1}$$

genügen, gilt:

$$\|u\|_{\mathcal{C}^{\tilde{\beta}/2-1/q}([0,T];B_{\tau,\tau}^\alpha(\mathcal{O}))} < \infty \qquad \mathbb{P}\text{-fast sicher.}$$

Notation

We collect here frequently used notations from this thesis. The number in the right column refers to the page where the symbol is introduced or where it appears first.

General mathematics

$\mathbb{N}$	set of positive integers $\{1, 2, \ldots\}$	
$\mathbb{N}_0$	set of non-negative integers $\{0, 1, 2, \ldots\}$	
$\mathbb{Z}$	set of integers	
$\mathbb{R}$	set of real numbers	
$\mathbb{R}_+$	set of positive real numbers $(0, \infty)$	
$\mathbb{C}$	set of complex numbers	
$d \in \mathbb{N}, d \geq 2$	dimension	
$\mathbb{R}^d$	d-dimensional Euclidian space $\{(x^1, \ldots, x^d) : x^1, \ldots, x^d \in \mathbb{R}\}$	
$\mathbb{R}_+^d$	half space in $\mathbb{R}^d$, $\{(x^1, \ldots, x^d) \in \mathbb{R}^d : x^1 > 0\}$	
$B_r(x)$	open ball with radius $r > 0$ centred at x, $\{y \in \mathbb{R}^d : \lvert y - x \rvert < r\}$	
A°	interior of a set $A \subseteq \mathbb{R}^d$	
A^B	set of all mappings from a set B to a set A	
$\lvert . \rvert$	absolute value, Euclidian norm on $\mathbb{R}^d$, or cardinality of a finite set; in Example 5.6 also used for the scale level (see p. 91)	
Σ_σ	the sector $\{z \in \mathbb{C} \setminus \{0\} : \lvert \arg(z) \rvert < \sigma\} \subset \mathbb{C}$	52
$\delta_{j,k}$	Kronecker symbol	
$\asymp$	norm equivalence	22
$\simeq$	isomorphic	22
$\cong$	isometrically isomorphic	22
Id	identity operator	
$\mathbb{1}_A$	indicator or characteristic function	
$u\vert_G$	restriction of u to G	
$\hookrightarrow$	continuously linearly embedded	22
$T > 0$	time horizon	20
$\overline{A}^{\lVert \cdot \rVert_B}, \overline{A}$	closure of $A \subseteq B$ in $(B, \lVert \cdot \rVert_B)$	
$[E_1, E_2]_\eta$	complex interpolation space of exponent $\eta \in (0, 1)$	22, 77
$E_1 \cap E_2$	intersection space for a compatible couple (E_1, E_2)	22
$E_1 \times E_2$	cartesian product	22
Δu	$\sum_{i=1}^d u_{x^i x^i}$, whenever it makes sense	
$\exists$	existential quantifier	
$A \subseteq B$	A is a subset of B	
$A \subset B$	A is a proper (or strict) subset of B, i.e., $A \subset B$ and $A \neq B$	
$A \subsetneq B$	same meaning as $A \subset B$, emphasizing that $A \neq B$	
$\bar{z}$	complex conjugate of a complex number $z \in \mathbb{C}$	
$\wedge, a \wedge b$	$\min\{a, b\}$	

Operators

$\mathcal{L}(E_1, E_2)$	vector space of all linear and bounded operators from E_1 to E_2	15
$\mathcal{L}(E)$	$\mathcal{L}(E, E)$	15
$\mathcal{L}_f(H, E)$	vector space of finite rank operators from H to E	25
$\mathcal{L}_1(H, U)$	vector space of nuclear operators from H to U	16
$\mathcal{L}_2(H, U)$	vector space of Hilbert-Schmidt operators from H to U	16, 27
E^*	dual space of E, i.e., $E^* := \mathcal{L}(E, \mathbb{R})$	15
$\langle x^*, x \rangle_{E^* \times E}$	dual pairing of $x^* \in E^*$ and $x \in E$	16
$\langle x^*, x \rangle$	dual pairing of $x^* \in E^*$ and $x \in E$	16
$\Gamma^\infty(H, E)$	vector space of γ-summing operators from H to E	25
$\|\cdot\|_{\Gamma_p^\infty(H,E)}$, $p \geq 1$	norm on $\Gamma^\infty(H, E)$, equivalent to $\|\cdot\|_{\Gamma^\infty(H,E)}$	25
$\Gamma(H, E)$	vector space of γ-radonifying operators	26
$\|\cdot\|_{\Gamma_p(H,E)}$, $p \geq 1$	norm on $\Gamma(H, E)$, equivalent to $\|\cdot\|_{\Gamma(H,E)}$	26
$h \otimes x$	rank one operator $\langle h, \cdot \rangle_H x \in \mathcal{L}(H, E)$	25
$\mathrm{ran}(S)$	range of an operator $S : E_1 \to E_2$, $\{Sx : x \in E_1\}$	
$D(A)$	domain of an (unbounded) operator A	51
$\rho(A)$	resolvent set of an operator A	51
$\sigma(A)$	spectrum of an operator A	51
$\Delta_{p,w}^D$	weak Dirichlet-Laplacian	123

Domains

G	arbitrary domain, i.e., an open and connected subset of $\mathbb{R}^d$	16
∂G	boundary of a domain $G \subseteq \mathbb{R}^d$	16
$\rho(x)$, $\rho_G(x)$	distance of a point $x \in G$ to the boundary ∂G	16
ψ	infinitely differentiable function on G, equivalent to ρ_G	37
$\mathcal{O}$	**bounded Lipschitz domain** in $\mathbb{R}^d$	16

Measure theory and probability

$(\mathfrak{M}, \mathcal{A}, \mu)$	σ-finite measure space	17
$L_p(\mathfrak{M}, \mathcal{A}, \mu; E)$	space of μ-equivalence classes of p-integrable strongly $\mathcal{A}$-measurable functions from $\mathfrak{M}$ to E, $p \in (0, \infty)$	17
$L_\infty(\mathfrak{M}, \mathcal{A}, \mu; E)$	space of μ-equivalence classes of strongly $\mathcal{A}$-measurable functions with a.e. finite norm	17
$L_p(\mathfrak{M}; E)$	shorthand for $L_p(\mathfrak{M}, \mathcal{A}, \mu; E)$, $p \in (0, \infty]$	17
$L_p(\mathfrak{M})$	shorthand for $L_p(\mathfrak{M}, \mathcal{A}, \mu; \mathbb{R})$, $p \in (0, \infty]$	17
L_p	$L_p(\mathbb{R}^d, \mathcal{B}(\mathbb{R}^d), \lambda^d; \mathbb{R})$	18
$\mathcal{B}(E)$	Borel σ-field on a quasi-normed space E, i.e., the σ-field generated by the standard topology on E	17
$\mathscr{P}(I)$	power set of I	18
δ_i	Dirac measure	18
$\ell_p(I)$	$L_p(I, \mathscr{P}(I), \sum_{i \in I} \delta_i; \mathbb{R})$	18
$\langle \cdot, \cdot \rangle_{\ell_2(I)}$	scalar product on $\ell_2(I)$	18
$|\cdot|_{\ell_2(I)}$	norm $\sqrt{\langle \cdot, \cdot \rangle_{\ell_2(I)}}$ on $\ell_2(I)$	18
ℓ_2	$\ell_2(\mathbb{N})$	18
λ^d	Lebesgue measure on $(\mathbb{R}^d, \mathcal{B}(\mathbb{R}^d))$ and restrictions on $(A, \mathcal{B}(A))$ for $A \in \mathcal{B}(\mathbb{R}^d)$	18
λ	λ^1	

$g\lambda^d$	measure with density g with respect to λ^d	18
$\mathrm{d}x$	shorthand for $\lambda^d(\mathrm{d}x)$	18
$\langle f, g \rangle$	$\int_G fg\,\mathrm{d}x$ for $fg \in L_1(G, \mathcal{B}(G), \lambda^d; \mathbb{R})$	18
$(\Omega, \mathcal{F}, \mathbb{P})$	complete probability space	18
$\mathbb{P}$	complete probability measure on $(\Omega, \mathcal{F})$	18
$\mathbb{E}[\cdot], \mathbb{E}$	expectation	18
$(\mathcal{F}_t)_{t\in[0,T]}$	normal filtration on $(\Omega, \mathcal{F}, \mathbb{P})$	20
$L_p^{\mathbb{F}}(\Omega; \ldots)$	closure of the finite rank $(\mathcal{F}_t)_{t\in[0,T]}$-adapted step processes in $L_p(\Omega; \Gamma(L_2([0,T]; H), E))$	31
$\{(w_t^k)_{t\in[0,T]}\}_{k\in\mathbb{N}}$	sequence of stochastically independent real-valued standard Brownian motions with respect to a normal filtration $(\mathcal{F}_t)_{t\in[0,T]}$	20
W_H	H-cylindrical Brownian motion	27
$\int_0^T \Phi(t)\,\mathrm{d}W_H(t)$	stochastic integral of Φ with respect to W_H	28–31
Ω_T	$\Omega \times [0,T]$	20
$\mathcal{P}_T$	predictable σ-field on Ω_T	20
$\mathbb{P}_T$	product measure $\mathbb{P} \times \lambda^1$ on $(\Omega_T, \mathcal{F} \otimes \mathcal{B}([0,T]))$ and on $(\Omega_T, \mathcal{P}_T)$	20
a.e., μ-a.e.	almost everywhere	
a.s., $\mathbb{P}$-a.s.	almost surely	

Distributions and derivatives

$\mathcal{C}_0^\infty(G)$	space of infinitely differentiable real-valued functions with compact support in the domain G	20
$\mathcal{D}'(G)$	space of real-valued distributions	21
$\mathcal{S}(\mathbb{R}^d)$	Schwartz space of rapidly decreasing real-valued functions on $\mathbb{R}^d$	21
$\mathcal{S}'(\mathbb{R}^d)$	space of real-valued tempered distributions	21
$\mathcal{S}'(\mathbb{R}^d; \mathbb{C})$	space of complex-valued tempered distributions	21
$\mathfrak{F}, \mathfrak{F}^{-1}$	Fourier transform on $\mathcal{S}'(\mathbb{R}^d; \mathbb{C})$ and its inverse	21
(u, φ)	application of $u \in \mathcal{D}'(G)$ ($u \in \mathcal{S}'(\mathbb{R}^d)$) to $\varphi \in \mathcal{C}_0^\infty(G)$ ($\varphi \in \mathcal{S}(\mathbb{R}^d)$); see also (2.24) and (2.38) for generalizations	21
$D^{(\alpha)}u, \alpha \in \mathbb{N}_0^d$	classical derivative	21
$D^\alpha u, \alpha \in \mathbb{N}_0^d$	generalized/weak/distributional derivative	21
$D^m u, m \in \mathbb{N}$	generalized/weak/distributional derivative of order m and the vector of all generalized/weak/distributional derivatives of order m	21
$u_x, u_{xx}, u_{x^i}, u_{x^ix^j}$	generalized/weak/distributional derivatives of first and second order	21
supp u	support of a distribution $u \in \mathcal{D}'(G)$	
$H^\infty(\Sigma_\sigma)$	set of all bounded analytic functions on the sector Σ_σ	52
$H_0^\infty(\Sigma_\sigma)$	subset $H^\infty(\Sigma_\sigma)$ consisting of all functions fulfilling (2.51)	52

Function spaces

$(*)$	non-linear approximation scale	2
$\mathcal{C}(G)$	space of real-valued continuous functions on a domain G	20
$\mathcal{C}^r(G), r \in \mathbb{N}$	space of real-valued r-times continuously differentiable functions	20
$\mathcal{C}(\overline{G})$	space of real-valued continuous functions on $\overline{G}$	21
$\mathcal{C}^r(\overline{G}), r \in \mathbb{N}$	space of real-valued r-times continuously differentiable functions with derivatives which can be extended to $\overline{G}$	21
$\mathcal{C}^\kappa([0,T]; E)$	space of Hölder continuous functions taking values in the quasi-Banach space E ($\kappa \in (0,1)$)	19

Spaces of stochastic processes and random variables

Wavelets

Semi-(Quasi-)Norms

Bibliography

[1] Felix Abramovich, Theofanis Sapatinas, and Bernard W. Silverman, *Wavelet thresholding via a Bayesian approach*, J. R. Stat. Soc. Ser. B Stat. Methodol. **60** (1998), no. 4, 725–749.

[2] Robert A. Adams and John J. F. Fournier, *Sobolev Spaces*, 2 ed., Pure Appl. Math. (Amst.), vol. 140, Academic Press, Amsterdam, 2003.

[3] Hugo Aimar and Ivana Gómez, *Parabolic Besov regularity for the heat equation*, Constr. Approx. **36** (2012), no. 1, 145–159.

[4] Hugo Aimar, Ivana Gómez, and Bibiana Iaffei, *Parabolic mean values and maximal estimates for gradients of temperatures*, J. Funct. Anal. **255** (2008), no. 8, 1939–1956.

[5] ———, *On Besov regularity of temperatures*, J. Fourier Anal. Appl. **16** (2010), no. 6, 1007–1020.

[6] David Albrecht, Xuan Duong, and Alan McIntosh, *Operator theory and harmonic analysis*, in: Instructional Workshop on Analysis and Geometry (Canberra) (Tim Cranny and John Hutchinson, eds.), Proceedings of the Centre for Mathematics and its Applications, vol. 34, ANU, 1996, pp. 77–136.

[7] Herbert Amann, *Ordinary Differential Equations. An Introduction to Nonlinear Analysis*, De Gruyter Studies in Mathematics, vol. 13, de Gruyter, Berlin–New York, 1990.

[8] ———, *Linear and Quasilinear Parabolic Problems. Volume I. Abstract Linear Theory*, Monographs in Mathematics, vol. 89, Birkhäuser, Basel–Boston–Berlin, 1995.

[9] Ludwig Arnold, *Mathematical models of chemical reactions*, in: Stochastic Systems: The Mathematics of Filtering and Identification and Applications (Michiel Hazewinkel and Jan C. Willems, eds.), NATO Advanced Study Institutes Series, Series C - Mathematical and Physical Sciences, vol. 78, Springer Netherlands, Dordrecht, 1981, pp. 111–134.

[10] Pascal Auscher, Steve Hofmann, Michael Lacey, Alan McIntosh, and Philippe Tchamitchian, *The solution of the Kato square root problem for second order elliptic operators on $\mathbb{R}^n$*, Ann. of Math. (2) **156** (2002), no. 2, 633–654.

[11] Pascal Auscher, Alan McIntosh, and Andrea R. Nahmod, *Holomorphic functional calculi of operators, quadratic estimates and interpolation*, Indiana Univ. Math. J. **46** (1997), no. 2, 375–403.

[12] Pascal Auscher and Philippe Tchamitchian, *Square roots of elliptic second order divergence operators on strongly Lipschitz domains: L^p theory*, Math. Ann. **320** (2001), no. 3, 577–623.

[13] Jöran Bergh and Jörgen Löfström, *Interpolation Spaces. An Introduction*, Grundlehren der mathematischen Wissenschaften, vol. 223, Springer, Berlin–Heidelberg–New York, 1976.

[14] Natalia Bochkina, *Besov regularity of functions with sparse random wavelet coefficients*, Preprint, http://www.bgx.org.uk/Natalia/Bochkina_BesovWavelets.pdf, 2006, last accessed: 17 December 2013.

[15] Zdzisław Brzeźniak, *Stochastic partial differential equations in M-type 2 Banach spaces*, Potential Anal. **4** (1995), no. 1, 1–45.

[16] ———, *On stochastic convolution in Banach spaces and applications*, Stochastics Stochastics Rep. **61** (1997), no. 3, 245–295.

[17] Zdzisław Brzeźniak, Jan M.A.M. van Neerven, Mark C. Veraar, and Lutz Weis, *Itô's formula in UMD Banach spaces and regularity of solutions of the Zakai equation*, J. Differential Equations **245** (2008), no. 1, 30–58.

[18] Donald L. Burkholder, *Martingales and singular integrals in Banach spaces*, in: Handbook of the Geometry of Banach Spaces (William B. Johnson and Joram Lindenstrauss, eds.), vol. 1, Elsevier, Amsterdam, 2001–2003, pp. 233–269.

[19] Claudio Canuto, Anita Tabacco, and Karsten Urban, *The wavelet element method. Part I. Construction and analysis*, Appl. Comput. Harmon. Anal. **6** (1999), no. 1, 1–52.

[20] ———, *The wavelet element method. Part II. Realization and additional features in 2D and 3D*, Appl. Comput. Harmon. Anal. **8** (2000), no. 2, 123–165.

[21] Philippe G. Ciarlet, *The Finite Element Method for Elliptic Problems*, Studies in Mathematics and its Applications, vol. 4, North-Holland, Amsterdam–New York–Oxford, 1978.

[22] Petru A. Cioica and Stephan Dahlke, *Spatial Besov regularity for semilinear stochastic partial differential equations on bounded Lipschitz domains*, Int. J. Comput. Math. **89** (2012), no. 18, 2443–2459.

[23] Petru A. Cioica, Stephan Dahlke, Nicolas Döhring, Ulrich Friedrich, Stefan Kinzel, Felix Lindner, Thorsten Raasch, Klaus Ritter, and René L. Schilling, *On the convergence analysis of Rothe's method*, Preprint, DFG-SPP 1324 Preprint Series, no. 124, http://www.dfg-spp1324.de/download/preprints/preprint124.pdf, 2012, last acessed: 17 December 2013.

[24] Petru A. Cioica, Stephan Dahlke, Nicolas Döhring, Stefan Kinzel, Felix Lindner, Thorsten Raasch, Klaus Ritter, and René L. Schilling, *Adaptive wavelet methods for the stochastic Poisson equation*, BIT **52** (2012), no. 3, 589–614.

[25] Petru A. Cioica, Stephan Dahlke, Stefan Kinzel, Felix Lindner, Thorsten Raasch, Klaus Ritter, and René L. Schilling, *Spatial Besov regularity for stochastic partial differential equations on Lipschitz domains*, Studia Math. **207** (2011), no. 3, 197–234.

[26] Petru A. Cioica, Kyeong-Hun Kim, Kijung Lee, and Felix Lindner, *On the $L_q(L_p)$-regularity and Besov smoothness of stochastic parabolic equations on bounded Lipschitz domains*, Electron. J. Probab. **18** (2013), no. 82, 1–41.

[27] Albert Cohen, *Numerical Analysis of Wavelet Methods*, 1st ed., Studies in Mathematics and its Applications, vol. 32, Elsevier, Amsterdam, 2003.

[28] Albert Cohen, Wolfgang Dahmen, and Ronald A. DeVore, *Adaptive wavelet methods for elliptic operator equations: Convergence rates*, Math. Comp. **70** (2001), no. 233, 27–75.

[29] _____, *Adaptive wavelet methods II–beyond the elliptic case*, Found. Comput. Math. **2** (2002), no. 3, 203–245.

[30] Albert Cohen, Ingrid Daubechies, and Jean-Christophe Feauveau, *Biorthogonal bases of compactly supported wavelets*, Comm. Pure Appl. Math. **45** (1992), no. 5, 485–560.

[31] Sonja G. Cox, *Stochastic Differential Equations in Banach Spaces. Decoupling, Delay Equations, and Approximations in Space and Time*, PhD thesis, Technische Universiteit Delft, 2012.

[32] Giuseppe Da Prato and Jerzy Zabczyk, *Stochastic Equations in Infinite Dimensions*, Encyclopedia Math. Appl., vol. 45, Cambridge Univ. Press, Cambridge, 1998.

[33] Stephan Dahlke, *Wavelets: Construction Principles and Applications to the Numerical Treatment of Operator Equations*, Habilitation thesis, RWTH Aachen, Berichte aus der Mathematik, Shaker, Aachen, 1997.

[34] _____, *Besov regularity for second order elliptic boundary value problems with variable coefficients*, Manuscripta Math. **95** (1998), no. 1, 59–77.

[35] _____, *Besov regularity for elliptic boundary value problems in polygonal domains*, Appl. Math. Lett. **12** (1999), no. 6, 31–36.

[36] _____, *Besov regularity of edge singularities for the Poisson equation in polyhedral domains*, Numer. Linear Algebra Appl. **9** (2002), no. 6-7, 457–466.

[37] Stephan Dahlke, Wolfgang Dahmen, and Ronald A. DeVore, *Nonlinear approximation and adaptive techniques for solving elliptic operator equations*, in: Multiscale Wavelet Methods for Partial Differential Equations (Wolfgang Dahmen, Andrew J. Kurdila, and Peter Oswald, eds.), Wavelet Analysis and Its Applications, vol. 6, Academic Press, San Diego, 1997, pp. 237–283.

[38] Stephan Dahlke and Ronald A. DeVore, *Besov regularity for elliptic boundary value problems*, Comm. Partial Differential Equations **22** (1997), no. 1-2, 1–16.

[39] Stephan Dahlke, Massimo Fornasier, Thorsten Raasch, Rob Stevenson, and Manuel Werner, *Adaptive frame methods for elliptic operator equations: the steepest descent approach*, IMA J. Numer. Anal. **27** (2007), no. 4, 717–740.

[40] Stephan Dahlke and Winfried Sickel, *Besov regularity for the Poisson equation in smooth and polyhedral cones*, in: Sobolev Spaces in Mathematics II, Applications to Partial Differential Equations (Vladimir G. Maz'ya, ed.), International Mathematical Series 9, Springer, jointly published with Tamara Rozhkovskaya Publisher, Novosibirsk, 2008, pp. 123–145.

[41] Wolfgang Dahmen, *Wavelet and multiscale methods for operator equations*, Acta Numer. **6** (1997), 55–228.

[42] Wolfgang Dahmen and Reinhold Schneider, *Wavelets with complementary boundary conditions – function spaces on the cube*, Results Math. **34** (1998), no. 3-4, 255–293.

[43] _____, *Composite wavelet bases for operator equations*, Math. Comp. **68** (1999), no. 228, 1533–1567.

[44] _____, *Wavelets on manifolds I: Construction and domain decomposition*, SIAM J. Math. Anal. **31** (1999), no. 1, 184–230.

[45] Ingrid Daubechies, *Ten Lectures on Wavelets*, CBMS-NSF Regional Conference Series in Applied Mathematics, vol. 61, SIAM, Philadelphia PA, 1992.

[46] Ronald A. DeVore, *Nonlinear approximation*, Acta Numer. **7** (1998), 51–150.

[47] Ronald A. DeVore, Björn Jawerth, and Vasil Popov, *Compression of wavelet decompositions*, Amer. J. Math. **114** (1992), no. 4, 737–785.

[48] Ronald A. DeVore and Robert C. Sharpley, *Maximal Functions Measuring Smoothness*, Mem. Amer. Math. Soc., vol. 47, no. 293, AMS, Providence RI, 1984.

[49] Joe Diestel, Hans Jarchow, and Andrew Tonge, *Absolutely Summing Operators*, Cambridge Stud. Adv. Math., vol. 43, Cambridge Univ. Press, Cambridge, 1995.

[50] Sophie Dispa, *Intrinsic characterizations of Besov spaces on Lipschitz domains*, Math. Nachr. **260** (2003), no. 1, 21–33.

[51] Manfred Dobrowolski, *Angewandte Funktionalanalysis. Funktionalanalysis, Sobolev-Räume und Elliptische Differentialgleichungen*, Springer, Berlin, 2006.

[52] Klaus-Jochen Engel and Rainer Nagel, *One-Parameter Semigroups for Linear Evolution Equations*, Grad. Texts in Math., vol. 194, Springer, New York, 2000.

[53] Lawrence C. Evans, *Partial Differential Equations*, Grad. Stud. Math., vol. 19, AMS, Providence, RI, 2002.

[54] Lawrence C. Evans and Ronald F. Gariepy, *Measure Theory and Fine Properties of Functions*, Stud. Adv. Math., CRC Press, Boca Raton, FL, 1992.

[55] Franco Flandoli, *Dirichlet boundary value problem for stochastic parabolic equations: compatibility relations and regularity of solutions*, Stochastics **29** (1990), no. 3, 331–357.

[56] Michael Frazier and Björn Jawerth, *A discrete transform and decompositions of distribution spaces*, J. Funct. Anal. **93** (1990), no. 1, 34–170.

[57] Pierre Grisvard, *Elliptic Problems in Nonsmooth Domains*, Monographs and Studies in Mathematics, vol. 24, Pitman, Boston–London–Melbourne, 1985.

[58] ――――, *Singularities in boundary value problems*, Recherches en mathématiques appliquées, vol. 22, Masson, Paris, Springer, Berlin, 1992.

[59] István Gyöngy and Annie Millet, *Rate of convergence of space time approximations for stochastic evolution equations*, Potential Anal. **30** (2009), no. 1, 29–64.

[60] Markus Haase, *The Functional Calculus for Sectorial Operators*, Oper. Theory Adv. Appl., vol. 169, Birkhäuser, Basel–Boston–Berlin, 2006.

[61] Wolfgang Hackbusch, *Elliptic Differential Equations. Theory and Numerical Treatment*, Springer Ser. Comput. Math., vol. 18, Springer, Berlin–Heidelberg–New York, 1992.

[62] Piotr Hajłasz, *Change of variables formula under minimal assumptions*, Colloq. Math. **64** (1993), no. 1, 93–101.

[63] Markus Hansen, *n-term approximation rates and Besov regularity for elliptic PDEs on polyhedral domains*, Preprint, DFG-SPP 1324 Preprint Series, no. 131, http://www.dfg-spp1324.de/download/preprints/preprint131.pdf, 2012, last accessed: 17 December 2013.

[64] Lars I. Hedberg and Yuri Netrusov, *An Axiomatic Approach to Function Spaces, Spectral Synthesis, and Luzin Approximation*, Mem. Amer. Math. Soc., vol. 188, AMS, Providence, RI, 2007.

[65] Jørgen Hoffmann-Jørgensen, *Sums of independent Banach space valued random variables*, Studia Math. **52** (1974), no. 2, 159–186.

[66] Vasile I. Istrățescu, *Fixed Point Theory: An Introduction*, Mathematics and its Applications, vol. 7, Kluwer, Dordrecht, 1981.

[67] David Jerison and Carlos E. Kenig, *The inhomogeneous Dirichlet problem in Lipschitz domains*, J. Funct. Anal. **130** (1995), no. 1, 161–219.

[68] Hans Johnen and Karl Scherer, *On the equivalence of the K-functional and moduli of continuity and some applications*, in: Constructive Theory of Functions of Several Variables (Walter Schempp and Karl Zeller, eds.), Lecture Notes in Math., vol. 571, Springer, Heidelberg, 1977, pp. 119–140.

[69] Alf Jonsson and Hans Wallin, *Function Spaces on Subsets of $\mathbb{R}^n$*, Mathematical Reports, vol. 2, pt. 1, Harwood Academic Publishers, Chur, 1984.

[70] Nigel J. Kalton, Svitlana Mayboroda, and Marius Mitrea, *Interpolation of Hardy-Sobolev-Besov-Triebel-Lizorkin spaces and applications to problems in partial differential equations*, in: Interpolation Theory and Applications (Laura De Carli and Mario Milman, eds.), Contemporary Mathematics, vol. 445, AMS, Providence, RI, 2007, pp. 121–177.

[71] Nigel J. Kalton and Lutz Weis, *The H^∞-calculus and sums of closed operators*, Math. Ann. **321** (2001), no. 2, 319–345.

[72] Kyeong-Hun Kim, *On stochastic partial differential equations with variable coefficients in C^1 domains*, Stochastic Process. Appl. **112** (2004), no. 2, 261–283.

[73] _____, *An L_p-theory of SPDEs on Lipschitz domains*, Potential Anal. **29** (2008), no. 3, 303–326.

[74] _____, *Sobolev space theory of SPDEs with continuous or measurable leading coefficients*, Stochastic Process. Appl. **119** (2009), no. 1, 16–44.

[75] _____, *A weighted Sobolev space theory of parabolic stochastic PDEs on non-smooth domains*, J. Theoret. Probab. **27** (2012), no. 1, 107–136.

[76] Kyeong-Hun Kim and Nicolai V. Krylov, *On the Sobolev space theory of parabolic and elliptic equations in C^1 domains*, SIAM J. Math. Anal. **36** (2004), no. 2, 618–642.

[77] Mihály Kovács, Stig Larsson, and Karsten Urban, *On wavelet-Galerkin methods for semilinear parabolic equations with additive noise*, Preprint, arXiv:1208.0433v1, 2012.

[78] Nicolai V. Krylov, *A W_2^n-theory of the Dirichlet problem for SPDEs in general smooth domains*, Probab. Theory Related Fields **98** (1994), no. 3, 389–421.

[79] _____, *On L_p-theory of stochastic partial differential equations in the whole space*, SIAM J. Math. Anal. **27** (1996), no. 2, 313–340.

[80] _____, *An analytic approach to SPDEs*, in: Stochastic Partial Differential Equations: Six Perspectives (René A. Carmona and Boris L. Rozovskii, eds.), Math. Surveys Monogr., vol. 64, AMS, Providence, RI, 1999, pp. 185–242.

[81] _____, *Some properties of weighted Sobolev spaces in* $\mathbb{R}^d_+$, Ann. Sc. Norm. Super. Pisa, Cl. Sci., IV. Ser. **28** (1999), no. 4, 675–693.

[82] _____, *SPDEs in* $L_q((0,\tau], L_p)$ *spaces*, Electron. J. Probab. **5** (2000), no. 13, 1–29.

[83] _____, *Some properties of traces for stochastic and deterministic parabolic weighted Sobolev spaces*, J. Funct. Anal. **183** (2001), no. 1, 1–41.

[84] _____, *Lectures on Elliptic and Parabolic Equations in Sobolev Spaces*, Grad. Stud. Math., vol. 96, AMS, Providence, RI, 2008.

[85] Nicolai V. Krylov and Sergey V. Lototsky, *A Sobolev space theory of SPDE with constant coefficients on a half line*, SIAM J. Math. Anal. **30** (1999), no. 2, 298–325.

[86] _____, *A Sobolev space theory of SPDEs with constant coefficients in a half space*, SIAM J. Math. Anal. **31** (1999), no. 1, 19–33.

[87] Alois Kufner, *Weighted Sobolev Spaces*, Teubner-Texte zur Mathematik, vol. 31, BSB B. G. Teubner Verlagsgesellschaft, Leipzig, 1980.

[88] Stanisław Kwapień, *On Banach spaces containing* c_0. *A supplement to the paper by J. Hoffmann-Jørgensen "Sums of independent Banach space valued random variables"*, Studia Math. **52** (1974), no. 2, 187–188.

[89] George C. Kyriazis, *Wavelet coefficients measuring smoothness in* $H_p(\mathbb{R}^d)$, Appl. Comput. Harmon. Anal. **3** (1996), no. 2, 100–119.

[90] Werner Linde and Albrecht Pietsch, *Mappings of Gaussian cylindrical measures in Banach spaces*, Theory Probab. Appl. **19** (1974), no. 3, 445–460.

[91] Felix Lindner, *Approximation and Regularity of Stochastic PDEs*, PhD thesis, TU Dresden, Shaker, Aachen, 2011.

[92] _____, *Singular behavior of the solution to the stochastic heat equation on a polygonal domain*, Preprint, DFG-SPP 1324 Preprint Series, no. 139, `http://www.dfg-spp1324.de/download/preprints/preprint139.pdf`, 2013, last accessed: 17 December 2013.

[93] Sergey V. Lototsky, *Sobolev spaces with weights in domains and boundary value problems for degenerate elliptic equations*, Methods Appl. Anal. **7** (2000), no. 1, 195–204.

[94] Alessandra Lunardi, *Analytic Semigroups and Optimal Regularity in Parabolic Problems*, Progr. Nonlinear Differential Equations Appl., vol. 16, Birkhäuser, Basel–Boston–Berlin, 1995.

[95] Alan McIntosh, *Operators which have an* H^∞ *functional calculus*, in: Miniconference on Operator Theory and Partial Differential Equations (Canberra) (Brian Jefferies, Alan McIntosh, and Werner J. Ricker, eds.), Proceedings of the Centre for Mathematical Analysis, vol. 14, ANU, 1987, pp. 210–231.

[96] Osvaldo Mendez and Marius Mitrea, *The Banach envelopes of Besov and Triebel-Lizorkin spaces and applications to partial differential equations*, J. Fourier Anal. Appl. **6** (2000), no. 5, 503–531.

[97] Yves Meyer, *Wavelets and Operators. Translated by D.H. Salinger*, Cambridge Stud. Adv. Math., vol. 37, Cambridge Univ. Press, Cambridge, 1992.

[98] Gustavo A. Muñoz, Yannis Sarantopoulos, and Andrew Tonge, *Complexifications of real Banach spaces, polynomials and multilinear maps*, Studia Math. **134** (1999), no. 1, 1–33.

[99] Jindřich Nečas, *Sur une méthode pour résoudre les équations aux dérivées partielles du type elliptique, voisine de la variationnelle*, Ann. Sc. Norm. Super. Pisa, Cl. Sci., III. Ser. **16** (1962), no. 4, 305–326.

[100] Peter Oswald, *Multilevel Finite Element Approximation. Theory and Applications*, Teubner-Skripten zur Numerik, Teubner, Stuttgart, 1994.

[101] Etienne Pardoux, *Équations aux dérivées partielles stochastiques non linéaires monotones. Étude de solutions fortes de type Itô*, Thèse, Univ. Paris Sud, 1975.

[102] Amnon Pazy, *Semigroups of Linear Operators and Applications to Partial Differential Equations*, Appl. Math. Sci., vol. 44, Springer, New York, 1983.

[103] Gilles Pisier, *Probabilistic methods in the geometry of Banach spaces*, in: Probability and Analysis (Giorgio Letta and Maurizio Pratelli, eds.), Lecture Notes in Math., vol. 1206, Springer, Berlin–Heidelberg–New York, 1986, pp. 167–241.

[104] Claudia Prévôt and Michael Röckner, *A Concise Course on Stochastic Partial Differential Equations*, Lecture Notes in Math., vol. 1905, Springer, Berlin–Heidelberg, 2007.

[105] Thorsten Raasch, *Adaptive Wavelet and Frame Schemes for Elliptic and Parabolic Equations*, PhD thesis, Philipps-Universität Marburg, Logos, Berlin, 2007, .

[106] Jan Rosiński and Zdisław Suchanecki, *On the space of vector-valued functions integrable with respect to the white noise*, Colloq. Math. **43** (1980), 183–201.

[107] Boris L. Rozovskii, *Stochastic Evolution Systems. Linear Theory and Applications to Non-Linear Filtering*, Mathematics and its Applications (Soviet Series), vol. 35, Kluwer, Dordrecht, 1990.

[108] Walter Rudin, *Functional Analysis*, 2nd ed., McGraw-Hill, New York, 1991.

[109] Thomas Runst and Winfried Sickel, *Sobolev Spaces of Fractional Order, Nemytskij Operators, and Nonlinear Partial Differential Equations*, De Gruyter Ser. Nonlinear Anal. Appl., vol. 3, de Gruyter, Berlin, 1996.

[110] Vyacheslav S. Rychkov, *On restrictions and extensions of the Besov and Triebel-Lizorkin spaces with respect to Lipschitz domains*, J. Lond. Math. Soc. (2) **60** (1999), no. 1, 237–257.

[111] René L. Schilling, *Measures, Integrals and Martingales*, Cambridge Univ. Press, Cambridge, 2005.

[112] Christoph Schwab and Rob Stevenson, *Space-time adaptive wavelet methods for parabolic evolution problems*, Math. Comp. **78** (2009), no. 267, 1293–1318.

[113] Elias M. Stein, *Singular Integrals and Differentiability Properties of Functions*, Princeton Math. Ser., vol. 30, Princeton Univ. Press, Princeton, NJ, 1970.

[114] Jesús Suárez and Lutz Weis, *Addendum to "Interpolation of Banach spaces by the γ-method"*, Extracta Math. **24** (2009), no. 3, 265–269.

[115] Hans Triebel, *Theory of Function Spaces*, Monog. Math., vol. 78, Birkhäuser, Basel–Boston–Stuttgart, 1983.

[116] _____, *Interpolation Theory, Function Spaces, Differential Operators*, 2nd ed., Barth, Heidelberg–Leipzig, 1995.

[117] _____, *Theory of Function Spaces III*, Monogr. Math., vol. 100, Birkhäuser, Basel–Boston–Berlin, 2006.

[118] Jan M.A.M. van Neerven, *Stochastic Evolution Equations*, ISEM Lectrue Notes, http://ocw.tudelft.nl/courses/mathematics/stochastic-evolution-equations/lectures, 2008, last accessed: 17 December 2013.

[119] _____, *γ-radonifying operators – a survey*, in: The AMSI-ANU Workshop on Spectral Theory and Harmonic Analysis (Andrew Hassel, Alan McIntosh, and Robert Taggart, eds.), Proceedings of the Centre for Mathematics and its Applications, vol. 44, 2010, pp. 1–62.

[120] Jan M.A.M. van Neerven, Mark C. Veraar, and Lutz Weis, *Stochastic integration in UMD Banach spaces*, Ann. Probab. **35** (2007), no. 4, 1438–1478.

[121] _____, *Maximal L^p-regularity for stochastic evolution equations*, SIAM J. Math. Anal. **44** (2012), no. 3, 1372–1414.

[122] _____, *Stochastic maximal L^p-regularity*, Ann. Probab. **40** (2012), no. 2, 788–812.

[123] Jan M.A.M. van Neerven and Lutz Weis, *Stochastic integration of functions with values in a Banach space*, Studia Math. **166** (2005), no. 2, 131–170.

[124] _____, *Weak limits and integrals of Gaussian covariances in Banach spaces*, Probab. Math. Statist. **25** (2005), no. 1, 55–74.

[125] Manuel Werner, *Adaptive Wavelet Frame Domain Decomposition Methods for Elliptic Operator Equations*, PhD thesis, Philipps-Universität Marburg, Logos, Berlin, 2009.

[126] Ian Wood, *Maximal L_p-regularity for the Laplacian on Lipschitz domains*, Math. Z. **255** (2007), no. 4, 855–875.

Index